CRC SERIES IN AGRICULTURE

Editor-in-Chief

Angus A. Hanson, Ph.D.
Vice President-Research
W-L Research, Inc.
Highland, Maryland

HANDBOOK OF SOILS AND CLIMATE IN AGRICULTURE

Editor
Victor J. Kilmer (Deceased)
Chief
Soils and Fertilizer Research Branch
National Fertilizer Development Center
Tennessee Valley Authority
Muscle Shoals, Alabama

HANDBOOK OF PLANT SCIENCE IN AGRICULTURE

Editor
B. R. Christie, Ph.D.
Professor
Department of Crop Science
Ontario Agricultural College
University of Guelph
Guelph, Ontario, Canada

HANDBOOK OF PEST MANAGEMENT IN AGRICULTURE

Editor
David Pimentel, Ph.D.
Professor
Department of Entomology
New York College of Agricultural
and Life Sciences
Cornell University
Ithaca, New York

HANDBOOK OF ENGINEERING IN AGRICULTURE

Editor
R. H. Brown, Ph.D.
Chairman
Division of Agricultural Engineering
Agricultural Engineering Center
University of Georgia
Athens, Georgia

HANDBOOK OF TRANSPORTATION AND MARKETING IN AGRICULTURE

Editor
Essex E. Finney, Jr., Ph.D.
Assistant Center Director
Agricultural Research Center
U.S. Department of Agriculture
Beltsville, Maryland

HANDBOOK OF PROCESSING AND UTILIZATION IN AGRICULTURE

Editor
Ivan A. Wolff, Ph.D. (Retired)
Director
Eastern Regional Research Center
Science and Education Administration
U.S. Department of Agriculture
Philadelphia, Pennsylvania

CRC Handbook of Plant Science in Agriculture

Volume II

Editor

B. R. Christie
Professor
Department of Crop Science
Ontario Agricultural College
University of Guelph
Guelph, Ontario
Canada

CRC Series in Agriculture

A. A. Hanson, Editor-in-Chief
Vice President-Research
W-L Research, Inc.
Highland, Maryland

CRC Press, Inc.
Boca Raton, Florida

Library of Congress Cataloging-in-Publication Data

Handbook of plant science in agriculture.

   (CRC series in agriculture)
   Bibliography: p.
   Includes index.
   1. Crops. 2. Agriculture. I. Christie, B. R.
(Bertram R.), 1933-   II. Series.
SB91.H36 1987   631   86-12937
ISBN 0-8493-3821-2

   This book represents information obtained from authentic and highly regarded sources. Reprinted material is quoted with permission, and sources are indicated. A wide variety of references are listed. Every reasonable effort has been made to give reliable data and information, but the author and the publisher cannot assume responsibility for the validity of all materials or for the consequences of their use.

   All rights reserved. This book, or any parts thereof, may not be reproduced in any form without written consent from the publisher.

   Direct all inquiries to CRC Press, Inc., 2000 Corporate Blvd., N.W., Boca Raton, Florida, 33431.

© 1987 by CRC Press, Inc.

International Standard Book Number 0-8493-3821-2 (set)
International Standard Book Number 0-8493-3822-0 (v. 1)
International Standard Book Number 0-8493-3823-9 (v. 2)

Library of Congress Card Number 86-12937
Printed in the United States

EDITOR-IN-CHIEF

Angus A. Hanson, Ph.D., is Vice President-Research, W-L Research, Inc., Highland, Maryland, and has had broad experience in agricultural research and development. He is a graduate of the University of British Columbia, Vancouver, and McGill University, Quebec, and received the Ph.D. degree from the Pennsylvania State University, University Park, in 1951.

An employee of the U.S. Department of Agriculture from 1949 to 1979, Dr. Hanson worked as a Research Geneticist at University Park, Pa., 1949 to 1952, and at Beltsville, Md., serving successively as Research Leader for Grass and Turf Investigations, 1953 to 1965, Chief of the Forage and Range Research Branch, 1965 to 1972, and Director of the Beltsville Agricultural Research Center, 1972 to 1979. He has been appointed to a number of national and regional task forces charged with assessing research needs and priorities, and has participated in reviewing agricultural needs and research programs in various foreign countries. As Director at Beltsville, he was directly responsible for programs that included most dimensions of agricultural research.

In his personal research, he has emphasized the improvement of forage crops, breeding and management of turfgrasses, and the breeding of alfalfa for multiple pest resistance, persistence, quality, and sustained yield. He is the author of over 100 technical and popular articles on forage crops and turfgrasses, and has served as Editor of *Crop Science* and the *Journal of Environmental Quality*.

PREFACE

Plants are the ultimate source of food, fiber, fuel, and many other products of importance to man. Thousands of years ago, people in various parts of the world began to plant, cultivate, and harvest those species which could provide for their needs. From this early cultivation, a vast array of cultivated plants has developed. Some species, such as wheat, cotton, and alfalfa are grown on every continent of the world, while others are grown only within a small geographical area. In the *Handbook,* we have attempted to provide information on all economically important crops, except those grown for timber or as ornamentals.

The study of crop plants enabled early man to learn their cultivation and processing. In this century, there has been a vast amount of information accumulated, as the study of these plant species still retains a fascination for us. Information on crop plants, e.g., on their genetics, growth, morphology, physiology, cultivation, and processing, is accumulating daily, and is reported in a vast array of publications and conferences and in many different languages. To assemble all of that information in one place would require more than one book. Therefore the aim of this *Handbook* is to assemble, in a condensed form, as much information as possible. The contributors have attempted to collect and to present the most recent information available on their assigned topics. Not every topic is included, nor are all crops covered under every topic. Limitations of time and space necessitated some selection. Volume I presents information on the genetics, botany, and growth of crop plants, while Volume II covers the production of crops and their utilization.

This *Handbook* was produced to serve as a guide and reference for all those involved with and interested in crop plants, such as producers, processors, researchers, students, and teachers. It has been made possible through the time and effort provided by the members of the Advisory Board and by our various contributors. Any comments or suggestions for future editions would be welcomed.

I want to thank all those involved for their efforts, their cooperation, and for their patience. It has been a privilege to work with all of them.

B. R. Christie
Editor

THE EDITOR

Bertram R. Christie, Ph.D., is Professor of Crop Science, Ontario Agricultural College, University of Guelph, Guelph, Ontario.

Professor Christie obtained his B.S.A. and M.S.A. degrees from the Ontario Agricultural College (University of Toronto) in 1955 and 1956, respectively, and the Ph.D. Degree from Iowa State University, in 1959. Since then, he has been a member of the faculty of the Crop Science Department at the University of Guelph, where he is involved in undergraduate and graduate education and in research on forage crops. He was one of the early recipients of the O.A.C. Alumni Distinguished Teaching Award. In 1985, he was invited to the People's Republic of China to present a series of lectures on quantitative genetics and on forage crop breeding to Chinese scientists.

Professor Christie is a member of the Editorial Board of the *Canadian Journal of Genetics and Cytology* and has served in a similar capacity with the *Canadian Journal of Plant Science*. He is a Director of the American Forage and Grassland Council and the Canada Committee on Forage Crops. He has served as President of the Canadian Society of Agronomy and also of the Ontario Institute of Professional Agrologists. Professor Christie is also a member of the American Society of Agronomy, the Canadian Society of Genetics and Cytology, and the Agricultural Insitute of Canada.

Professor Christie has been the author or co-author of numerous articles and has developed several cultivars of forage crops.

ADVISORY BOARD

Charles O. Gardner, Ph.D.
Professor
Department of Agronomy
University of Nebraska
Lincoln, Nebraska

Jules Janick, Ph.D.
Professor
Department of Horticulture
Purdue University
West Lafayette, Indiana

Donald G. Hanway, Ph.D.
Professor Emeritus
Department of Agronomy
University of Nebraska
Lincoln, Nebraska

William C. Kennard, Ph.D.
Professor of Plant Physiology
Department of Plant Science
University of Connecticut
Storrs, Connecticut

Allan K. Stoner, Ph.D.
Director
Plant Genetics and Germplasm Institute
Agricultural Research Center
Beltsville, Maryland

CONTRIBUTORS

W. Powell Anderson
Associate Professor
Department of Agronomy
New Mexico State University
Las Cruces, New Mexico

K. C. Armstrong, Ph.D.
Research Cytogeneticist
Research Branch
Agriculture Canada
Ottawa, Ontario
Canada

Donald K. Barnes, Ph.D.
U.S.D.A., A.R.S. Plant Science
  Research Unit
Department of Agronomy and Plant
  Genetics
University of Minnesota
St. Paul, Minnesota

Stephen R. Bowley, Ph.D.
Assistant Professor
Department of Crop Science
University of Guelph
Guelph, Ontario
Canada

Calvin Chong, Ph.D.
Research Scientist
Horticultural Research Institute
Ontario Ministry of Agriculture and Food
Vineland Station, Ontario
Canada

Meryl N. Christiansen, Ph.D.
Director
Plant Physiology Institute
Agricultural Research Service
U.S. Department of Agriculture
Beltsville, Maryland

F. W. Cope, Ph.D.
Professor, Emeritus
Biological Science Department
University of the West Indies
St. Augustine
Trinidad

Bruce E. Coulman, Ph.D.
Associate Professor
Department of Plant Science
MacDonald College
McGill University
Ste. Anne de Bellevue, Quebec
Canada

S. K. A. Danso, Ph.D.
Joint FAO/IAEA Division
International Atomic Energy Agency
Vienna
Austria

Arnel R. Hallauer, Ph.D.
Research Geneticist, USDA, ARS
Department of Agronomy
Iowa State University
Ames, Iowa

Gudni Hardarson, Ph.D.
Soil Fertility Section
FAO/IAEA Program
Seibersdorf Laboratory
Seibersdorf
Austria

Jack R. Harlan, Ph.D.
Professor Emeritus, Plant Genetics
University of Illinois
Urbana, Illinois

S. B. Helgason, Ph.D.
Professor Emeritus
Department of Plant Science
Faculty of Agriculture
University of Manitoba
Winnipeg, Manitoba
Canada

Robert J. Hilton, Ph.D.
Professor Emeritus, Horticulture
University of Guelph
Guelph, Ontario
Canada

D. L. Jennings, Ph.D.
Scottish Crop Research Institute
Invergowrie, Dundee
Scotland

Joshua A. Lee
Department of Crop Science
North Carolina State University
Raleigh, North Carolina

Robert C. Leffel, Ph.D.
Research Agronomist
Agricultural Research Service
U.S. Department of Agriculture
Beltsville, Maryland

F. W. Liu, Ph.D.
Associate Professor
Department of Pomology
Cornell University
Ithaca, New York

E. V. Maas
Research Leader
U.S. Salinity Laboratory
U.S. Department of Agriculture
Riverside, California

David J. Major, Ph.D.
Senior Research Scientist
Plant Science Section
Agriculture Canada
Research Station
Lethbridge, Alberta
Canada

Beverley H. Marie
Research Associate
Department of Horticultural Science
University of Guelph
Guelph, Ontario
Canada

Douglas P. Ormrod, Ph.D.
Professor of Horticultural Science
University of Guelph
Guelph, Ontario
Canada

Craig C. Sheaffer, Ph.D.
Professor
Department of Agronomy and Plant
  Genetics
University of Minnesota
St. Paul, Minnesota

Alfred E. Slinkard, Ph.D.
Senior Research Scientist
Crop Development Centre
University of Saskatchewan
Saskatoon, Saskatchewan
Canada

Garry A. Smith, Ph.D.
Research Geneticist
Crops Research Laboratory
Agriculture Research Service
U.S. Department of Agriculture
Colorado State University
Fort Collins, Colorado

Thomas M. Starling, Ph.D.
Professor
Department of Agronomy
Virginia Polytechnic Institute and State
  University
Blacksburg, Virginia

Anna K. Storgaard, Ph.D.
Department of Plant Science
Faculty of Agriculture
University of Manitoba
Winnipeg, Manitoba
Canada

Norman L. Taylor, Ph.D.
Professor
Department of Agronomy
University of Kentucky
Lexington, Kentucky

M. Tollenaar, Ph.D.
Assistant Professor
Crop Science Department
University of Guelph
Guelph, Ontario
Canada

Dan Wiersma, Ph.D.
Water Resources Research Center
 (Retired)
Department of Agronomy
Purdue University
West Lafayette, Indiana

James R. Wilcox, Ph.D.
Research Geneticist, USDA, ARS
Department of Agronomy
Purdue University
West Lafayette, Indiana

B. Young
Research Assistant
Department of Horticultural Science
University of Guelph
Guelph, Ontario
Canada

F. Zapata, Ph.D.
International Atomic Energy Agency
Seibersdorf Laboratory
Seibersdorf
Austria

TABLE OF CONTENTS

Volume I

GENETICS OF CROPS
Chromosome Numbers of Crop Species 3
K. C. Armstrong

Centers of Origin 15
Jack R. Harlan

Introgressive Hybridization 23
S. R. Bowley and Norman L. Taylor

Breeding Systems 61
Arnel R. Hallauer

BOTANY OF CROPS
Plant Propagation 91
Calvin Chong

Life Cycles 115
S. B. Helgason and Anna K. Storgaard

Biological Nitrogen Fixation 165
Gudni Hardarson, S. K. A. Danso, and F. Zapata

ENVIRONMENTAL FACTORS AND PLANT GROWTH
Light and Photoperiod 195
David J. Major

Plant Temperature Stress 217
Meryl N. Christiansen

Sensitivity of Crop Plants to Gaseous Pollution Stress 225
Douglas P. Ormrod, B. Young, and Beverley Marie

INDEX 275

Volume II

CROP PRODUCTION
Water and Agricultural Productivity 3
Dan Wiersma and B. R. Christie

Salt Tolerance of Plants 57
E. V. Maas

World Production and Distribution 77
Bruce E. Coulman

Dry Matter Production from Crops 89
M. Tollenaar

Weed Science As It Relates to Crop Production 99
W. Powell Anderson

UTILIZATION OF CROPS
Cereals 117
Thomas M. Starling

Sugar Crops 125
Garry A. Smith

Starch Crops 137
D. L. Jennings

Oilseed Crops 145
James R. Wilcox and Robert C. Leffel

Protein Crops 167
Alfred E. Slinkard

Plant Fibers 173
Joshua A. Lee

Vegetable Crops 183
Robert J. Hilton

Fruit Crops 195
F. W. Liu

Drug Crops 209
F. W. Cope

Forage Crops 217
Craig C. Sheaffer and Donald K. Barnes

INDEX 253

Crop Production

WATER AND AGRICULTURAL CROP PRODUCTIVITY

Dan Wiersma and B. R. Christie

INTRODUCTION

The ultimate objective of agronomic agriculture is to attain the highest potential yield per unit area of land. Several economic inputs are involved; however, the prime limiting factors are physiochemical, as illustrated in Figure 1. Each of these stress factors contribute to depressing yield, and to an extent may be interdependent; however, "lack of moisture" heads the list. The relative importance of moisture will vary geographically, dependent on the amount and timely seasonal distribution of precipitation.

Crop yields are a measure of that portion of the plant which has the priority economic value. For the cereal crops it is the grain, for forages the vegetative portion, and for sugar beets the sucrose. The highest yield of record can be an approximation of the potential yield of a particular crop.[7] In Table 1, there is listed the record yield as of 1975 for eight agricultural crops compared with average yield over a period of years. Although the stress factors of disease and insects can be devastating to individual localities and farmers, their contribution to decreasing yield is relatively small, 4.1 and 2.6 %, respectively. The "Other" category of environmental factors, which is primarily lack of moisture, contributes over 69% of the stress.

In addition to precipitation, the type of soil greatly influences the amount of water available to the plant. In Table 2, the two soils with plant moisture limitations are drought, which can be defined as soils with low available water, and shallowness resulting from an impediment relatively close to the surface affecting root penetration. Almost 45% (Table 2) of the U.S. area has soil with limiting moisture-supplying capabilities. From Table 3, insurance indemnities for crop losses shows that for the period 1939 to 1978, drought was the major cause for payment with almost 41%.

Importance of Water to Plants[54]

Since water is such a major factor in crop yield, what is its role in the life of a plant?

First, water is a major constituent of the cell protoplasm of actively growing plants. Plants vary as to their percentage of fresh weight in water. Table 4 lists the water content of some common crop plants. When a plant becomes dehydrated physiological processes within the protoplasm become impeded, and below a certain water content the protoplasm is killed.

Second, water is involved directly in a number of chemical reactions occurring in protoplasm, most notably the process of photosynthesis. This process involves the reduction of carbon dioxide into organic materials such as sugars, starches, proteins etc. This is a rather complex process, however simply stated the reaction is

$$CO_2 + H_2O \xrightarrow[\text{chloroplasts}]{\text{energy}} \text{Organic Substances}$$

The reverse of this reaction is respiration, a continuing process within the plant, and whose product includes water and carbon dioxide. Hydration and condensation reactions, in which water is added to or removed from organic molecules, are important in various metabolic processes, for example:

$$(C_6H_{10}O_5)_n + n\ H_2O \leftrightarrows n\ C_6H_{12}O_6$$

$$\text{starch} \qquad\qquad\qquad \text{glucose}$$

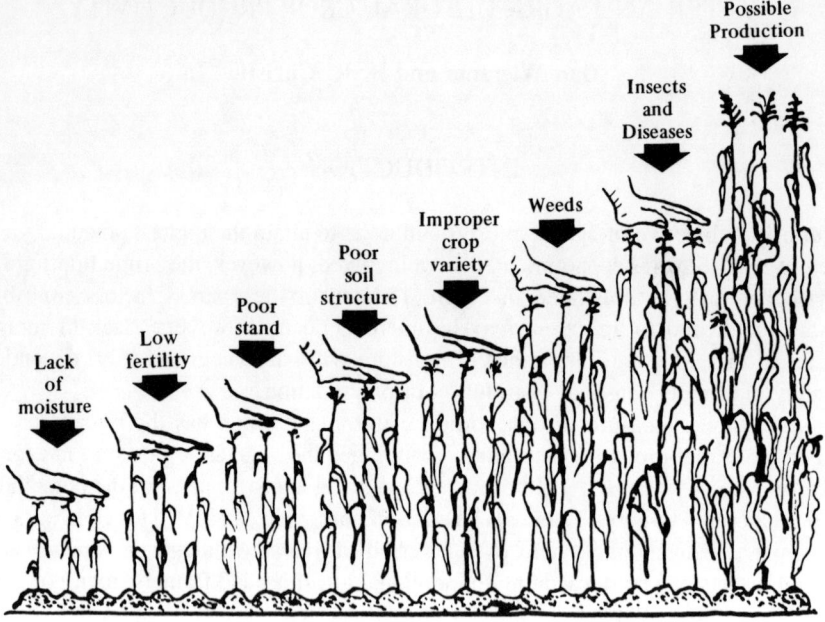

FIGURE 1. Factors which can limit crop production. The lack of moisture heads the list. (From U.S. Department of Agriculture, Soil Conservation Bulletin 199.)

Table 1
RECORD YIELDS, AVERAGE YIELDS, AND YIELD LOSSES DUE TO DISEASES, INSECTS, AND UNFAVORABLE PHYSIOCHEMICAL ENVIRONMENTS FOR MAJOR U.S. CROPS

			Average losses			
					Unfavorable environment[a]	
Crop	Record yield	Average yield	Diseases	Insects	Weeds	Other
Corn	19,300	4,600	750	691	511	12,700
Wheat	14,500	1,880	336	134	256	11,900
Soybeans	7,390	1,610	269	67	330	5,120
Sorghum	20,100	2,830	314	314	423	16,200
Oats	10,600	1,720	465	107	352	7,960
Barley	11,400	2,050	377	108	280	8,590
Potatoes	94,100	28,300	8,000	5,900	875	50,900
Sugar beets	121,000	42,600	6,700	6,700	3,700	61,300
Mean percentage of record yield		21.6	4.1	2.6	2.6	69.1

Note: Values given in kilograms per hectare.

[a] Calculated as follows: record yield − (average yield + disease loss + insect loss). The "unfavorable environment" includes moisture deficiency.

From Boyer, J. S., *Science*, 218, 443, 1982. With permission.

Table 2
AREA OF THE U.S. WITH SOILS SUBJECT TO ENVIRONMENTAL LIMITATIONS

Environmental limitation	Area of U.S. soil affected (%)
Drought[a]	25.3
Shallowness[a]	19.6
Cold	16.5
Wet	15.7
Alkaline salts	2.9
Saline or no soil	4.5
Other	3.4
None	12.1

[a] Relates to moisture deficiencies.

From Boyer, J. S., *Science*, 218, 443, 1982. With permission.

Table 3
DISTRIBUTIONS OF INSURANCE INDEMNITIES FOR CROP LOSSES IN THE U.S. FROM 1939 TO 1978[7]

Cause of crop loss	Proportion of payments (%)
Drought	40.8
Excess water	16.4
Cold	13.8
Hail	11.3
Wind	7.0
Insect	4.5
Disease	2.7
Flood	2.1
Other	1.5

From Boyer, J. S., *Science*, 218, 443, 1982. With permission.

Table 4
WATER CONTENT COMMONLY FOUND IN VARIOUS PLANT MATERIALS

	Water contents expressed as %	
	Fresh weight	Dry weight
Cucumber fruits	96	2400
Lettuce, leafy head	94	1567
Cabbage, leafy head	90	900
Potato, tuber (white)	79	376
Tomato, leaves	85—95	567—1900
Nasturtium, leaves	80—85	400—567
Grape, leaves	72—80	257—400
Peach, leaves	61—70	156—233
Apple, leaves	59—62	144—163
Norway maple, leaves	54—58	117—138
Oak, leaves	54	117
Corn, leaves	65—82	186—456
Sorghum, leaves	58—79	138—376
Cotton, leaves	70—78	233—354
Sugarcane, blades	65—74	186—285

From Curtis, O. F. and Clark, D. G., An *Introduction to Plant Physiology*, McGraw Hill, New York, 1950, 141. With permission.

Third, water is the solvent in which many other substances are dissolved, and in which they undergo chemical reactions.

Fourth, much of the water in plants occurs in large vacuoles within the protoplasts, where it is largely responsible for maintaining the turgidity of the cells and hence of the plant as a whole.

Fifth, there is a thin layer of water surrounding each cell of a plant and this permeates the microspaces between solid materials in the cell wall. These surface films, which are continuous from cell to cell and form a network throughout the plant, are important in the entry and movement of dissolved substances.

And, finally, water fulfills a variety of additional functions in plants; for example, it provides a medium for the movement of dissolved substances in the xylem and phloem. It is the medium through which motile gametes effect fertilization, and it assists in various ways in the dissemination of spores, fruits, and seeds.

Plants and the Hydrologic Cycle

Plants are an integral segment of the hydrologic cycle. In the cycle, the transformation of water from the liquid directly into the vapor phase is through the processes of evaporation. Transpiration is a similar process whereby water which moves through the plant is evaporated from the plant surfaces. To change water from a liquid to a vapor requires almost 600 cal of energy per gram of water at 20°C. The source of this energy is the sun. Since it is difficult to separate the amounts of water involved in evaporation and transpiration, and essentially the same energy transformation system is involved, the terms are frequently combined and the process is referred to as evapotranspiration.

The ratio of water evaporating from a bare surface and that transpired by the plant is dependent on the vegetative cover. For a crop with a leaf-area index (that is, the ratio of total leaf surface to ground surface) greater than 3, the major portion of the process is transpiration. Under certain conditions as much as a liter of water can pass through a single maize (corn) plant during a warm, clear mid-July day. About 65 to 70% of the precipitation received in the U.S. is evapotranspired.

The amount of annual rainfall varies widely over the earth; numerous stations record averages less than 13 cm while there are some which exceed 1000 cm. Although the total annual average rainfall usually receives the most attention, for agricultural crops the seasonal distribution is very important. For example, a large portion of the North American mid-continent rainfall occurs during the months of May through August, the crop-growing season, and a very small percentage from November to February when crops are dormant.

Geographic regions have been classified from arid to humid according to their precipitation patterns. One such classification is Thornwaite's, which is a ratio of the total monthly precipitation to the total monthly evaporation (P/E) and a sum of the 12-month P/E ratios is considered the P/E index. According to this classification, a P/E index of over 128 is a wet region, 64 to 127 is humid, 32 to 63 is subhumid, 16 to 31 is semiarid, and less than 16 is arid.[57]

Some of the rainfall is intercepted by the plant canopy; however, the percentage is small depending on the rainfall intensity and duration and the type of vegetative cover. Condensation or dew can also contribute to the water economy of the plant, but again the amount is relatively small and variable.

Looking further to the hydrologic cycle, the major source of water for the plant is that which is infiltrated and stored in the soil profile. Infiltration is defined as the downward entry of water into the soil and infiltration rate is a soil characteristic determining or describing the maximum rate at which water can enter the soil under specified condition.

The Soil as Plant Water Storage

Soil serves as a water-storage reservoir for plant usage during periods of nonprecipitation. Soil is made up of various size particles, with the generalized classification of sand, silt, and clay, ranging from coarse to fine, respectively. The arrangement of the soil particles make up pores where the water is held and stored. Several layers of water molecules are strongly adsorbed to the surface of the soil particles, and is referred to as adhesive water

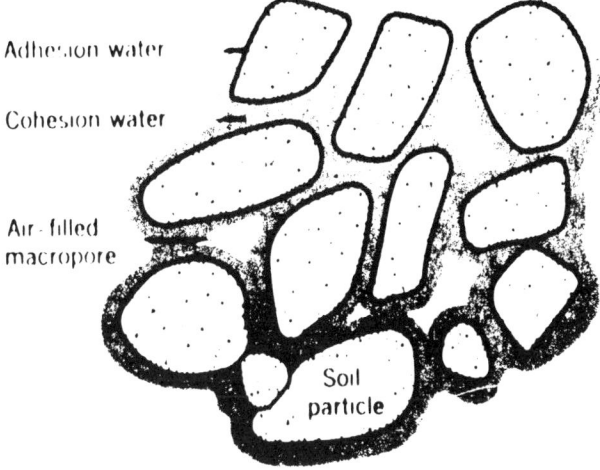

FIGURE 2. Schematic drawing showing the relationship of adhesion and cohesion water with respect to soil particles. (From Foth, H. D., *Fundamentals of Soil Science*, 6th ed., John Wiley & Sons, New York, 1978. With permission).

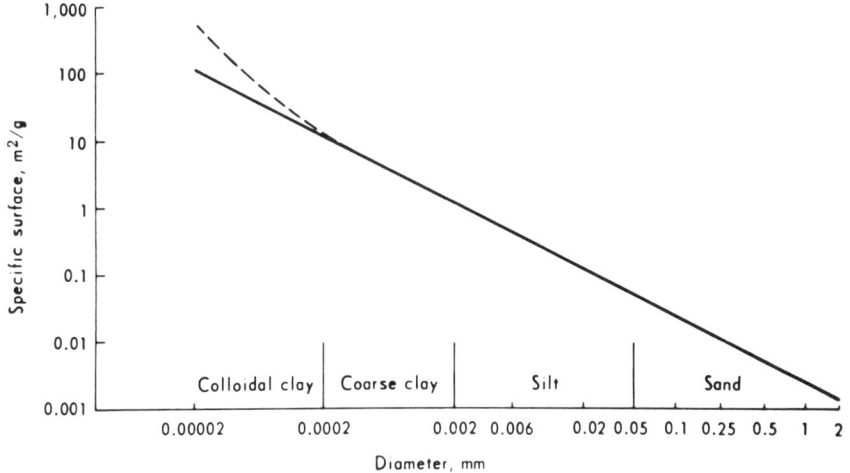

FIGURE 3. Relationship between particle size and soil fractions and the approximate specific surfaces. (From Kohnke, H., *Soil Physics*, McGraw-Hill, New York, 1968, 87. With permission.)

(Figure 2). This strongly adsorbed water is not readily available to the plants. Further from the particle surface, water is held with less and less attraction, is much more mobile, and is referred to as cohesive water (Figure 2). The outer approximately two thirds of the film can be considered available to plants and constitutes the major source of water for plants.

The amount of water stored in soil is thus dependent on the total surface area per unit volume of soil. A grain of fine colloidal clay has about 10,000 times as much surface area as the same weight of medium-size sand. The specific surface area (area per unit weight) of colloidal clay ranges from about 10 to 1000 m^2/g. The same figures for the smallest silt particles and for fine sand are 0.1 and 0.01 m^2/g, respectively (Figure 3 and Table 5).

The attractive force referred to above is the energy of retention of the soil surface for the water. In common terminology, this is suction, pressure, tension, or potentials. A frequently

Table 5
RANGES OF SPECIFIC SURFACE OF CLAYS AND SOILS IN m²/g

Clay, montmorillonite	500—800
Clay, illite	60—120
Clay, kaolinite (soil)	20—40
Clay, kaolinite (ceramic)	10—20
Clay, soil-clay fraction in Middle West	300—400
Clay, from glacial calcareous till	150—200
Clay soil	150—250
Silty clay loam	120—200
Silt loam	50—150
Loam	50—100
Sandy loam	10—40
Silt soil	5—20

From Kohnke, H., *Soil Physics*, McGraw-Hill, New York, 1968, 88. With permission.

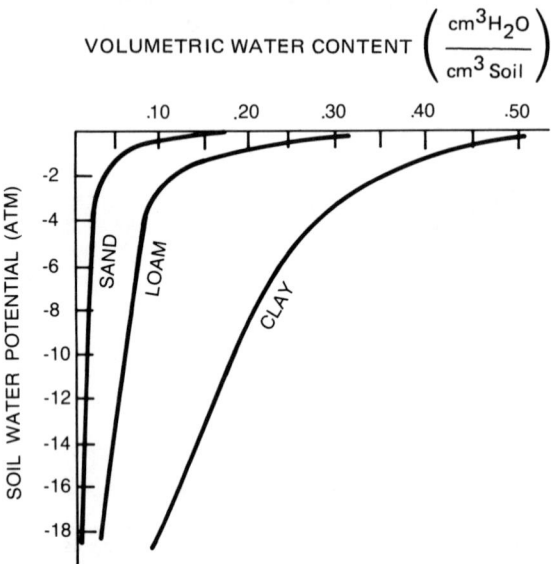

FIGURE 4. Soil water potential as a function of volumetric water content for typical sand, loam, and clay soils. 1 atm. = 1.013 bars. (From Merva, G. E., *Physioengineering Principles*, AVI, Westport, Conn., 1975. With permission.

used unit is "bar" which is 1 million dyn/cm² or approximately 14.9 lb/in.² Moisture tension or pressure is equal to the equivalent pressure that must be applied to the soil water to bring it to hydraulic equilibrium through a porous permeable membrane, with a pool of water of the same composition. It is a common laboratory measurement. Soils vary as to the amount of water in relation to the energy of retention (Figure 4). A soil moisture-retention constant which has been quite commonly used is termed "field capacity". This is defined as the percentage of water remaining in a soil 2 or 3 days after saturation when free drainage has practically ceased. This concept has proven useful in plant-soil-water relations as it has been considered the upper limit of plant available stored water. The constant as such is rather ill defined in any specific soil, as can be noted from the moisture-retention curves of Figure 4. There is no discontinuity which would indicate a "field capacity" constant. The one

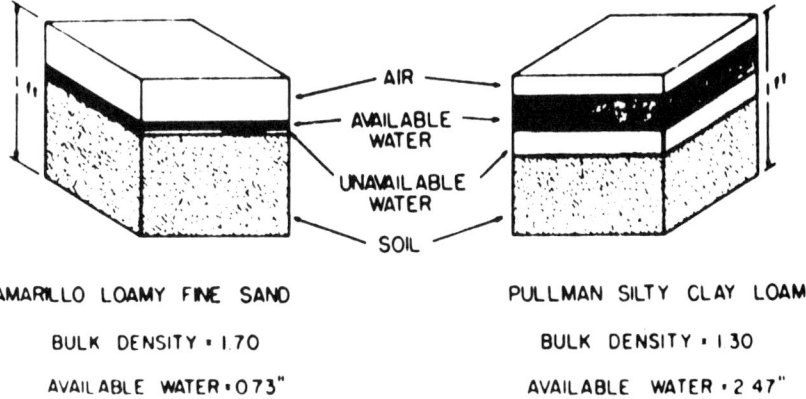

FIGURE 5. The water-holding capacity of two typical soils. (Reproduced from Wadleigh, C. H., Raney, W. H., and Herschfield, D. M., *Plant Environment and Efficient Water Use*, 1966, pages 1—19 and 73—94, by permission of the American Society of Agronomy and Soil Science Society of America.

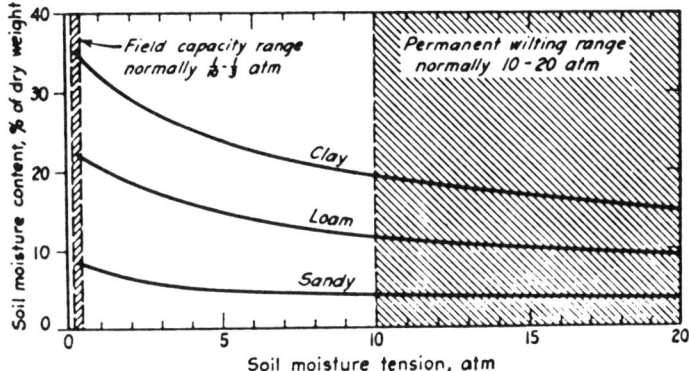

FIGURE 6. Typical soil moisture characteristic curves indicating field capacity and permanent wilting point range. (From Hansen, V. E., Israelson, O. W., and Stringham, G. E., *Irrigation Principles and Practices*, 4th ed., John Wiley & Sons, New York, 1980. With permission.)

tenth or one third bar tension is usually used as the estimate for field capacity; however, better measurement can be made from field sampling.

The constant used as the lower limit of plant water availability is termed either "wilting percentage" or "permanent wilting percentage". This is defined as the moisture content of a soil, on an oven-dry basis, at which plants (specifically sunflower) wilt and fail to recover their turgidity when placed in a dark humid atmosphere. Again this is not a real constant on the retention curve; however, because the curve becomes flatter as the tension increases, a more unique value can be defined. The usual tension used to estimate the wilting percentage of a soil is 15 bars. The water content of the soil between the upper limit, field capacity, and the lower limit, wilting percentage, is considered available to plants. The amount varies with the soil type (Figures 5 through 8).

Tables 6 and 7 show moisture characteristics of some typical soil types.

Figure 9 depicts three hypotheses as to the availability of soil water to the plant which have been postulated and accepted by some scientists at some time. However all these are

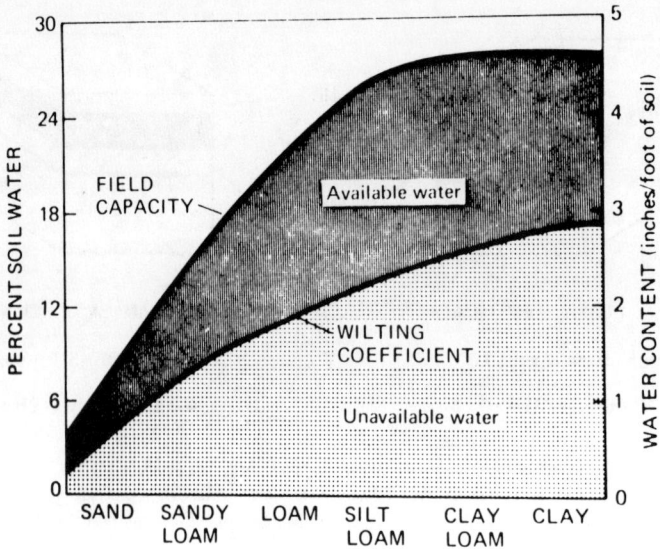

FIGURE 7. General relationship between soil moisture characteristics and soil texture. Note field capacity increases with finer texture until the silt loams, then levels off, whereas the wilting coefficient tends to continue to increase. (Reprinted with permission of the Macmillan Publishing Co., Inc. from Brady, N. C., *The Nature and Properties of Soils,* 8th ed., Copyright © 1974 by Macmillan Publishing Co., Inc.)

generalized conclusions, and in reality none are correct, since the plant is subjected not only to the water supply in the soil, but also to the evaporative demand of the atmosphere.

Plant Roots as They Affect the Soil Water Supply

Plant roots penetrating through the soil absorb the water and transport it to the xylem of the plant. The depth and extent to which roots penetrate into the soil profile will in part determine the supply of water available to the plant and the effectiveness with which this supply can be utilized. A realization as to the extent of this root penetration can be gained from the fact that total root length of individual plants can be measured in kilometers, and the number of root hairs counted in millions or billions. A single wheat plant has been found to possess as much as 71 km of roots, while a single rye plant, growing in a container 30.5 cm square and 56 cm deep produced 623 km in 4 months — an average of 5 km/day. The total root length was divided among some 13 million roots; in addition, there were some 14 billion root hairs providing surface for water absorption (Table 8).

Plant species vary in their rooting characteristics, ranging from a deep tap root to a shallower fibrous system (Figure 10). Root density (root length per unit volume) is a factor relating to the effectiveness of a plant in utilizing the soil water supply (Figures 11 and 12). Soils may have an impeding layer limiting the penetration of roots and thus decrease the total water supply. A shallow water table can also limit the depth of rooting. The moisture content of the soil affects the depth of rooting (Figure 13), as does its fertility status. With annual crops such as maize (corn) the roots develop quite rapidly until tasselling, or the reproductive stage, then almost cease further extension (Table 9). The proportion of total water extracted decreases with depth (Figure 14).

In Table 10 are generalized data on rooting depths for some typical crops as well as the fraction of available soil water, and the amount of water readily available for different soil types when evapotranspiration is 5 to 6 mm/day.

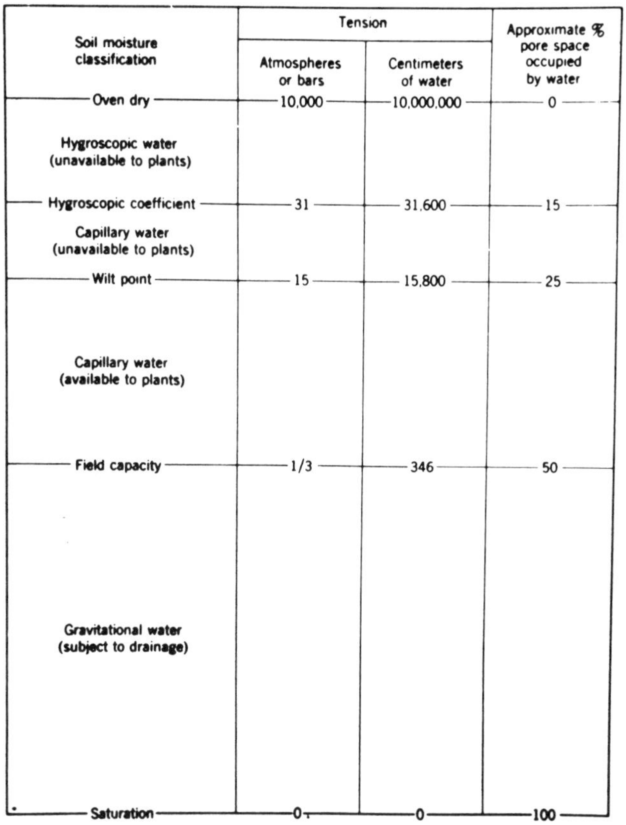

Soil moisture classification	Tension - Atmospheres or bars	Tension - Centimeters of water	Approximate % pore space occupied by water
Oven dry	10,000	10,000,000	0
Hygroscopic water (unavailable to plants)			
Hygroscopic coefficient	31	31,600	15
Capillary water (unavailable to plants)			
Wilt point	15	15,800	25
Capillary water (available to plants)			
Field capacity	1/3	346	50
Gravitational water (subject to drainage)			
Saturation	0	0	100

FIGURE 8. Diagram showing soil moisture classification, soil moisture tension equivalents, and the approximate percentage of the soil pore space occupied by water at various tensions. (From Foth, H. D., *Fundamentals of Soil Science*, 6th ed., John Wiley & Sons, New York, 1978. With permission.)

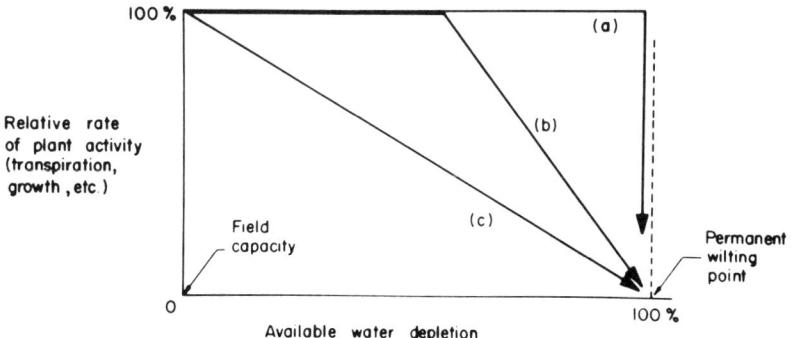

FIGURE 9. Three classical hypotheses regarding the availability of soil water to plants: (a) equal availability from field capacity to wilting point, (b) equal availability from field capacity to a ''critical moisture'' beyond which availability decreases, and (c) availability decreases gradually as soil moisture content decreases. (From Hillel, D., *Applications of Soil Physics*, Academic Press, New York, 1980. With permission.)

Table 6
REPRESENTATIVE PHYSICAL PROPERTIES OF SOILS

Soil texture	Infiltration[a] and permeability (cm/hr) I_f	Total pore space (%) N	Apparent specific gravity A_s	Field capacity (%) FC	Permanent wilting (%) PW	Total available moisture[b] Dry weight (%) P_w = FC − PW	Volume (%) $P_v = P_w A_s$	cm/m d = P_w /100 A_sD
Sandy	5 (2.5—25)	38 (32—42)	1.65 (1.55—1.80)	6 (6—12)	4 (2—6)	5 (4—6)	8 (6—10)	8 (7—10)
Sandy loam	2.5 (1.3—7.6)	43 (40—47)	1.50 (1.40—1.60)	14 (10—18)	6 (4—8)	8 (6—10)	12 (9—15)	12 (9—15)
Loam	1.3 (0.8—2.0)	47 (43—49)	1.40 (1.35—1.50)	22 (18—26)	10 (8—12)	12 (10—14)	17 (14—20)	17 (14—19)
Clay loam	0.8 (0.25—1.5)	49 (47—51)	1.35 (1.30—1.40)	27 (23—31)	13 (11—15)	14 (12—16)	19 (16—22)	19 (17—22)
Silty clay	.25 (.03—0.5)	51 (49—53)	1.30 (1.30—1.40)	31 (27—35)	15 (13—17)	16 (14—18)	21 (18—23)	21 (18—23)
Clay	0.5 (.01—0.1)	53 (51—55)	1.25 (1.20—1.30)	35 (31—39)	17 (15—19)	18 (16—20)	23 (20—25)	23 (20—25)

Note: Normal ranges are shown in parentheses.

[a] Intake rates vary greatly with soil structure and structural stability, even beyond the normal ranges shown above.
[b] Readily available moisture is approximately 75% of the total available moisture.

From Hansen, V. E., Israelsen, O. W., and Stringham, G. E., *Irrigation Principles and Practices*, 4th ed., John Wiley & Sons, New York, 1980. With permission.

Table 7
RELATION BETWEEN SOIL-WATER TENSION IN BARS AND AVAILABLE SOIL WATER IN mm/m SOIL DEPTH FOR SOILS OF VARIOUS TEXTURES

	\multicolumn{4}{c}{Soil-water tension in bars}
	0.2	0.5	2.5	15
Soil texture	\multicolumn{4}{c}{Available soil water (mm/m)}
Heavy clay	180	150	80	0
Silty clay	190	170	100	0
Loam	200	150	70	0
Silt loam	250	190	50	0
Silty clay loam	160	120	70	0
Fine-textured soils	200	150	70	0
Sandy clay loam	140	110	60	0
Sandy loam	130	80	30	0
Loamy fine sand	140	110	50	0
Medium-textured soils	140	100	50	0
Medium fine sand	60	30	20	0
Coarse-textured soils	60	30	20	0

From Doorenbos, J. and Pruitt, W. O., Crop Water Requirements, Irrigation Drainage Paper No. 24, Food and Agriculture Organization, Rome, 1977. With permission of the Food and Agriculture Organization of the United Nations.

Table 8
TOTAL LENGTH OF ROOTS AND ROOT HAIRS, AND TOTAL SURFACE OF ROOT OF A LARGE 4-MONTH-OLD RYE PLANT GROWN IN A BOX OF SOIL 30.5 cm SQUARE AND 56 cm DEEP

	Miles	Meters	Calculated volumes (cc)
Total lengths of roots	387.17	623,089.4	7529.7
Total lengths of root hairs	6603.86	10,627,751.9	1180.8
Roots plus root hairs	6991.03	16,850,841.3	8710.5

	ft²	m²	Ratio total surface underground/ total surface aboveground
Total root surface	2554.09	237.375	
Total hair surface	4321.31	401.450	
Total roots plus hairs	6875.40	638.825	
Total leaf surface (including stems)	50.0	4.645	137.5:1

From Curtis, O. F. and Clark, D. G., *An Introduction to Plant Physiology*, McGraw-Hill, New York, 1960, 212. With permission.

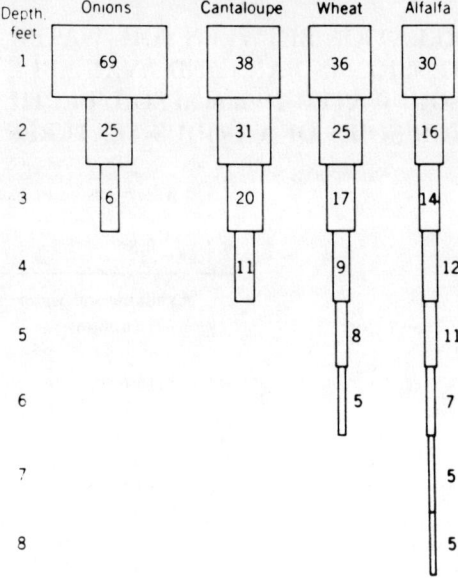

FIGURE 10. Moisture extraction patterns of various crops with different rooting systems. (From Technical Bulletin 69, Arizona Agricultural Experiment Station, Tucson, 1965.

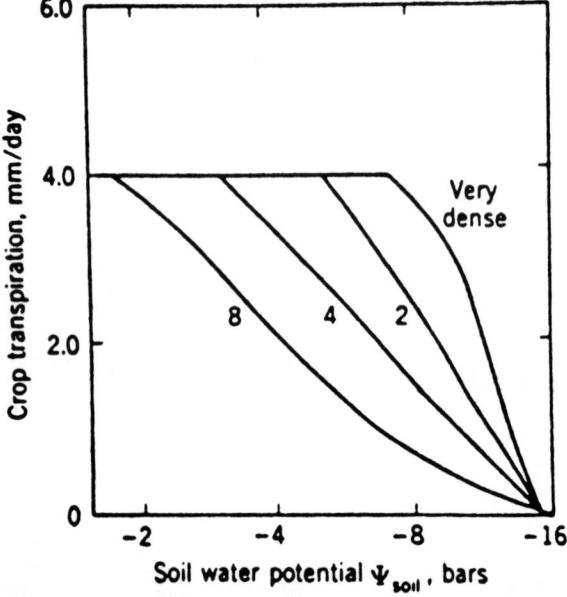

FIGURE 11. Effects of root density on the expected relationship between daily transpiration of a crop and soil water potential, assuming an intermediate rate of potential transpiration. Soil water potential is less limiting at high root densities than at low densities. Root densities are 8, 4, and 2 cm³ of soil per centimeter of root length. (From Crown, I. R., *J. Appl. Ecol.*, 2, 221, 1965. With permission.)

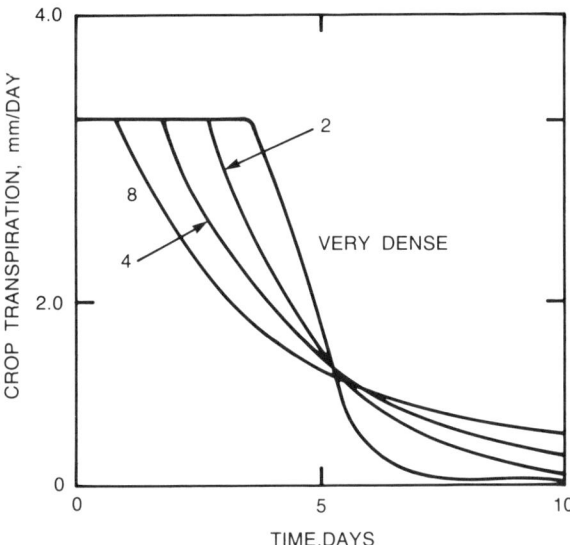

FIGURE 12. The expected decrease with time in transpiration of a crop having the root densities of 8, 4, and 2 cm³ of soil per centimeter of root length. At high root densities transpiration is maintained at a potential rate for a longer period, but it then falls more rapidly than for plants with low root density which remove soil water more slowly. (From Crown, I. R., *J. Appl. Ecol.*, 2, 221, 1965. With permission.)

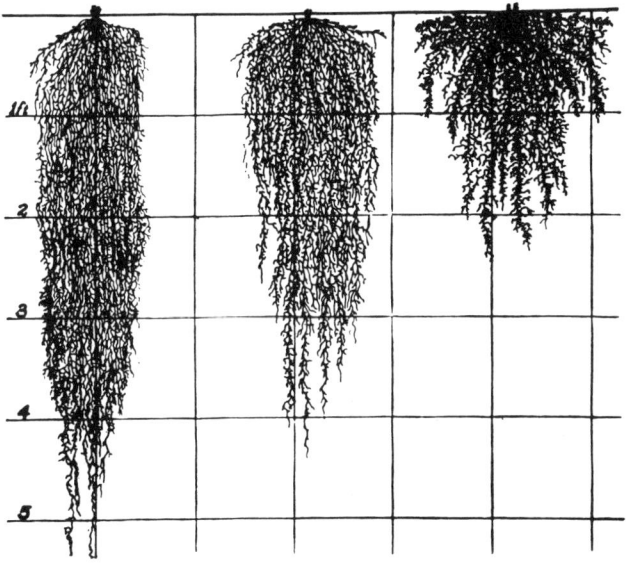

FIGURE 13. Effect of soil moisture on root development. Winter wheat roots in a fertile silt loam under a precipitation of 66 to 81 cm (left), 53 to 61 cm (center), and 41 to 48 cm (right). (From Weaver, J. E. and Clements, F. E., *Plant Ecology*, 2nd ed., McGraw-Hill, New York, 1938, 299. With permission.)

Table 9
WATER EXTRACTION FROM THE SOIL PROFILE AT DIFFERENT DEPTHS DURING THE GROWING SEASON. VALUES FOR EACH DATE ARE GIVEN AS THE PERCENTAGE OF EVAPORATION (E) OR EVAPOTRANSPIRATION (ET) THAT OCCURS FROM EACH OF THE DEPTHS LISTED

Dates	% of E or ET which comes from respective depths	Depths from which water was extracted
To June 7	100	1st 6 in.
June 8 to 14	100	1st ft (equally from each 6 in.)
June 15 to 27	67.7, 33.3	1st, 2nd ft
June 28 to July 4	60, 20, 20	1st, 2nd, and top half of 3rd ft
July 5 to 11	60, 20, 20	1st—3rd ft
July 12 to 18	60, 15, 15, 10	1st—3rd and top half of 4th ft
July 19 to 25	60, 15, 15, 10	1st—4th ft
July 26 to August 1	60, 10, 10, 10, 10[a]	1st—4th and upper half of 5th ft
After August 1	60, 15, 15, 10[b]	1st—4th ft
	60, 10, 10, 10, 10[a]	1st—5th ft
	60, 15, 15, 10[b]	1st—4th ft

[a] Used only if first 4 ft all have <50% available water.
[b] Used if any of first 4 ft have >50% available water; however, after August 1, the percent available is always computed on the total available water in the 5-ft profile.

Reproduced from Shaw, R. H. and Laing, D. R., in *Plant Environment and Efficient Water Use*, Pierre, W. H., Ed., 1966, pages 1—19 and 93—94, by permission of the American Society of Agronomy and the Soil Science Society of America.

In addition to the roots permeating the soil as a factor in supplying water to the plant, as water is absorbed by the roots it moves along an energy gradient from the moist soil toward the roots. The rate of this movement is referred to as hydraulic conductivity and is an expression of the readiness with which a liquid such as water flows through the soil in response to a given potential gradient. This hydraulic conductivity is also dependent on the soil texture and moisture content (Figure 15). At saturation the rate of movement is highest in sand, then silt, and finally clay; however, as the water content decreases to unsaturated flow the order reverses to clay, silt, and sand from highest to lowest. In Figure 16, the relationships between soil water potential, hydraulic conductivity, and volumetric water content for a clay soil is shown.

The Evaporative Demand, or Atmospheric Sink

The driving force for the hydrologic cycle is the energy from the sun. It requires almost 600 cal of energy to evaporate 1 g of water at 20°C. Evaporation plays an important role in the energy balance (Figure 17). In addition to energy necessary to accomplish the change in state of water from a liquid to a vapor, for evaporation to occur there must be a vapor pressure gradient with water moving from high to lower. Vapor pressure of water increases with temperature (Table 11). For example, the vapor pressure of water at 20°C is 17.54 mm of mercury, whereas at 30°C it is 31.82. This temperature-vapor pressure relationship is important to the evaporation process. The amount of moisture in the air is often expressed as "relative humidity", which is the ratio of the actual vapor pressure to that at saturation. The vapor pressure at a free water surface is at or near saturation. For a 100% relative humidity the vapor pressure at a free water surface and that of the air is the same; however, as air temperature increases the vapor pressure for saturation increases thus the relative

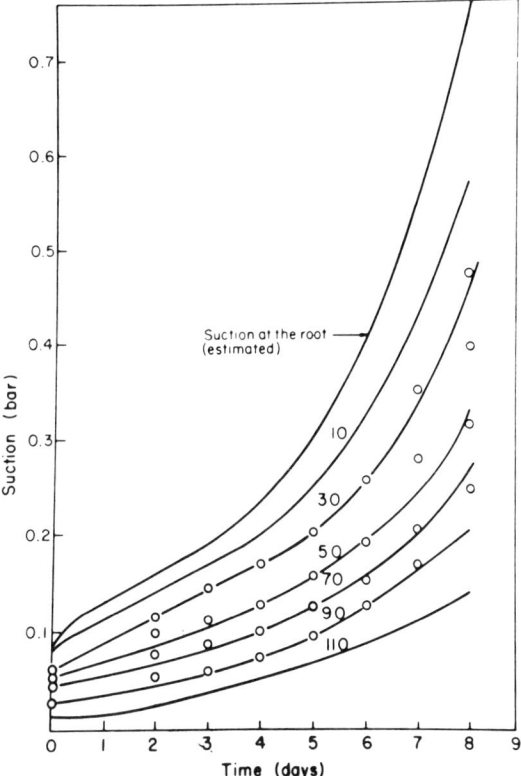

FIGURE 14. The increase of average soil-water suction with time at different depths for alfalfa. The numbers on the curves represent depths in centimeters within the rooting zone. (Reproduced from Gardner, W. R., *Agronomy Journal*, 56, 35, 1964, by permission of the American Society of Agronomy, Inc.)

Table 10
GENERALIZED DATA OF ROOTING DEPTH OF FULL-GROWN CROPS, FRACTION OF AVAILABLE SOIL WATER, AND READILY AVAILABLE SOIL WATER FOR DIFFERENT SOIL TYPES (mm/m SOIL DEPTH) WHEN EVAPOTRANSPIRATION(ET) OF THE CROP IS 5 TO 6 mm/DAY

Crop	Rooting depth (m)	Fraction of available soil water	Readily available soil water (mm/m)		
			Fine	Medium	Coarse
Alfalfa	1.0—2.0	0.55	110	75	35
Bananas	0.5—0.9	0.35	70	50	20
Beans	0.5—0.7	0.45	90	65	30
Carrots	0.5—1.0	0.35	70	50	20
Citrus	1.2—1.5	0.5	100	70	30
Clover	0.6—0.9	0.35	70	50	20
Cotton	1.0—1.7	0.65	130	90	40
Deciduous orchards	1.0—2.0	0.5	100	70	30
Grains, small	0.9—1.5	0.6	120	80	40
Grains, winter	1.5—2.0	0.6	120	80	40
Grass	0.5—1.5	0.5	100	70	30
Groundnuts	0.5—1.0	0.4	80	55	25

Table 10 (continued)
GENERALIZED DATA OF ROOTING DEPTH OF FULL-GROWN CROPS, FRACTION OF AVAILABLE SOIL WATER, AND READILY AVAILABLE SOIL WATER FOR DIFFERENT SOIL TYPES (mm/m SOIL DEPTH) WHEN EVAPOTRANSPIRATION(ET) OF THE CROP IS 5 TO 6 mm/DAY

Crop	Rooting depth (m)	Fraction of available soil water	Readily available soil water (mm/m) Fine	Medium	Coarse
Maize	1.0—1.7	0.6	120	80	40
Melons	1.0—1.5	0.35	70	50	25
Palm trees	0.7—1.1	0.65	130	90	40
Peppers	0.5—1.0	0.25	50	35	15
Potatoes	0.4—0.6	0.25	50	30	15
Safflower	1.0—2.0	0.6	120	80	40
Sorghum	1.0—2.0	0.55	110	75	35
Soybeans	0.6—1.3	0.5	100	75	35
Sugar beet	0.7—1.2	0.5	100	70	30
Sugarcane	1.2—2.0	0.65	130	90	40
Sunflower	0.8—1.5	0.45	90	60	30
Sweet potatoes	1.0—1.5	0.65	130	90	40
Tomatoes	0.7—1.5	0.4	180	60	25
Vegetables	0.3—0.6	0.2	40	30	15
Wheat	1.0—1.5	0.55	105	70	35
Wheat (ripening)		0.9	180	130	55
Total available soil water			200	140	60

Note: When ET of the crop is 3 mm or smaller increase values by some 30%; when ET of the crop is 8 mm/day or more reduce values by some 30%, assuming nonsaline conditions (EC 2 mmhos/cm). Higher values than those shown apply during ripening.

From Doorenbos, J. and Pruitt, W. O., Crop Water Requirements, Irrigation Drainage Paper No. 24, Food and Agriculture Organization, Rome, 1977. With permission of the Food and Agriculture Organization of the United Nations.

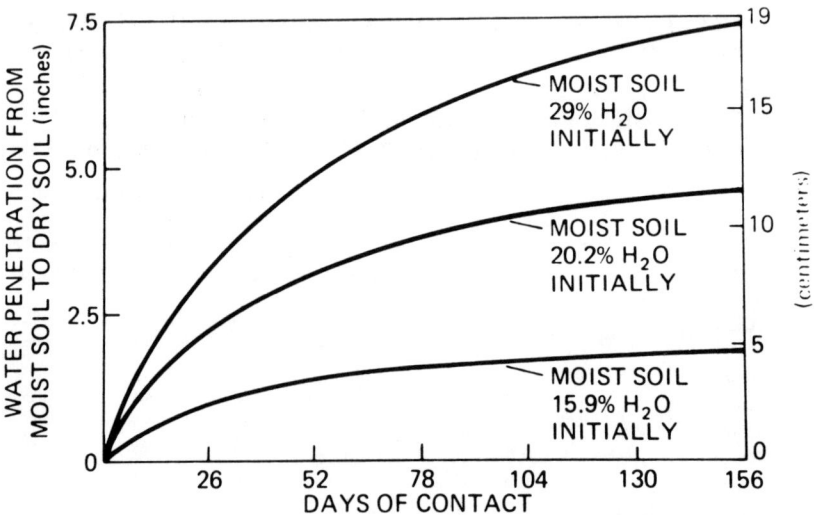

FIGURE 15. Rate of water movement from moist soils at three moisture levels to a drier soil. The higher the water content of the moist soil, the greater will be the tension gradient and the more rapid the delivery. (From Gardner, W. and Widtsol, J. A., *Soil Sci.*, 11, 230, 1921. With permission.)

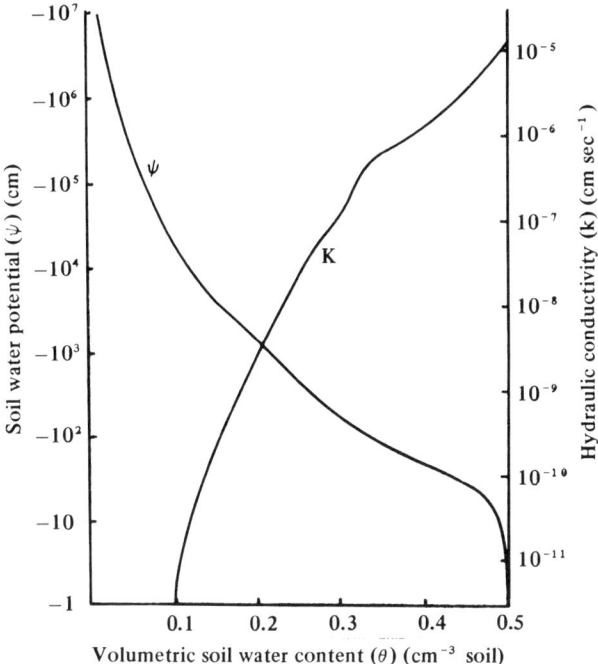

FIGURE 16. Soil water potential and hydraulic conductivity, K, plotted against volumetric water content for a Yolo light clay soil. (From Philip, J. R., *J. Meteorol.*, 14, 354, 1957. With permission of the American Meteorological Society.)

humidity decreases, and a vapor pressure gradient develops from the free water surface to the atmosphere. Further, the energy from the sun absorbed by the soil, free water, or a plant canopy, will cause a rise in their temperature to the extent it will be somewhat higher than the surrounding air. As air temperature rises from early morning to midday, the vapor presure of the air does not change significantly. However, the vapor pressure at the free water surface will remain at saturation plus the increase in vapor pressure caused by the increase in temperature. The resulting vapor pressure gradient promotes evaporation. The effect of this difference in temperature on vapor pressure gradient is shown graphically in Figure 18. Figure 19 shows the disposition of energy with varying leaf-air temperature differences.

Wind has an affect on the rate of evaporation and transpiration as does the roughness of the plant canopy by a mixing action of the air thus providing a continual vapor pressure gradient. However this effect is essentially dissipated at relatively low wind velocities as shown in Figure 20. There is some evidence that winds of higher velocity, 16 to 20 km/hr tend to decrease the transpiration rate.

Transpiration rates vary with weather conditions, such as a clear or cloudy, hot or cool, humid or dry day, and also with soil moisture content as shown in Figures 21 and 22.

The potential evapotranspiration rate is attained only when the soil moisture supply is not limiting. Table 12 gives some average water uses per day for maize (corn) through the growing season at various temperatures. Table 13 is a generalized estimate of daily evapotranspiration rates considering latitude, month, and different weather conditions. Evapotranspiration rates increase through the season to approximately the reproductive stage of a crop then levels and decreases toward maturity.

Thus the plant is subjected to two environments: a slow and sluggish water supply provided by the soil and the very turbulent demanding environment of the atmosphere. Water is first

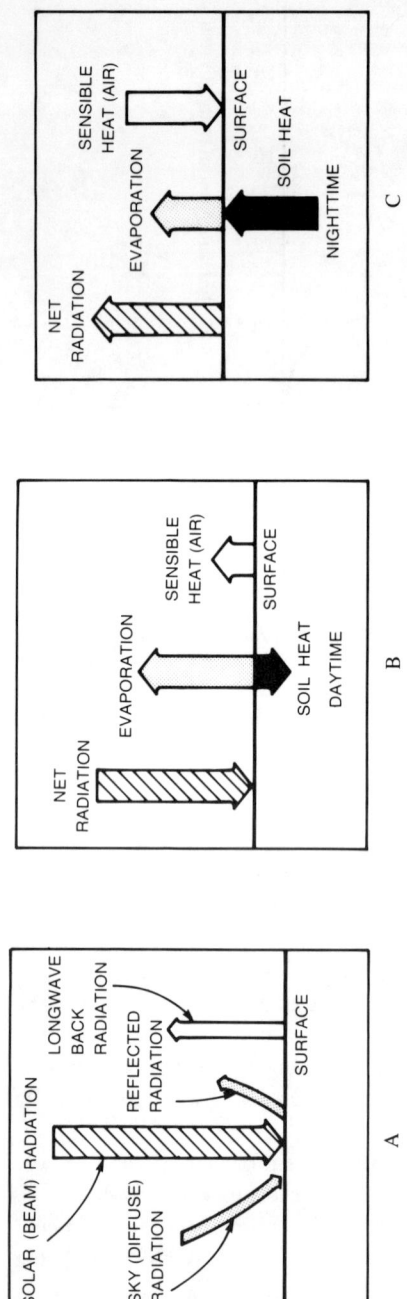

FIGURE 17. Schematic representation of (A) the radiation balance and (B) the daytime and (C) the nighttime energy balance. (Net radiation = [solar radiation + sky radiation] − [reflected radiation + back radiation]). (From Tanner, C. B., in *Water Deficits and Plant Growth*, Vol. 1, Kozlowski, T. T., Ed., Academic Press, New York, 1968, 84. With permission.)

Table 11
RELATION BETWEEN VAPOR PRESSURE AND RELATIVE HUMIDITY AT DIFFERENT TEMPERATURES[17]

Vapor pressure (mm Hg) at different values of relative humidity

Temp. (°C)	0	10%	20%	30%	40%	50%	60%	70%	80%	90%	100%
0	0	0.46	0.92	1.37	1.83	2.29	2.75	3.21	3.66	4.12	4.58
10	0	0.92	1.84	2.76	3.68	4.60	5.53	6.45	7.37	8.29	9.21
20	0	1.75	3.51	5.26	7.02	8.77	10.52	12.28	14.03	15.79	17.54
30	0	3.18	6.36	9.55	12.73	15.91	19.09	22.27	25.46	28.64	31.82
40	0	5.53	11.06	16.60	22.13	27.66	33.19	38.72	44.25	49.79	55.32

From Devlin, R. M., *Plant Physiology*, Reinhold, New York, 1968. With permission.

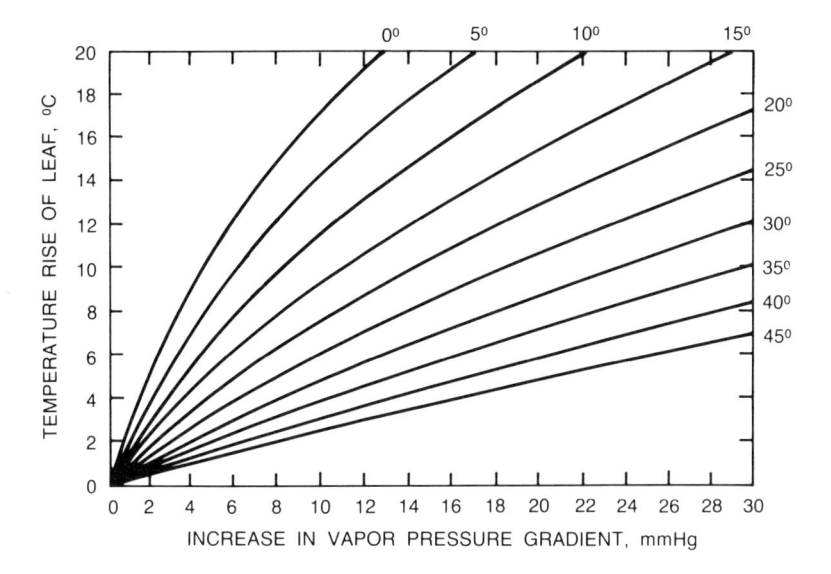

FIGURE 18. Effect of raising leaf temperature above air temperature on increase in vapor pressure difference between leaf and air, expressed as millimeters of mercury. (From Curtis, O. F. and Clark, D. G., *An Introduction to Plant Physiology*, McGraw-Hill, New York, 1950, 197. With permission.)

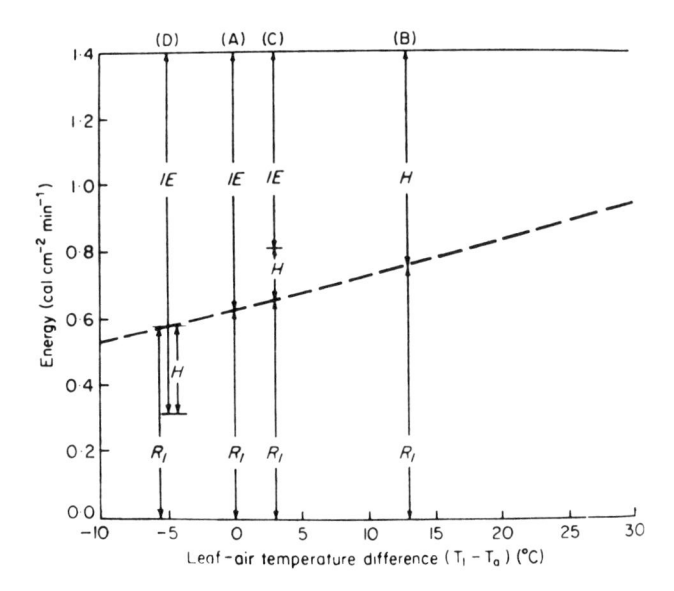

FIGURE 19. Diagram showing estimated energy exchange by transpiration (IE), radiation (R), and by sensible heat transfer (H) for a leaf 10 cm wide at 25°C with wind velocity of 200 cm/sec and energy absorption of 1.4 cal/cm^2/min. centimeter per minute. Data are for one surface only of a leaf exposed to intense radiation. (From Slatyer, R. O., *Plant-Water Relationships*, Academic Press, New York, 1967. With permission.)

absorbed from the soil by root hairs and other epidermal cells in or near the root hair zone. It then moves through the cortex tissue and across the endodermis and pericycle of the roots and finally into the xylem ducts. The xylem tissue of the roots connects directly with the xylem tissue of the stem, thus allowing water to move out of the root and into the stem.

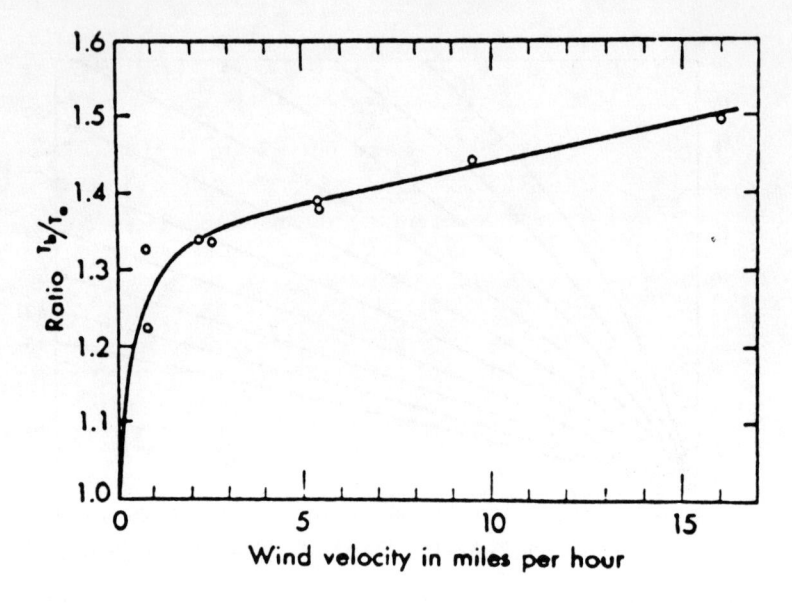

FIGURE 20. The effect of wind velocity on transpiration. T_b, actual transpiration rate; T_a, potential or rate at saturated vapor pressure. (From Martin, E. V. and Clements, F. E., *Plant Physiol.*, 10, 613, 1935. With permission.)

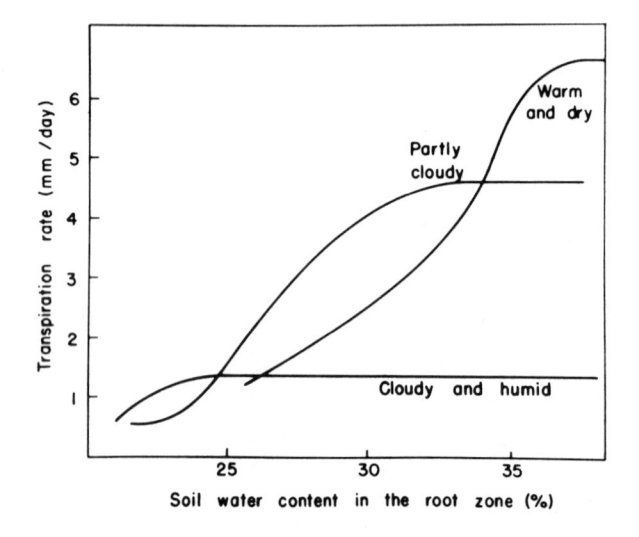

FIGURE 21. Relation of actual transpiration rate to soil-water content under different meteorological conditions. (Reproduced from Denmead, O. T. and Shaw, R. H., *Agronomy Journal*, 54, 385—390, 1962, by permission of the American Society of Agronomy, Inc.)

The xylem of the stem is divided and subdivided many times to form a complex network of water-conducting tissues, finally ending in the fine veins of the leaf. Water moves from the leaf veins into the mesophyll cells, is evaporated from their surfaces, and finally moves as water vapor through the stomates into the surrounding atmosphere.

The overall significance of evapotranspiration in the water balance for a humid, semiarid, and arid region is portrayed in Figure 23.

For the arid region irrigation is introduced which increases the water use almost to its potential.

MOISTURE STRESS

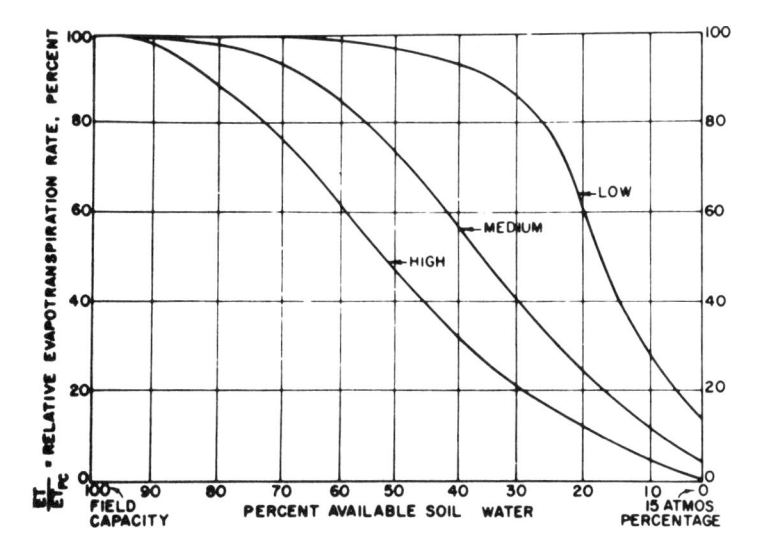

FIGURE 22. Ratio of actual evapotranspiration to evapotranspiration at field capacity at different levels of soil water for three atmospheric demand conditions. (From Shaw, R. H., Estimation of Soil Moisture Under Corn, Bull. 520, Iowa Agricultural and Home Economics Experiment Station, 1963.

The Soil-Plant-Atmosphere Continuum

Formerly, the status of the water supply (that is, the soil moisture) was used as the prime criteria in evaluating plant response to water deficits. In more recent years with the development of instruments capable of measuring water potentials in the plant, the concept of a soil-plant-atmosphere continuum was visualized. This takes into account the entire system and recognizes plant responses to a limited water supply and a limitless atmospheric sink or demand.

In the literature it has been frequently implied that only when water loss by transpiration exceeds water absorption by the roots does a water deficit occur in the plant. This does not take into account that during the period when the plant is undergoing wilting or a water deficit is being established the transpiration is greater than absorption, while during a recovery period absorption is greater than transpiration, and once a deficit develops and is maintained transpiration will nearly equal absorption. However, whenever the atmospheric demand or sink is greater than the water supply capabilities, a deficit will persist. A plant may experience some degree of water deficit in relatively moist soil simply because of the excessive evaporative demand and the limited capabilities of the root to absorb sufficient water to maintain cell turgidity.

The Energy System in the Plant

As water is transported from the soil to the atmosphere through the plant it moves along an energy gradient, from higher to lower. The total energy status within the plant is comprised of osmotic pressure, turgor pressure, and matrix potential. The osmotic pressure is due to solutes, both organic and inorganic substances, in solution, while turgor pressure is the pressure exerted by the cell walls. The matrix potential has been considered only recently by scientists as a component of the energy status of water in plants, and consists of the effect of colloidal hydration, adsorption, and capillary phenomena which is the matrix affinity of the tissue for water. These energy components are not necessarily independent of one another and hence not additive. A change in one brings about a change in another; for

Table 12
AVERAGE WATER USE FOR MAIZE (CORN) DURING THE GROWING SEASON CONSIDERING DIFFERENT TEMPERATURE RANGES; CENTIMETERS PER DAY

Week after emergence	4	5	6	7	8	9	10	11	12	13	14	15	16	17
Date														
Temperature														
10—15°C	0.10	0.13	0.18	0.20	0.23	0.23	0.23	0.23	0.23	0.23	0.20	0.20	0.15	0.13
(50—59°F)	(0.04)	(0.05)	(0.07)	(0.08)	(0.09)	(0.09)	(0.09)	(0.09)	(0.09)	(0.09)	(0.08)	(0.08)	(0.06)	(0.05)
15.6—20.6°C	0.18	0.23	0.28	0.33	0.38	0.38	0.38	0.36	0.36	0.36	0.33	0.33	0.23	0.18
(60—69°F)	(0.04)	(0.09)	(0.11)	(0.13)	(0.15)	(0.15)	(0.15)	(0.14)	(0.14)	(0.14)	(0.13)	(0.13)	(0.09)	(0.07)
21.1—26.1°C	0.25	0.30	0.38	0.48	0.56	0.56	0.56	0.51	0.51	0.51	0.46	0.43	0.28	0.23
(70—79°F)	(0.10)	(0.12)	(0.15)	(0.19)	(0.22)	(0.22)	(0.22)	(0.20)	(0.20)	(0.20)	(0.18)	(0.17)	(0.11)	(0.09)
26.7—31.7°C	0.30	0.38	0.46	0.58	0.64	0.66	0.66	0.61	0.61	0.61	0.56	0.53	0.36	0.30
(80—89°F)	(0.12)	(0.15)	(0.18)	(0.23)	(0.25)	(0.26)	(0.26)	(0.24)	(0.24)	(0.24)	(0.22)	(0.22)	(0.14)	(0.12)
32.2—37.2°C	0.33	0.43	0.53	0.66	0.74	0.76	0.76	0.74	0.74	0.74	0.66	0.66	0.63	0.36
(90—99°F)	(0.13)	(0.17)	(0.21)	(0.26)	(0.29)	(0.30)	(0.30)	(0.29)	(0.29)	(0.29)	(0.26)	(0.25)	(0.17)	(0.14)

Note: Values in cm/day; values in parentheses, in./day.

From Lundstrom, D. R., Stegman, E. C., and Werner, H. C., in *Irrigation Scheduling for Water Energy Conservation in the 80s, Proc. Agricultural Engineers Irrigation Scheduling Conf.*, American Society of Agricultural Engineers, St. Joseph, Mich., 1981, 187. With permission.

Table 13
ESTIMATED DAILY EVAPOTRANSPIRATION RATES FOR MAIZE (CORN) BY LATITUDE, MONTH OF THE GROWING SEASON, AND DIFFERENT WEATHER CONDITIONS

	Daily evapotranspiration rate (mm)		
Latitude and month	Dull cloudy weather	Normal weather	Bright weather
Between 48° and 40°N			
April and September	1.5	2.3	3.3
May and August	1.8	3.0	4.6
June and July	3.0	4.2	5.6
Between 40° and 34°N			
April and September	2.0	2.8	3.6
May and August	2.8	3.6	4.8
June and July	3.6	4.3	5.8
Between 34° and 30°N			
April and September	2.3	3.3	4.1
May and August	3.3	4.1	5.6
June and July	3.6	4.3	5.8

From Jamison, V. C. and Beale, O. W., *U.S. Dep. Agric. Farmers' Bull.*, No. 2143, 1959.

example, an increase in turgor pressure is generally the consequence of the stretching of cell walls due to osmotic absorption of water, which reduces osmotic pressure through dilution of the cell sap. The interrelationship of osmotic pressure, turgor pressure, and the overall energy status of water in the plant is depicted in Figure 24. In Figure 25 this relationship is shown for a sunflower plant.

Plants respond to the energy status within their cells. In only recent years has instrumentation been available to make these energy measurements. The total water potential can be measured with a psychometer, hygrometer, or pressure chamber. To measure the osmotic potential, there are the thermocouple psychrometer, vapor pressure osmometer, cryoscopes, and, most commonly, the pressure chamber. The pressure chamber or thermocouple psychrometer is used for the measurement of matric potential. There are no direct measurements for turgor potential, hence this is usually calculated by difference.

The water potentials can be measured in each of the various plant components in its pathway through the plant. The potential gradients from the soil to the atmosphere are depicted in Figure 26. Note that by far the greatest "jump" in potential is from the leaf to the atmosphere.

Table 14 shows the relationship of negative energy potentials of atmospheric water at various relative humidities.

Assuming that the energy status of the water in leaf cells is at or near saturation, as the relative humidity of the air decreases from 100%, the energy gradient increases very rapidly. On a warm summer day, the relative humidity of the air is frequently 50% or lower. This terrific "jump" in negative potential has been used to support the concept that soil water could be equally available from "field capacity" to the "wilting percentage" in that 15 bars, the wilting percentage, is only a small percentage of the "drop" occurring in the entire system. The total drop in potential from soil to leaves is only about 10 to 30 bars.

Figure 27 illustrates the general distribution of water potential in the soil-plant-air continuum under different conditions of soil moisture and atmospheric demand. Curve 1 has a low value of water suction in the soil (AB) and hence at the root surfaces (B). In the mesophyll cells of plant leaves (DE), the negative potential is below the critical value at which leaves will lose their turgidity; hence no wilting. E represents the stomatal cavity. In

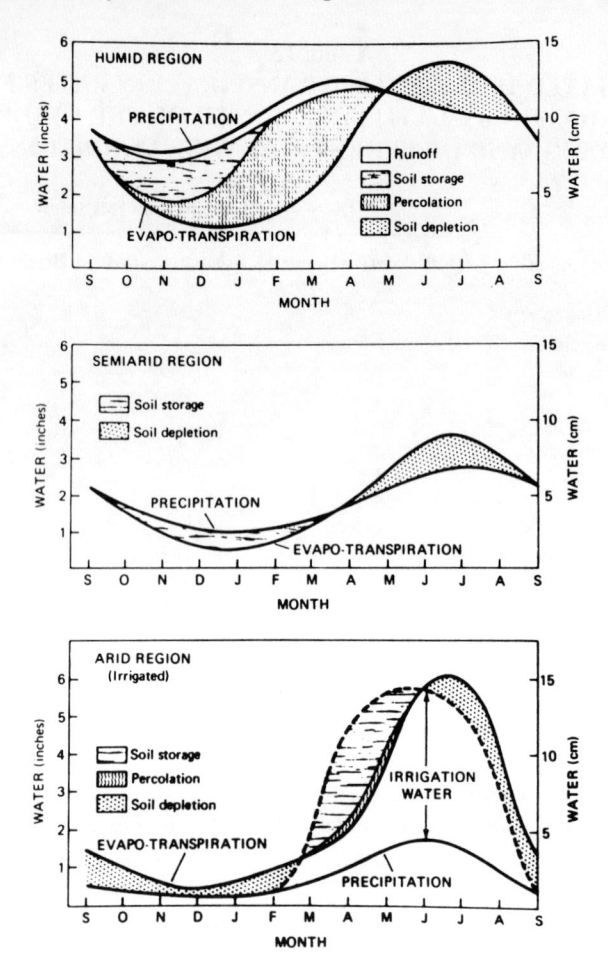

FIGURE 23. Generalized curves for precipitation and evapotranspiration for a humid region (upper), a semiarid region (middle), and an irrigated arid region (lower). Note the absence of percolation through the soil in the semiarid region. In each case water is stored in the soil. This moisture is released later when evapotranspiration demands exceed precipitation. In the semiarid region evapotranspiration would likely be much higher if ample soil moisture were available. In the irrigated arid region the very high evapotranspiration needs are supplied by irrigation. (Reprinted with permission of Macmillan Publishing Co., Inc. from Brady, N. C., *The Nature and Properties of Soils,* 8th ed. Copyright ©1974 by Macmillan Publishing Co., Inc.)

Curve 2, soil-water suction is also low, but the transpiration rate is higher and the negative potential in the leaf mesophyll cells approaches the critical wilting; for example, 30 bars. Curve 3 has a soil suction which is relatively high but the transpiration rate is low while Curve 4 demonstrates the extreme conditions where soil-water suction and transpiration are both high, and the leaf-water potential goes below the critical value, and thus the plant wilts.

The leaf-water potential decreases as the soil-water supply decreases. This relationship is shown in Figure 28. It also shows the diurnal effect with plant water content partially recovering during the dark or at night. Figure 29 shows the lowering of the leaf water potential with desiccation.

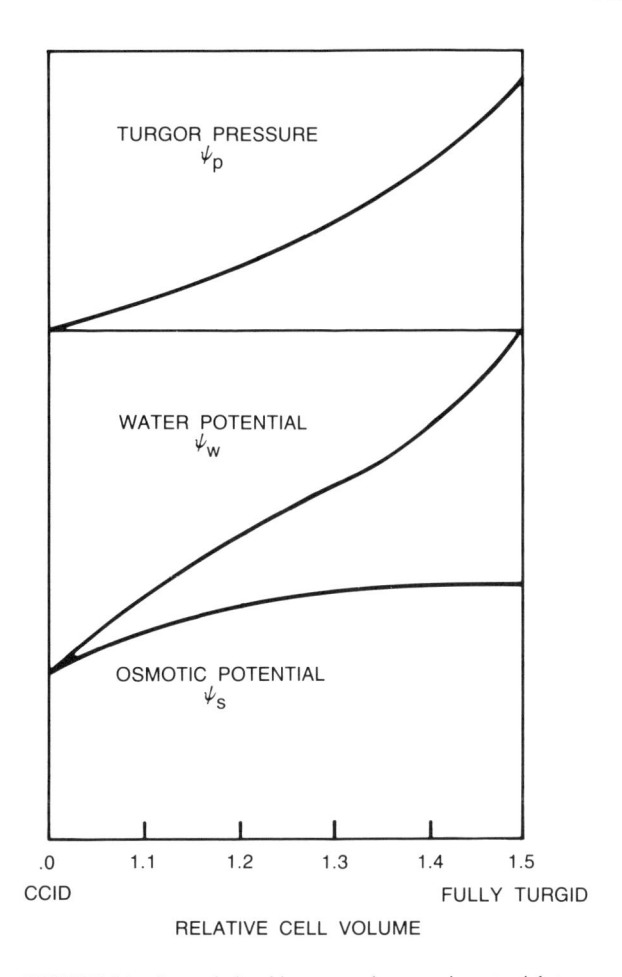

FIGURE 24. Interrelationship among the osmotic potential, turgor pressure, water potential, and cell volume of a plant cell. (From Hofler, K., *Ber. Dtsch. Bot. Ges.*, 38, 288, 1920. With permission.)

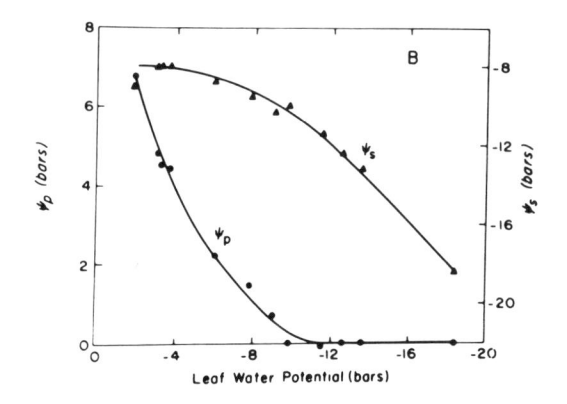

FIGURE 25. The relationship between leaf osmotic potential and leaf turgor in a sunflower having various leaf water potentials. (From Boyer, J. S., in *Water Deficits and Plant Growth, Vol. 4*, Kozlowski, T. T., Ed., Academic Press, New York, 1976, 169. With permission.)

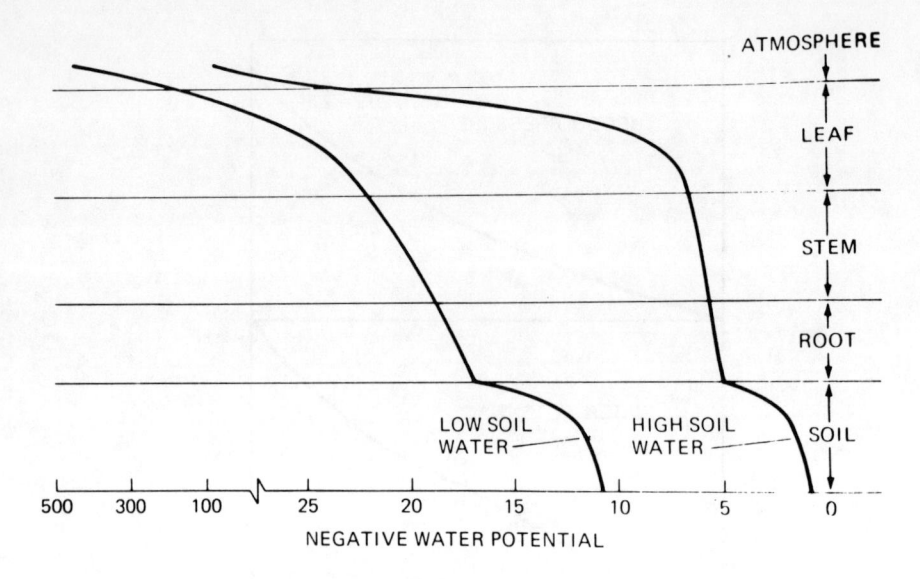

FIGURE 26. Changes in moisture potential as water moves from the soil through the root, stem, and leaf to the atmosphere. (From Hillel, D., *Soil and Water*, Academic Press, New York, 1971, 208. With permission.)

Table 14
RELATIONSHIP OF RELATIVE HUMIDITY, WATER VAPOR PRESSURE, AND EQUIVALENT POTENTIAL ENERGY

Relative humidity of air (%)	Absolute vapor pressure of water in air (mm Hg)	Energy potential of atmospheric water (in equilibrium bars)
100	17.54	0
99	17.36	13.6
98	17.19	27.2
97	17.01	41.1
95	16.66	69.3
90	15.79	142
80	14.03	302
50	4.60	1013
20	3.51	

As in the soil complex, as water moves into the roots through the xylem and into the leaves it encounters resistance to flow which is termed "hydraulic resistance". This, in theory, is the ratio of the flow-path length to its conductivity and its reciprocal is "conductance".

PLANT RESPONSES TO WATER DEFICITS

Stomatal Closures

The leaves of mesophytic plants are uniquely structured providing for small openings or pores called stomates which form a pathway from the mesophyll cells of the leaf to the atmosphere. Through these small openings carbon dioxide from the atmosphere diffuses into the leaf where it is combined with water in the photosynthetic process. The photosynthetic

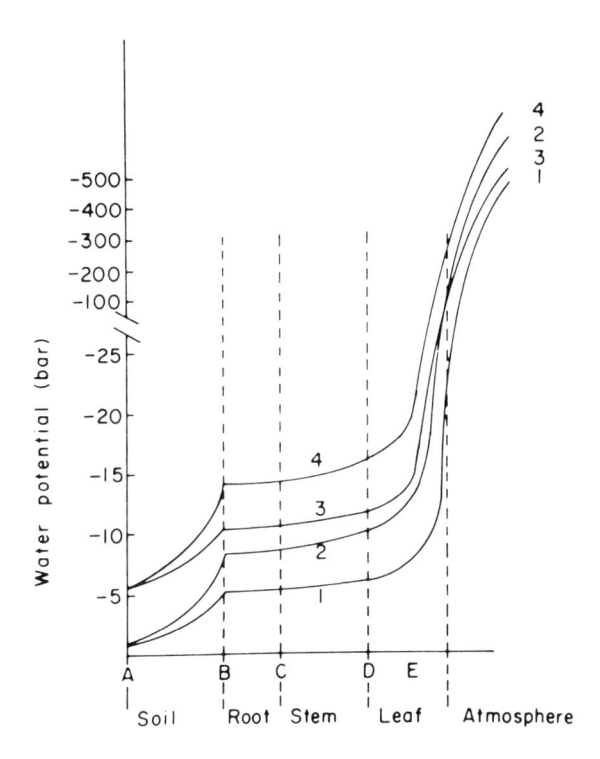

FIGURE 27. The distribution of potentials in the soil-plant-atmosphere continuum under different conditions of soil moisture and atmosphere evaporativity. (From Hillel, D., *Applications of Soil Physics*, Academic Press, New York, 1980. With permission.)

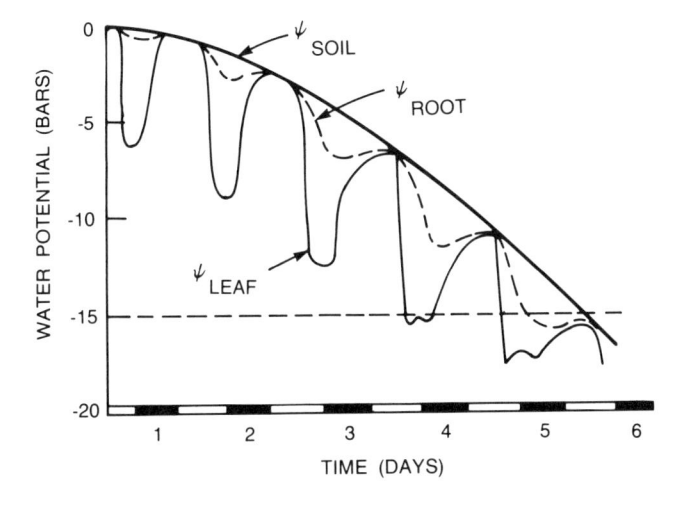

FIGURE 28. Schematic representation of changes in leaf water potential, root water potential, and soil water potential as transpiration proceeds during a drying cycle. The same evaporative conditions prevailed each day. The dark portion of the horizontal time line indicated night period and the dashed line at −15 bars is the potential at which wilting occurs. (From Slatyer, R. O., *Plant-Water Relationships*, Academic Press, New York, 1967. With permission.)

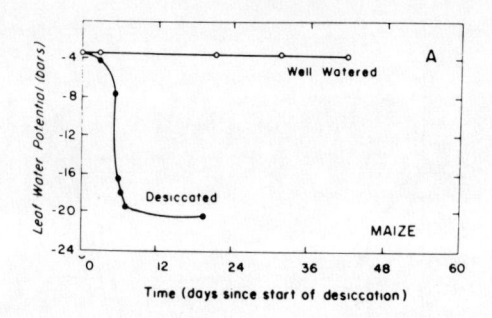

FIGURE 29. The leaf water potential of maize (corn) during a drying cycle in the grain filling period. (From Boyer, J. S., in *Water Deficits and Plant Growth,* Vol. 4, Kozlowski, T T., Ed., Academic Press, New York, 1976, 158. With permission.)

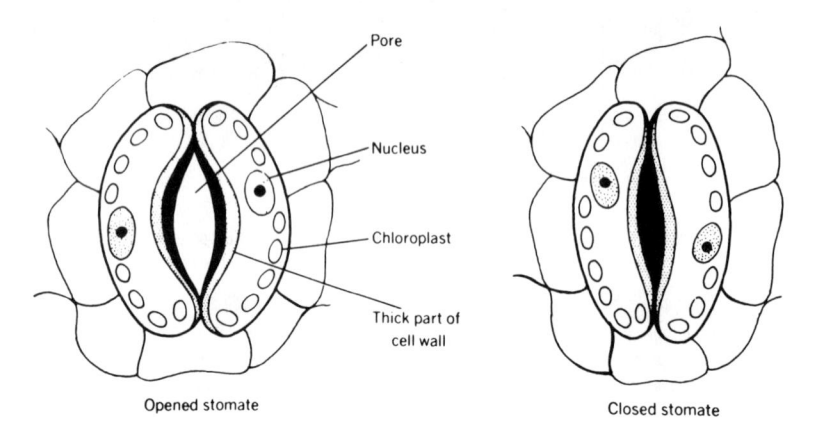

FIGURE 30. An open and closed stomatal pore showing the structure of the guard cells. (From Devlin, R. M., *Plant Physiology,* Reinhold, New York, 1968, 50. With permission.)

process is basically what the production of agronomic crops is all about. However as the carbon dioxide diffuses into the leaves, water which has been transported from the soil to the leaves literally escapes into the atmosphere. Figure 30 diagrammatically shows an open and a closed stomate; and Figure 31, the structure of the leaf in relation to a stomate. The stomates or pores are formed by "guard cells" which have enough flex to either open or close the stomates as shown in Figure 30. One of the factors determining the position of the guard cells is their moisture status, and they in turn respond to the overall moisture status within the plant. The plant thus has some control over the flow or loss of water by the opening or closing of its stomates. As the moisture supply becomes limited and/or the atmospheric demand becomes excessive, the leaf stomata will react accordingly. The action of the stomata tends to conserve the water in the plant, and coordinates the water-supplying capacity of the soil and the demands of the atmosphere. However, loss of turgor or wilting may occur despite the control mechanism.

Physiologically, it has been observed that the levels of abscissic acid and prolene in the guard cells increase by as much as sevenfold when plants are subjected to a moisture stress; however, their role in the closure of the stomates is not clearly understood.

There appears to be no unique value of leaf water potential for stomatal closure in any particular species or cultivar, but varies with position in the leaf canopy and the plant age. Another phenomenon which is not clearly understood is that stomatal closure tends to occur at a lower leaf water potential (more negative) if the plant has undergone a series of stress cycles. This has been attributed by some to osmotic adjustment.

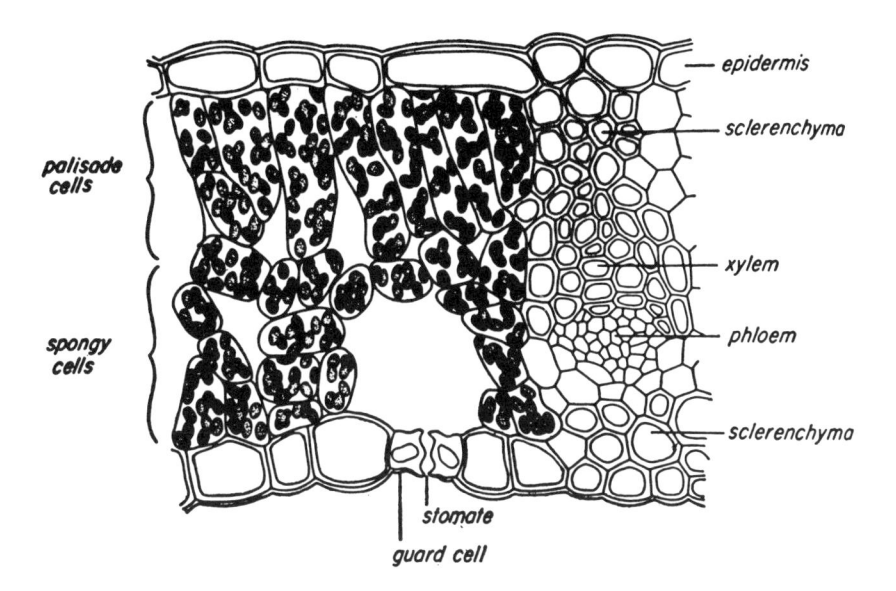

FIGURE 31. Cross-sectional portion of a leaf showing a stomate. (From Meyer, B. S., Anderson, D. B., Bohning, R. H., and Fratianne, D. G., *Introduction to Plant Physiology,* 2nd ed., D Van Nostrand, New York, 1973. With permission.)

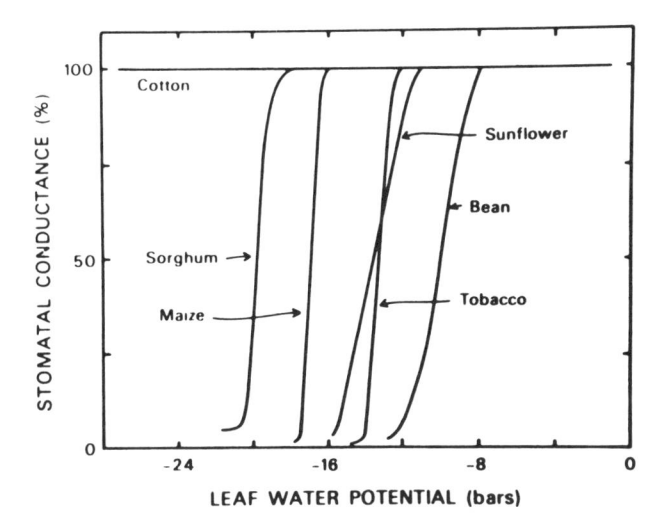

FIGURE 32. Relationship between stomatal conductance as a percentage of the maximum conductance, and leaf water potential for several crop species. (From Begg, J. E. and Turner, N. C., *Adv. Agron.*, 28, 161, 1976. With permission.)

Turgor pressure appears to be a better indicator of initiation of stomatal closures than total leaf water potential. It is generally recognized that stomata do not respond to changes in leaf water potential until a critical threshold, which may vary from 3 to 10 bars. As the stomates close, resistance to the flow of water increases, thus decreasing the rate of conductance.

Typical stomatal response for a range of field crops is shown in Figure 32. Although there have been a number of comparisons of stomatal response to stress among species, there have been few among cultivars and those which have been reported are mainly with the sorghums.

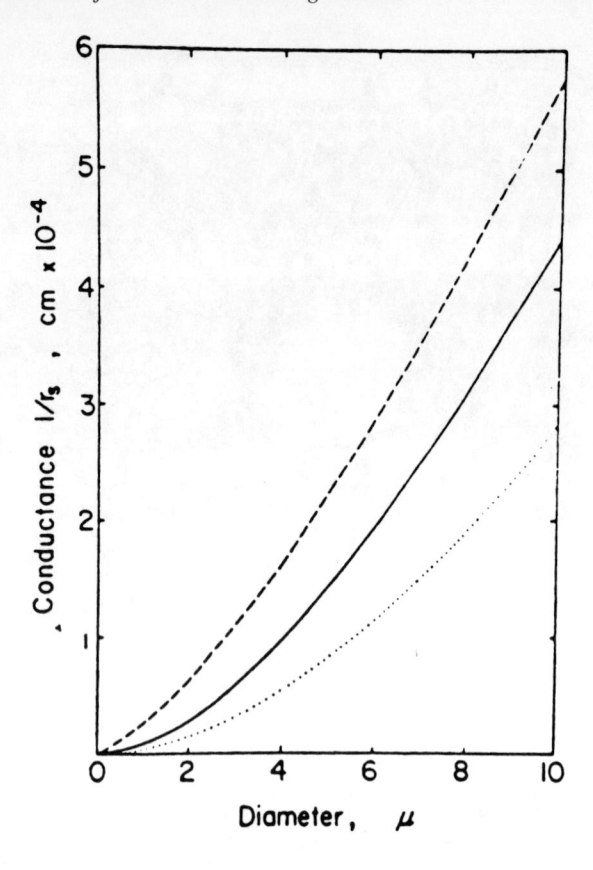

FIGURE 33. The conductance of closing pores. Cylindrical pores: $\ell = 10\ \mu$ (solid line); $\ell = 20\ \mu$ (dotted line). Elliptical pore: $\ell = 10\ \mu$ (dashed line); ℓ is the pore length or thickness of the leaf epidermis. (From Lee, R., *Water Resourc. Res.*, 3, 737, 1967. Copyright by the American Geophysical Union.)

The size and shape of the stomata temper the resistance to water movement as does the thickness of the leaf epidermis (Figure 33). The external resistance (that is, the adhering air layer next to the leaf surface) can also be of significance (Figure 34). A light wind will minimize the external resistance, with the resulting increased conductance of water vapor from the leaf surface (Figure 35).

Table 15 gives an average number of stomates per square centimeter of leaf surface for some typical crops and plants, while Tables 16 and 17 give, in addition to numbers, some stomatal dimensions. It is interesting that most plants have the greater number of stomata on the lower surface, while there are some which have the greater number on the upper.

Osmotic Adjustment

Plant physiological processes are generally recognized to be most responsive to the turgor pressure of the cells. Turgor potential at a particular water potential depends on the osmotic potential and the elasticity of the cell tissues. The osmotic potential is dependent on solute accumulation, cell size, osmotic volume, and cell wall thickness. A mechanism whereby plants tend to maintain turgor when water loss occurs is by osmotic adjustment. A lowering of the water content in cells concentrates the cell solution resulting in a decrease in osmotic potential. This decrease in the osmotic potential tends to maintain full or partial turgor in the plant. This lowering of the solute potential apparently arises from the alteration of the existing pathways and translocation patterns in the plant. Osmotic adjustment has been

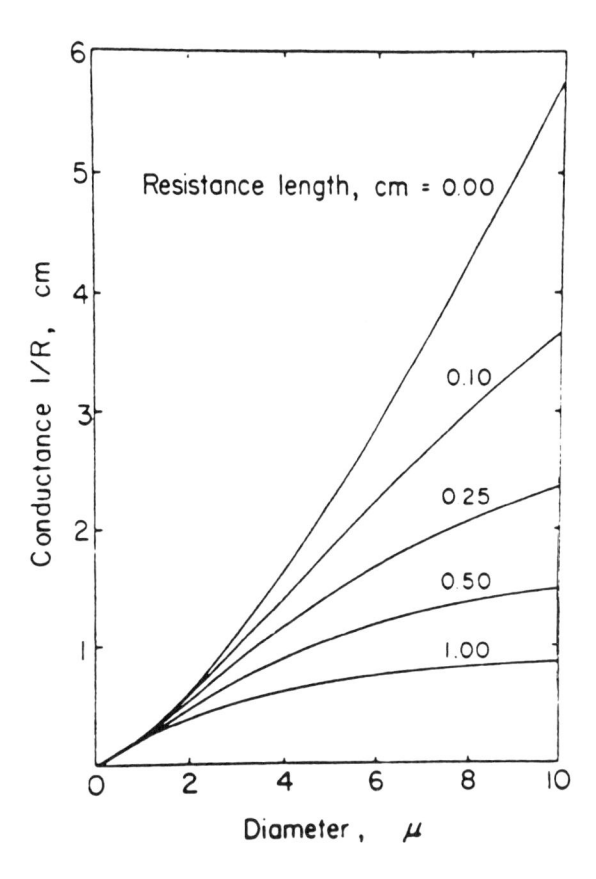

FIGURE 34. Leaf conductance (reciprocal of resistance of 1/R) changes with closing pores at various external resistance lengths. External resistance is that formed by still air at the leaf surface. (From Lee, R., *Water Resourc. Res.*, 3, 737, 1967. Copyright by the American Geophysical Union.)

attributed as a factor to the maintenance of growth, photosynthesis, and the increase in the root/shoot ratio of plants under moderate water stress. The degree and presence or absence of osmotic adjustment varies with species and cultivars. It is generally considered that plants which have the greater sensitivity to osmotic adjustment are most capable of withstanding water stress.

Plants that have undergone a series of moisture stress cycles appear to be more sensitive to cell osmotic adjustment. Further, the threshold water potential for stomatal closure to be initiated tends to decrease after such a series of moisture stresses. This change in threshold potential has been postulated as being a result of osmotic adjustment.

Solutes that tend to concentrate in the plant cells and contribute to the decrease in osmotic potential are the soluble carbohydrates, potassium, organic acids, chlorides, and free amino acids.

Photosynthesis

The photosynthetic process in the plant is the combining of carbon dioxide, water, and radiant energy to produce organic materials and is inhibited by a moisture stress. It is generally accepted that the initial reduction in photosynthesis arises primarily from a restriction of carbon dioxide diffusion into the leaves caused by stomatal closures; however, prolonged and severe stress can depress the chloroplast and enzyme activity in the plant cells.

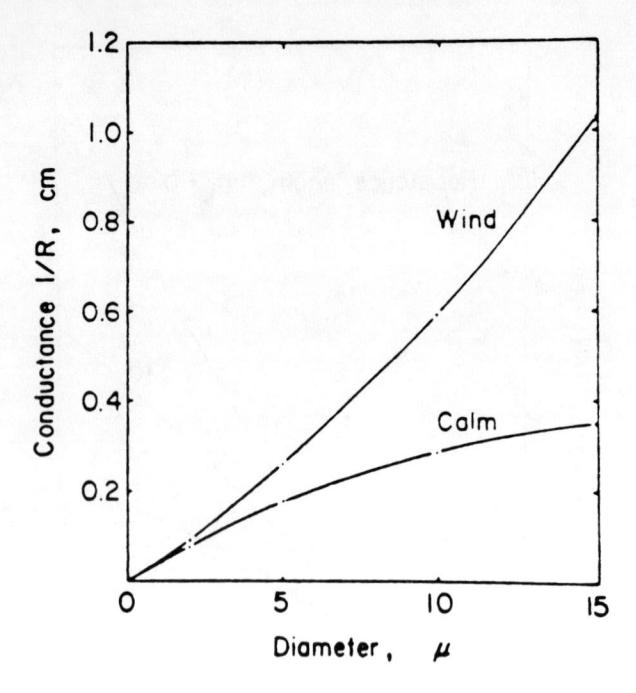

FIGURE 35. Conductance (reciprocal of resistance) curves in still air and in wind. (From Lee, R., *Water Resourc. Res.*, 3, 737, 1967. Copyright by the American Geophysical Union.)

Table 15
AVERAGE NUMBER OF STOMATES PER SQUARE CENTIMETER OF LEAF SURFACE

Species	Upper epidermis	Lower epidermis
Corn	5,200	6,800
Wheat	3,300	1,400
Oat	2,500	2,300
Bean	4,000	28,100
Potato	5,100	16,100
Tomato	1,200	13,000
Cabbage	14,100	22,600
Apple	0	29,400
Peach	0	22,500
Cherry	0	24,900
Black walnut	0	46,100
Willow oak	0	72,300
Coleus	0	14,100
Geranium	1,900	5,900

From *Fundamentals of Plant Physiology*, by James F. Ferry and Henry S. Ward. © Copyright Macmillan Publishing Company, 1959. Reprinted with permission of the publisher.

Net photosynthesis is the amount of photosynthate less that used by the plant in respiration. Respiration is essentially the reverse of photosynthesis in that it is the breakdown of the photosynthetic organic materials with the release of carbon dioxide, water, and energy.

Table 16
SIZE AND DISTRIBUTION OF STOMATA IN LEAVES OF VARIOUS PLANTS

Species	Mean stomatal number per cm^2 Upper epidermis	Mean stomatal number per cm^2 Lower epidermis	Mean size (μm)	Spacing on lower epidermis (μm)
Bean (*Phaseolus vulgaris*)	4,000	28,100	7 × 3	67.5
Ivy (*Hedera helix*)	0	15,800	11 × 4	90.0
Maize (*Zea mays*)	5,200	6,800	19 × 5	137
Tomato (*Lycopersicon esculentum*)	1,200	13,000	13 × 6	99.2
Wheat (*Triticum sativum*)	3,300	1,400	18 × 7	302
Sunflower (*Helianthus annuus*)	8,500	15,600	38 × 7	90.5
Oat (*Avena sativa*)	2,500	2,300	38 × 8	235.8
Geranium (*Pelargonium* sp.)	1,900	5,900	24 × 9	246.0
Wandering Jew (*Zebrina pendula*)	0	1,400	31 × 12	302

From Sutcliffe, J., *Plants and Water*, No. 14, The Institute of Biology's Studies in Biology, Institute of Biology, St. Martins Press, New York, 1968. With permission.

Table 17
LEAF STOMATAL CHARACTERISTICS OF SOME COMMON PLANTS

Species	Fully open pore (μ) Ellipse 2a × 2b	Fully open pore (μ) Circle (equiv.) 2 sec	Density (cm^{-2})	Open area (% = πs^{2}n(10^{-6}))	Spacing diameters = 5373/sn$^{1/2}$
Red pine (*Pinus resinosa*)	8 × 4	5.7	5,000	0.13	27
Wheat (*Triticum aestivum*)	38 × 7	16.3	1,400	0.29	18
Alfalfa (*Medicago sativa*)	9 × 4	6.0	15,350	0.43	15
Spanish oak (*Quercus triloba*)	5 × 1	2.2	119,200	0.47	14
Corn (*Zea mays*)	19 × 5	9.7	6,800	0.51	13
English ivy (*Hedera helix*)	11 × 4	6.6	15,800	0.55	13
Oat (*Avena sativa*)	38 × 8	17.4	2,300	0.55	13
Tomato (*Lycopersicon esculentum*)	13 × 6	8.8	13,000	0.80	11
American holly (*Ilex opaca*)	12 × 6	8.5	17,000	0.96	10
Sunflower (*Helianthus annuus*)	22 × 8	13.3	15,600	2.16	6

From Lee, R., *Water Resourc. Res.*, 3, 737, 1967. With permission. Copyright by the American Geophysical Union.

Current evidence indicates that in crop species dark respiration is depressed whenever the water stress in the plant is sufficient to cause stomatal closure and decrease photosynthesis, but the decrease in dark respiration is less than that of net photosynthesis.

Figure 36 indicates that below a threshold water potential, at about −11 bars, the photosynthetic activity in soybeans rapidly decreases. This is similarly shown for sunflowers in Figure 37, although in this case the threshold leaf water potential appears to be slightly higher. The significant aspect however is the rapid decline in the photosynthetic rate presumably after stomatal closure. This is further demonstrated in Figures 38, 39, and 40.

Recovery of the photosynthetic rate after the relief of a moisture stress is dependent somewhat on the severity and duration of the deficit. There is some evidence to indicate

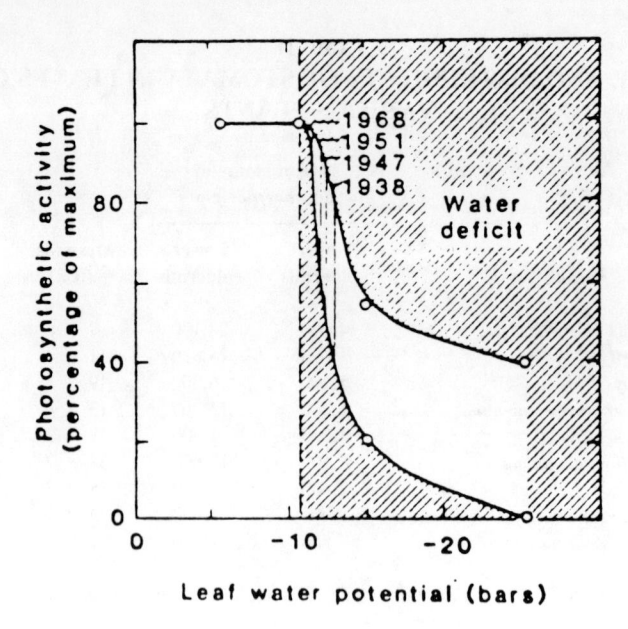

FIGURE 36. Response of leaf photosynthetic activity to low leaf water potential in soybeans. (With permission from Boyer, J. S., *Science*, 218, 443, 1982. Copyright 1982 by the AAAS.)

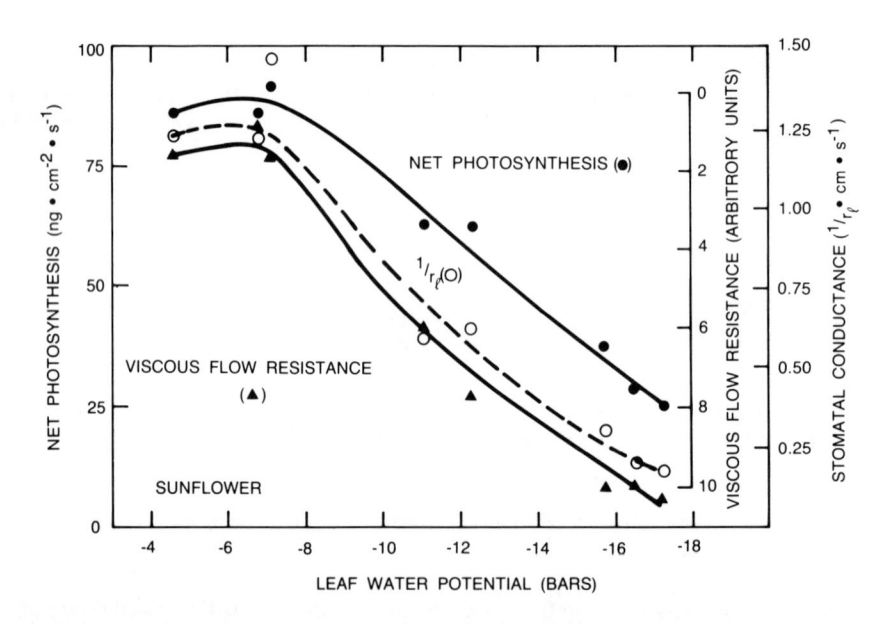

FIGURE 37. Net photosynthesis at saturated irradiance, stomatal conductance (1/r), and resistance to viscous flow of air through leaves of sunflower having various leaf water potentials. (From Boyer, J. S. and Bowen, B. L., *Plant Physiol.*, 45, 613, 1970. With permission.)

that after a short, moderate deficit, upon recovery plants tend to have an enhanced photosynthetic rate. However, if the cell chloroplasts have deteriorated because of dehydration the process will never fully recover.

The photosynthate is *translocated* in the plant to its various organs. However, during maturity of a cereal crop this process is primarily to the grain portion. It has been generally

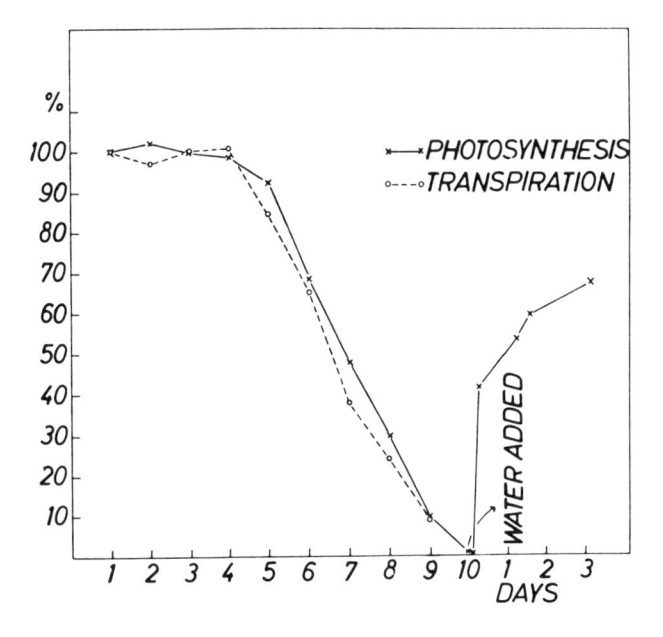

FIGURE 38. Decline of relative rates of photosynthesis (●) and transpiration (○) of tomato plants with decreasing soil water over a period of 10 days. (From Brix, H., *Physiol. Plant.*, 15, 10, 1962. With permission.)

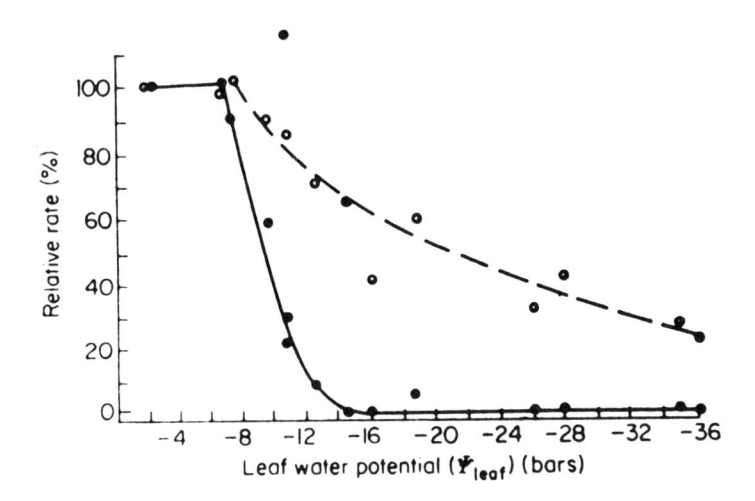

FIGURE 39. Effect of decreasing leaf water potential on relative rates of photosynthesis (●) and respiration (○) of tomato plants. (From Brix, H., *Physiol. Planta*, 15, 10, 1962. With permission.)

concluded that a reduction in translocation by a plant under moisture stress is primarily a result of reduced photosynthesis or growth of the sink, rather than any effect on the conducting system itself.

PLANT MORPHOLOGICAL RESPONSES TO MOISTURE STRESS

Growth

Growth of a plant is a result of a combination of cell division and cell elongation or expansion. Differentiation of cells into the various functional organs of the plant is also

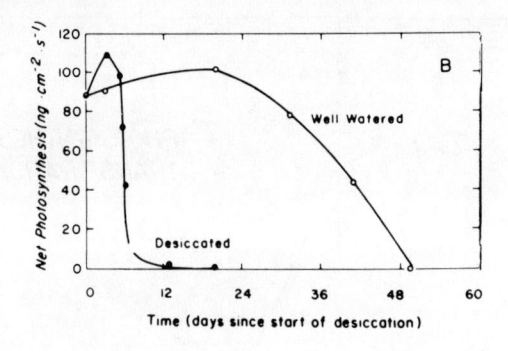

FIGURE 40. The net photosynthesis (photosynthesis less respiration) of maize (corn) during a drying cycle in the grain filling period. (From Boyer, J. S., in *Water Deficits and Plant Growth*, Vol. 4, Kozlowski, T. T., Ed., Academic Press, New York, 1976, 158. With permission.)

involved. In most cases cell division is less sensitive to a moisture stress than is cell enlargement. Cell enlargement is largely dependent on turgor pressure. There is a rapid and then a more gradual decline in the rate of cell enlargement as water stress develops. Cell enlargement can stop with turgor pressure still positive — as large as 6 to 8 bars in sunflower and maize (corn).

The growing tip or apical meristem is very sensitive to water stress. The apical meristem is not fully differentiated and is partially protected by the more developed leaves; hence it is able to survive a more severe stress better than most plant tissue. The sensitivity of sunflower growth to leaf-water potential is demonstrated in Figure 41. The relationship of leaf-water potential to the growth of bean leaves is shown in Figure 42.

A reduction in plant growth likewise reduces the leaf surface area, which contributes to the decrease in total photosynthetic product of a plant. Also during a prolonged water deficit leaves tend to senesce or die off, particularly the older leaves, which again lessens the total leaf area. Figure 43 shows how the leaf area of a maize (corn) plant can be affected by desiccation.

A diagrammatic scheme of the effects of soil moisture is given in Figure 44, while a schematic of the overall effect of water stress is given in Figure 45.

Leaf Orientation

How the leaves of a plant are oriented to the horizontal will influence the amount of incident radiation striking them. There are species which have more erect leaves, while in some cases cultivars are bred that way. Leaves angled at 70° to the horizontal have been found to reduce leaf temperature by 2°C and reduce transpiration by 7%. Since photosynthesis is saturated at relatively low quantum flux densities, it is likely that the reduced absorption of energy and the angling of leaves would have little effect on the rate of photosynthesis.

As moisture stress is experienced some plants are able to rotate their leaves to make most effective use of the incident radiation and also to conserve water. For example, soybeans tend to invert their leaves when under moisture stress. Their lower leaf surfaces are more reflective than their upper; thus when inverted by a water stress they tend to maintain a higher relative level of carbon dioxide conductance. There are some species and cultivars whose leaves tend to roll when under water stress, thus exposing less of the total leaf surface to the atmosphere. This will in effect conserve plant water. It is a mechanism whereby the plant enhances its survival under drought. Sorghum is a good example of a crop with leaves that roll; in fact they will do so on a bright warm day even with good soil moisture.

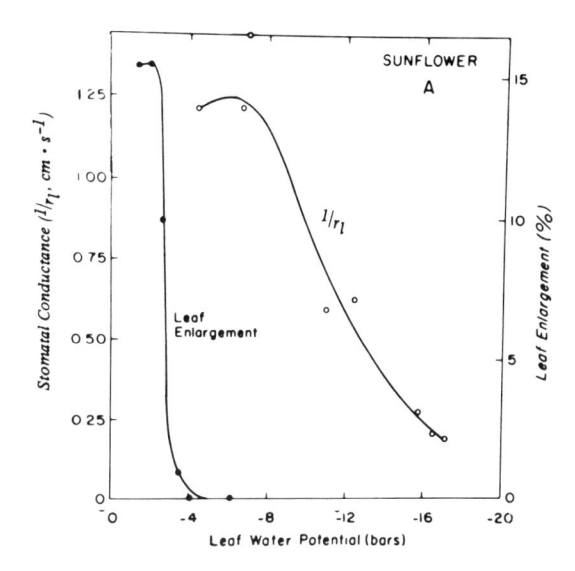

FIGURE 41. Leaf enlargement and stomatal conductance in sunflower having various leaf water potentials. (From Boyer, J. S., *Plant Physiol.*, 43, 1059, 1968. With permission.)

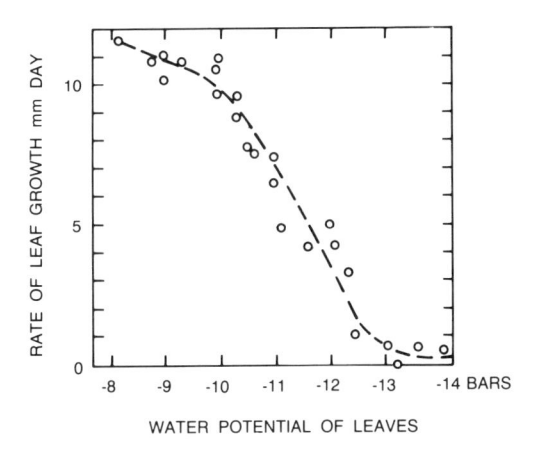

FIGURE 42. The rate of growth of bean leaves at various leaf water potentials. (From Brouwer, R., *Acta Bot. Neerl.*, 12, 248, 1963. With permission.)

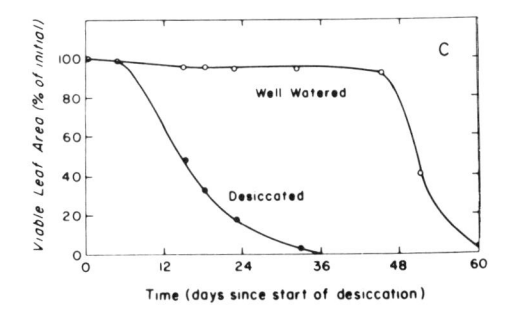

FIGURE 43. The viable leaf area of a drying maize (corn) plant compared with one well watered during the grain filling period. (From Boyer, J. S., in *Water Deficits and Plant Growth*, Vol. 4, Kozlowski, T. T., Ed., Academic Press, New York, 1976, 158. With permission.)

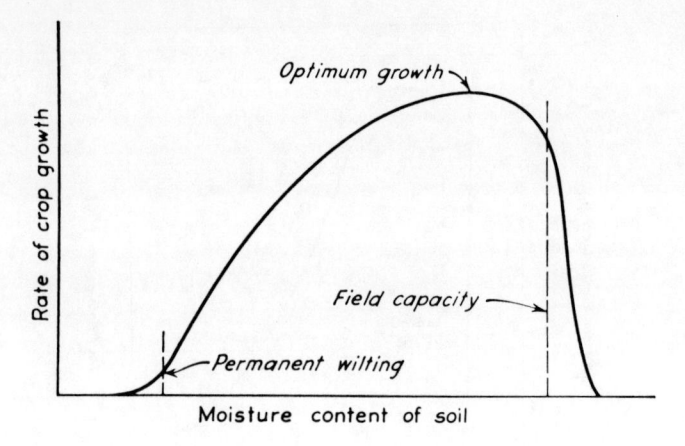

FIGURE 44. Relationship of rate of plant growth to moisture content of the soil; optimum growth varies somewhat with aeration, water-holding capacity of soil, and atmospheric conditions. (From Hansen, V. E., Israelsen, O. W., and Stringham, G. E., *Irrigation Principles and Practices*, 4th ed., John Wiley & Sons, New York, 1980. With permission.)

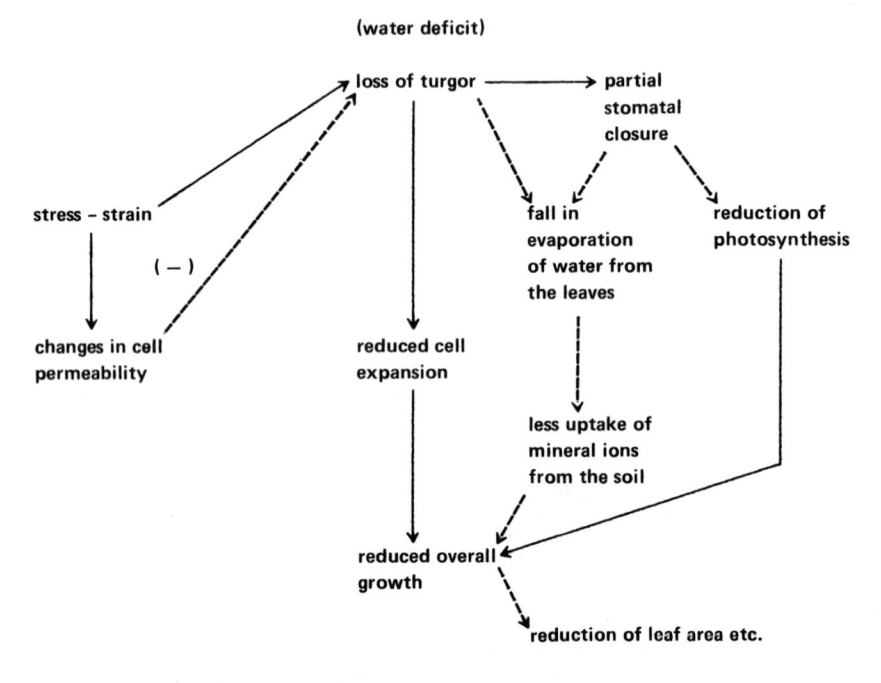

FIGURE 45. A schematic representing the effects of water stress on plants. (From Merva, G. E., *Physioengineering Principles*, AVI, Westport, Conn., 1975. With permission.)

Reflectance is another mechanism that enhances survival. Some plants have silvery white hairs on their leaf surface while others develop them when under water stress (for example, certain sunflower cultivars). There are sorghum cultivars which have or develop a shiny waxy bloom which enhances the reflection of energy.

Roots

The roots of plants under a moisture stress have been observed to continue their growth after shoot growth has ceased. The root/shoot ratio is therefore increased when comparing a plant which is well watered with one that is undergoing a temporary stress. This would appear to be somewhat contradictory to the previous statement that is roots do not effectively extend into dry soil; however, the conditions differ. In one case the soil was initially moist and roots proliferated whereas in the other case moisture was never available, and hence roots did not grow into the dry soil.

The increased root/shoot ratio would tend to enhance the water economy of the plant; however, it would be beneficial for only short-term drought periods. An explanation suggested for the increased root/shoot ratio of moisture-stressed plants is that some photosynthesis may continue to take place even after shoot growth is retarded and the resulting assimilate is translocated to the root promoting their growth. There is also the possibility that osmotic adjustment is more effective in roots than in shoots.

MOISTURE STRESS AND CROP YIELD

Plant Factors

Plants can be characterized according to the number of carbon atoms in the molecule of the first carboxylation product of photosynthesis. There are then either C_3 or C_4 species. The C_4 plants generally have higher photosynthetic and growth rates, particularly under high light and temperature. They have higher stomatal resistance resulting in a relatively greater reduction in transpiration as compared to photosynthesis and therefore make more efficient utilization of water. The C_4 crops should be grown in high-energy regions or seasons while C_3 crops can better be grown in temperate humid regions or seasons. Examples of C_4 crops include maize (corn), sorghum, and sugarcane, while for C_3 there are tobacco, spinach, and some cool-season grasses.

Determinate vs. Indeterminate Plants

Plants vary in their flowering habits which can affect crop yield when grown under moisture stress. There are plants with determinate growth characteristic in which the axis is limited by the development of the terminal flower bud and growth proceeds only during the vegetative season and then ceases. For plants which are indeterminate the growth axis is not limited by development of the terminal flower bud and continues to elongate indefinitely. Determinate crops such as maize (corn) have a distinct flowering period, whereas indeterminate crops such as cotton have more extended flowering periods. With soybeans some cultivars are determinate and some indeterminate.

A moisture stress at flowering of a plant is most detrimental to crops where economic yield is dependent on its reproductive organ. A short-term drought at flowering of a determinate plant can reduce yield significantly; in fact, if severe, to almost zero. With indeterminate plants with flowering occurring over a longer period of time, the crop has a better chance of escaping the effects of a short-term drought. However, a long-term or seasonal drought would have the same effect on both types of plants.

Further, roots of the determinate plants tend to stop growing at flowering, whereas with the indeterminates growth continues, although more slowly. Thus the rooting characteristics of these two types of plants could have an effect on the water economy of the crops.

Critical Periods of Plant Growth

The forage crops having the vegetative portion of the plant used as the economic yield have a different moisture regime than crops utilizing the reproductive segment, as, for example, maize (corn). For optimal vegetative growth of a forage it would be necessary to have a good supply of moisture the entire season, while for the cereal grains something less than optimal moisture during some of the growth stages may produce a good yield.

A yield-depressing critical period for a moisture stress to occur on several agronomic

Table 18
CRITICAL SOIL WATER STRESS PERIODS FOR DIFFERENT CROPS

Crop	Critical periods
Cereal crops	
Wheat	Booting, flowering, and grain formation early growth
Barley	Early booting, tillering, soft dough
Oats	Beginning of ear emergence to flowering
Maize	Flowering and early seed development prior to tasseling, grain-filling period
Sorghum	Tillering to booting, heading, flowering, and grain development
Pearl millet	Heading and flowering
Legume crops	
Alfalfa	Immediately after cutting for hay, start of flowering for seed production
Peas	Flowering and pod formation
Beans	Flowering and pod setting, earlier, ripening period
Groundnuts	Flowering and seed development, germination to flowering
Broad beans	Flowering
Snap beans	Flowering and fruit development
Oil crops	
Castor beans	Requires good moisture during full growing period
Sunflower	Flowering and seeding
Soybeans	Flowering and fruiting, period of maximum vegetative growth
Miscellaneous crops	
Sugar beets	3 to 4 weeks after emergence
Sugarcane	Period of maximum vegetative growth
Tobacco	Knee high to blossoming
Tomatoes	Flowering and period of rapid fruit development
Watermelon	Flowering to harvest
Broccoli	During head formation and enlargement
Cabbage	
Tree fruits	
Apricots	Flowering and bud development
Cherries	Period of rapid fruit growth
Citrus	Flowering and fruit setting
Peaches	Period of rapid fruit growth
Mangoes	Flowering, some water stress before flowering considered desirable
Other	
Bananas	Requires good moisture throughout growth and fruit-development periods, particularly during periods of hot weather
Coffee	Flowering and fruit development, some water stress before flowering generally held desirable
Olives	Just before flowering and during fruit enlargement

From Doorenbos, J. and Pruitt, W. O., Crop Water Requirements, Irrigation Drainage Paper 24, Food and Agriculture Organization, Rome, 1977. With permission of the Food and Agriculture Organization of the United Nations.

crops is given in Table 18. The majority of the critical periods occur at or near the flowering stage of the plant. The sugar beet is a notable exception. Although the vegetative and root growth is affected by moisture stress, the total sugar harvested per unit area is not, even though the plants may undergo afternoon wilting. Table 19 also shows critical water-deficit periods with additional information on the appearance of the crop under stress.

The economic yield of a grain crop is determined by the number of ears or spikelets per unit area, the number of kernels per ear or spikelet, and the weight per kernel. Moisture

Table 19
MOISTURE STRESS SYMPTOMS AND CRITICAL GROWTH PERIODS FOR SOME IRRIGATED CROPS

Crop	Moisture stress symptoms	Critical period	Comments
Alfalfa	Bluish-green color, then wilting	Seedling and immediately after cuttings	Soil kept moist in upper 5 ft; avoid over-irrigation; fall irrigation is desirable
Corn	Leaf curl by 10 a.m.	Tasseling stage until grain becomes firms	Sensitive to over-irrigation from seedling stage to 3 ft in height; needs adequate moisture from germination to dent stage
Pasture grass	Dull green color, then wilting	Seedling stage, for seed production boot to head formation	Late-fall irrigation is necessary; frequent, light applications; irrigate at end of grazing period in a rotation system
Dry beans and soybeans	Dull color, then wilting	Early bloom, seed forming	Very sensitive to over-irrigation; last irrigation at time of first set of pod maturity
Potatoes	Darkening of leaves, then wilting	Tuber swelling periods	Water should not stand around tubers; do not over-irrigate; lower tuber quality if plants go into serious moisture stress; last irrigation 3 to 4 weeks before harvest
Snap beans	Dull green color and wilting	During and immediately following blossoming	
Tomatoes	Darker green leaves, wilting and reduced growth; blossom end rot on fruit	During fruit accumulation and fruit-development period prior to ripening	Sensitive to over-irrigation; this causes delayed maturity, excessive vine growth, and reduces soluble solids content of fruit
Melon	Leaf wilting; blossom end rot and loss of small fruit	During fruit set and development	
Fruit trees	Decreased growth; fruit doesn't size; wilting of foliage	Postplanting on *young* trees; on mature trees immediate postbloom and 30-day period prior to harvest	Mature apples not usually very susceptible to drought except on sandy soils; trickle irrigation becoming popular; mature, close-spaced plantings will be more susceptible to drought, especially when crop is heavy; heavy crops increase water stress

From Irrigation Guide for Indiana, Soil Conservation Service, United States Department of Agriculture, Indianapolis, 1977.

stress may affect each of these yield components. For example, a stress during the vegetative stage of wheat may decrease the number of tillers, and thus the number of spikelets to emerge. For the determinate cereals — maize (corn), for example — with the critical moisture-stress period during flowering, the number of kernels per ear is most adversely affected.

WATER USE BY CROPS

Potential evapotranspiration of a crop can be defined as the volume of water vaporized per unit area per unit time given a sufficient soil water supply to satisfy the atmospheric

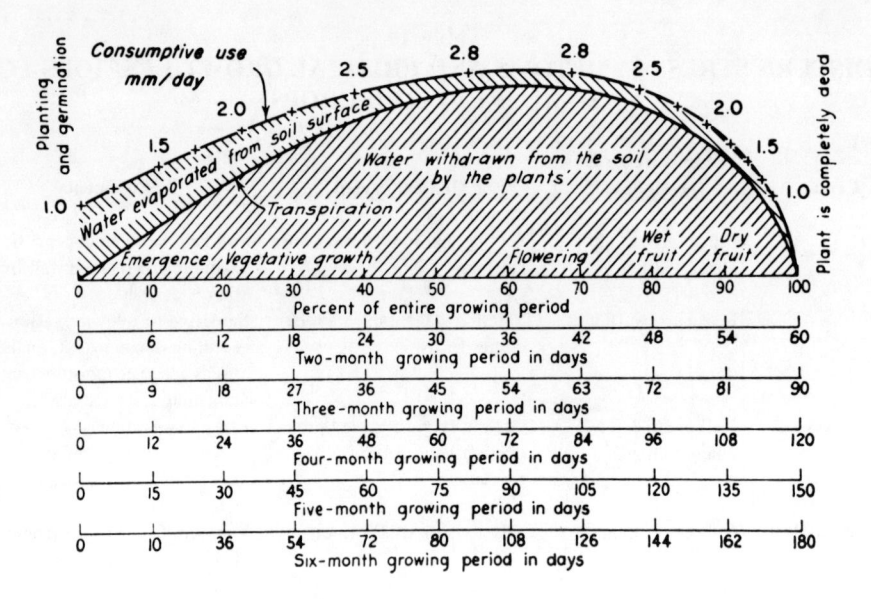

FIGURE 46. A generalized representation of water use by crops for various periods of growth. (From Hansen, V. E., Israelsen, O. W., and Stringham, G. E., *Irrigation Principles and Practices*, 4th ed., John Wiley & Sons, New York, 1980. With permission.)

demand. This condition only exists when the vapor pressure of the water at the soil and/or the leaf surface is at saturation. The evapotranspiration of a crop is something less than potential when the moisture supply is limiting, and where the crop vegetation does not completely cover the ground, as is usually the case where the leaf area index is less than 3. With incomplete vegetative cover, a portion of the soil surface is bare and thus on a warm, dry day the potential rate of vaporization occurring at the nonvegetatively covered soil surface would be greater than the unsaturated hydraulic conductivity of the soil, limiting the total evapotranspiration to something less than potential for a described area.

The concept of potential evapotranspiration as frequently used in the literature is defined as the rate of moisture conversion for a vegetative cover with short, densely spaced plants growing actively and having access to an unlimited soil-moisture supply. This does not allow for difference in plant species, internal resistance within the plant, or other changes under biological control.

Seasonal Use of Water by Crops

Crops differ in their seasonal use of water because of growth characteristics, date of planting, and so on. A perennial forage crop (for example, alfalfa) will begin growth and using water early in the season and will continue through the summer until fall, while an annual, such as maize (corn), will begin its growth cycle later in the season and thus have a different seasonal water-use pattern.

The graph shown in Figure 46 represents a generalized water-use picture for a growing season, not considering a specific crop but rather the length of growing season. When about 60 to 70% of the growing period has elapsed the evapotranspiration rate is the highest. This is also the flowering period which, as indicated above, is the most critical. Table 20 has similar information but integrates the water-use curve to give accumulated water use for the season.

As previously indicated, the potential evapotranspiration rate for a crop varies considerably with the climate and geographic area. For example, from Table 21, for a grass cover the highest rate in the northern latitudes occurs in June and July, and at Davis, Calif. is about

Table 20
ACCUMULATED USE OF WATER FOR CROPS WITH VARIOUS PERIODS FROM PLANTING TO COMPLETE MATURITY

Total period of growth (months)

Percentage of growing period[a]	Accumulated consumptive use of water in percentage of total use	2 Days since planting	2 Accumulated water use (cm)	3 Days since planting	3 Accumulated water use (cm)	4 Days since planting	4 Accumulated water use (cm)	5 Days since planting	5 Accumulated water use (cm)	6 Days since planting	6 Accumulated water use (cm)
10	6.0	6	0.76	9	1.0	12	1.5	15	2.0	18	2.3
20	14	12	1.80	18	2.5	24	3.6	30	4.3	36	5.1
30	24	18	2.80	27	4.3	36	5.8	45	7.4	54	8.9
40	35	24	4.30	36	6.6	48	8.6	60	10.9	72	13.0
50	47	30	5.80	45	8.6	60	11.7	75	14.5	90	17.3
60	59	36	7.40	54	11.2	72	14.7	90	18.5	108	22.1
70	72	42	8.90	63	13.5	84	18.0	105	22.6	126	26.9
80	84	48	10.40	72	15.7	96	21.1	120	26.2	144	31.5
90	95	54	11.90	81	17.8	108	23.6	135	29.5	162	35.3
100	100	60	12.40	90	18.8	120	24.9	150	31.2	180	37.3

[a] Growing period is used herein to refer to the entire time from planting to the time the plant dies, which is usually longer than the period from planting to harvesting. Flowering will occur at about 50 to 60% of the growing period and fruiting after 60%.

From Hansen, V. E., Israelson, O. W., and Stringham, G. E., *Irrigation Principles and Practices*, 4th ed., John Wiley & Sons, New York, 1980. With permission.

Table 21
POTENTIAL EVAPOTRANSPIRATION (mm/DAY) FOR VARIOUS LOCATIONS THROUGH THE YEAR FROM A GRASS OR GRASS-CLOVER COVER[42]

Location	J	F	M	A	M	J	J	A	S	O	N	D	Year
Seabrook, N.J. 4 years, 39°N lat.													
Mean	0.8	0.6	1.1	2.2	3.7	3.9	4.8	4.2	3.3	2.3	1.5	0.7	890
Std. error	0.2	0.2	0.3	0.1	0.2	0.2	0.3	0.2	0.3	0.2	0.3	0.1	
Raleigh, N.C. 3 years, 36°N lat.													
Mean	—	—	—	—	—	4.3	4.4	3.6	2.9	1.3	—	—	—
Std. Error	—	—	—	—	—	0.2	0.3	0.1	0.2	—	—	—	
Davis, Calif. 2 years, 39°N lat.													
Mean	0.8	1.8	2.8	4.0	5.2	7.1	7.0	5.7	4.2	2.9	1.5	0.8	—
Std. error	—	—	—	0.3	0.2	0.1	0.1	0.1	0.3	—	—	—	
Hong Kong 2 years, 22°N lat.													
Mean	2.3	2.6	2.8	3.2	4.0	4.0	4.3	3.7	4.2	4.5	3.7	2.7	1280
Std. error	0.1	0.2	0.2	0.3	0.3	0.4	0.3	0.3	0.4	0.1	0.2	0.1	
Aspendale, Vic. 3 years, 38°S lat.													
Mean	7.8	5.6	4.2	2.9	1.3	1.0	0.9	1.4	2.3	4.1	5.2	6.0	1310
Townsville, Qld. (Kalamia) 1 year, 19°S lat.													
Mean	6.4	4.5	5.2	4.3	4.0	4.3	5.3	6.8	5.3	6.8	8.7	11.0	1850

From Miller, D. H., *Water At the Surface of the Earth,* Academic Press, New York, 1977. With permission.

7.0 mm/day, whereas at Raleigh, N.C. for the same months it is 4.3 even though there is not a great difference in latitude between the two locations. In the southern latitudes the extreme high occurs in January. The range of seasonal water uses for several crops is given in Table 22.

Water-Use Efficiency

The yield of an agronomic crop is measured by the marketable produce per unit area. Crops vary, however, as to the amount of water transpired per unit of photosynthetic product. The classic expression of transpiration ratio was developed in 1914 by Briggs and Shantz.[10] Some of their data are given in Table 23.

More recently the concept of water-use efficiency has being expressed. Water-use efficiency is in essence the reciprocal of the transpiration ratio. Unfortunately there is no uniformity as to the units being used; for example, the units may be pounds or tons of dry weight produced or marketed per acre-inch of evapotranspiration or grams of dry matter per kilogram of water used. In the scientific literature it is sometimes given simply as grams of photosynthate per gram of water transpired.

Agronomic crop yields per unit area have increased dramatically during the past few decades. This has been accomplished without the concomitant increase in water consumption. The breeding of high-yielding adapted crop varieties and the adoption of cultural practices such as high fertility, proper tillage, control of weeds, diseases, and insects, and providing an adequate plant population have all contributed to the increased yields and thus water-use efficiency. In the broad classification of C_4 and C_3 species, the water-use efficiencies of the C_4 species are generally twice that of the C_3 species.

Table 22
APPROXIMATE RANGE OF SEASONAL EVAPOTRANSPIRATION (ET) FOR VARIOUS CROPS AND IN COMPARISON WITH GRASS

Crop	Seasonal crop ET mm	Grass (%)	Crop	Seasonal crop ET mm	Grass (%)
Alfalfa	600—1500	90—105	Oranges	600—950	60—75
Avocados	650—1000	65—75	Potatoes	350—625	25—40
Bananas	700—1700	90—105	Rice	500—950	45—65
Beans	250—400	20—25	Sisal	550—800	65—75
Cocoa	800—1200	95—110	Sorghum	300—650	30—45
Coffee	800—1200	95—110	Soybeans	450—825	30—45
Cotton	550—950	50—65	Sugar Beets	450—850	50—65
Dates	900—1300	35—100	Sugarcane	1000—1500	105—120
Deciduous trees	700—1050	60—70	Sweet potatoes	400—675	30—45
Flax	450—900	55—70	Tobacco	300—500	30—35
Grains (small)	300—450	25—30	Tomatoes	300—600	30—45
Grapefruit	650—1000	70—85	Vegetables	250—500	15—30
Maize	400—700	30—45	Vineyards	450—900	30—55
Oil seeds	300—600	25—40	Walnuts	700—1000	65—75
Onions	350—600	25—40			

From Doorenbos, J. and Pruitt, W. O., Crop Water Requirements, Irrigation Drainage Paper 24, Food and Agriculture Organization, Rome, 1977. With permission of the Food and Agriculture Organization of the United Nations.

Table 23
TRANSPIRATION RATIOS OF SEVERAL AGRONOMIC CROPS

Pounds of water transpired per pound of dry plant tissue produced

Crop	1911	1912	1913	Greatest variation	Average
Wheat	468	394	496	102	452.7
Oats	615	423	617	194	551.7
Corn	368	280	399	119	349.0
Sorghum	298	237	296	61	277.0
Alfalfa	1068	657	834	411	853.0

From Briggs, L. J. and Shantz, H. L., *Agric. Res.*, 3, 1, 1914. With permission.

Geographic location also has an effect on water-use efficiency as shown in Figure 47. The more arid the climate (greater atmospheric saturation deficit), the lower the water-use efficiency. (Figure 47 uses the evapotranspiration ratio which is the reciprocal of water-use efficiency.)

Water-Use Efficiency and Fertility

Figure 48 illustrates the relationship between nitrogen application, soil moisture, and yield, while Figure 49 shows the effect of phosphate fertilizer on water-use efficiency for a low and an adequate soil moisture.

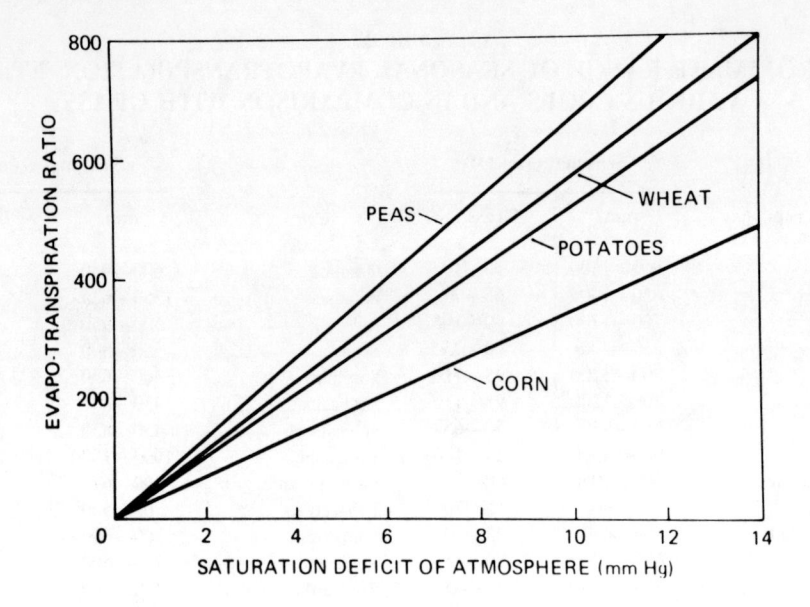

FIGURE 47. Evapotranspiration ratios of different crop plants grown at locations differing in the saturation deficit of the atmosphere. In the drier climates (high saturation deficit) maize (corn) uses less water per unit of dry matter produced, peas the most. (From Bierhuizen, J. F. and Slatyer, R. O., *Agric. Meteorol.*, 2, 265, 1965. With permission.)

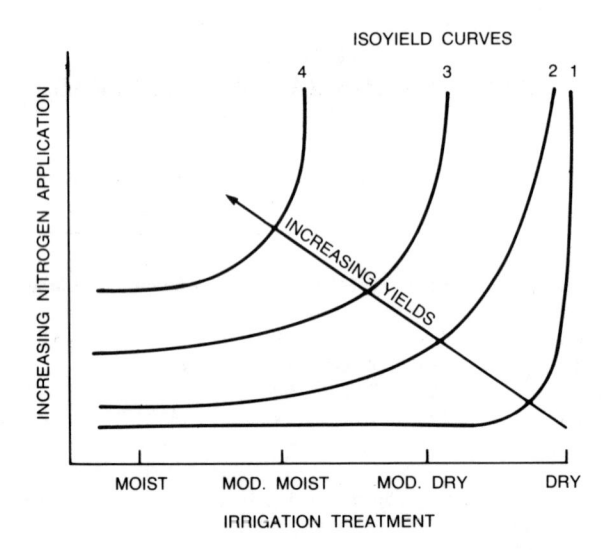

FIGURE 48. Isoyield curves illustrating the interaction between nitrogen application, irrigation, and yield. (From Hagan, R. M., Vaadia, Y., and Russell, M. B., *Adv. Agron.*, 11, 77, 1959. With permission.)

MANAGEMENT STRATEGIES AFFECTING PLANT DROUGHT RESISTANCE

Where drought or a limited moisture supply can be normally anticipated during a particular season of the year, strategies may be employed to at least partially overcome a yield-

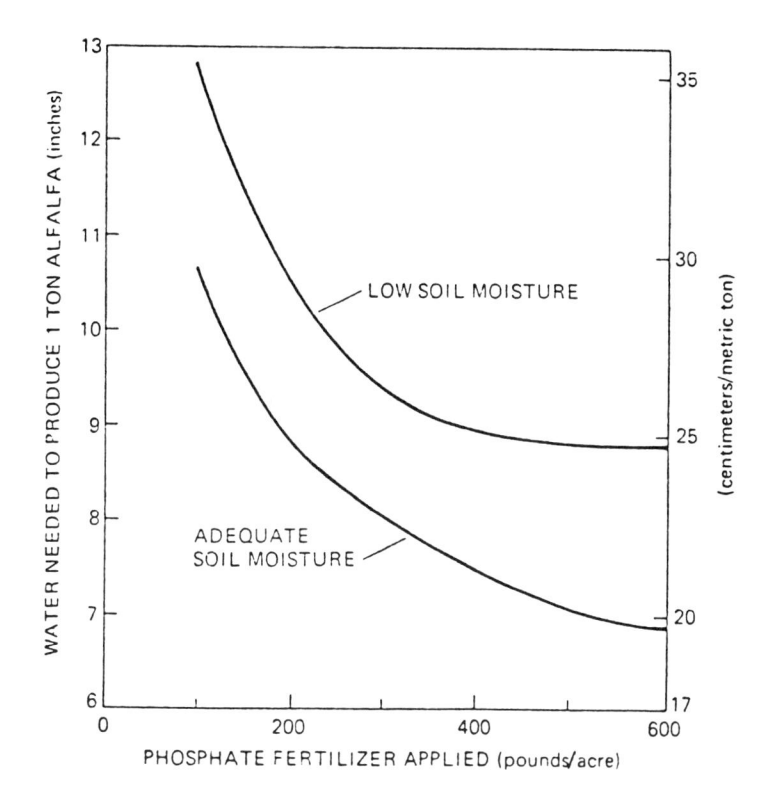

FIGURE 49. Effect of phosphate applications and soil moisture level on the amount of water required to produce 1 ton of alfalfa hay. Most efficient moisture utilization was obtained when the fertility level was high and the moisture supply adequate. (From Kelly, O. J., *Adv. Agron.*, 6, 67, 1954. With permission.)

depressing stress period. For example, using crops or cultivars which can tolerate cooler germination temperatures will permit earlier planting, and thus to an extent avoid the extreme water use or stress of the mid-season critical flowering period. Also, crops or cultivars which require fewer days for maturity will tend to avoid the possible low-moisture, high-stress period.

Species or cultivars which have the more erect leaves or can orient them during stress will reduce the amount of incident radiant energy striking the leaf surface, thus tending to conserve moisture without significantly reducing the photosynthetic activity of the plant.

Species or cultivars with deep and dense root systems can more effectively utilize the soil water supply and thereby may be able to overcome a moisture-stress period. Cultivars for areas of probable drought could be bred for better rooting systems. It has been noted that species or cultivars which have the better capability of osmotic adjustment in their plant cells tend to be the most drought tolerant.

Management practices which promote early vegetative growth may not be the most beneficial for the water economy of particularly determinate plants. In areas with a limited water supply later in the season, early vegetative growth would transpire more water during this stage and thus dwindle the supply for the more critical reproductive stage. High plant population, narrow rows, and high fertility are some of the practices which can promote such early growth.

Cultivars which produce the highest yield under excellent moisture conditions do not necessarily do the best under less favorable conditions. The converse is also in evidence, in that the highest-yielding cultivars under water-stress conditions are usually not the highest under more favorable conditions.

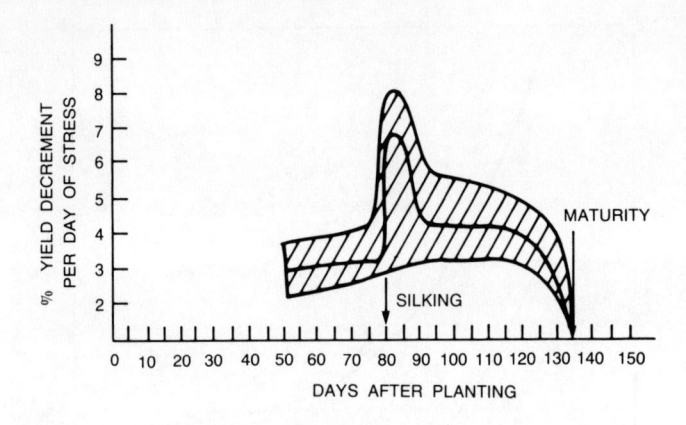

FIGURE 50. Schematic diagram showing effect of moisture stress on the yield of a maize (corn) crop. (From Shaw, R. H., in *Crop Reactions to Water and Temperature Stresses in Humid, Temperate Climates*, Ruper, C. D., Jr., and Kramer, P. J., Eds., Westville Press, Boulder, Colo., With permission.)

A practice which has not been used extensively with agricultural crops is foliar application of antitranspirants. This is the spraying of plants with certain chemicals to reduce transpiration. The objective is to increase stomatal resistance so that transpiration is reduced proportionately more than photosynthesis and thus decrease the water use and the likelihood of severe water deficits developing during a critical stage of growth. The kinds of materials that have been tried are either those which close the stomata, are film forming, or increase plant reflectivity. A problem with stomatal closure is that photosynthesis may be impeded as much as or more than transpiration. To date these materials have been used primarily on specialty crops.

Irrigation Strategies

A management technique that can replenish the soil water supply and alleviate potential plant moisture stress is irrigation. The question of when to irrigate and how much water to apply continually confronts the irrigation manager. In the past, practices have been quite inefficient in that frequently more water was applied than could effectively be utilized by the crop with the result that water was either lost by deep percolation or runoff. Inefficient irrigation practices are costly in that nutrients are lost by deep percolation, energy is used in water distribution, and there is possible salinization of the soil.

Irrigators with a limited water supply have an additional problem. When is the best time in the plant cycle to use the water available? Considering the critical moisture-stress period of a crop, it would appear that for a determinate plant, irrigation at the flowering stage would give the greatest response. An example is an experiment[46] which reported that corn wilted for 1 or 2 days during the tasseling or pollination stage had grain yield reduced as much as 22%. Extending the stress to 6 to 8 days resulted in a grain yield decline of 50%. The yield reduction due to stress during pollination was largely the result of pollination failure as grain filled only part of the ear. A schematic diagram summarizing the data of several experiments showing the relationship between age of a corn crop and percentage yield decrement due to 1 day of moisture stress is shown in Figure 50. The shaded area covers the range of most yield reduction, and the line through the shaded area represents the average yield reduction. There is experimental evidence,[51] however, that for wheat (given a limited water supply) it is more advantageous for yield to apply an irrigation amount equal to a fraction of the soil-water deficit each week, allowing stress to gradually and progressively intensify throughout the season. This would support the premise that plants adjust to cyclic moisture stresses.

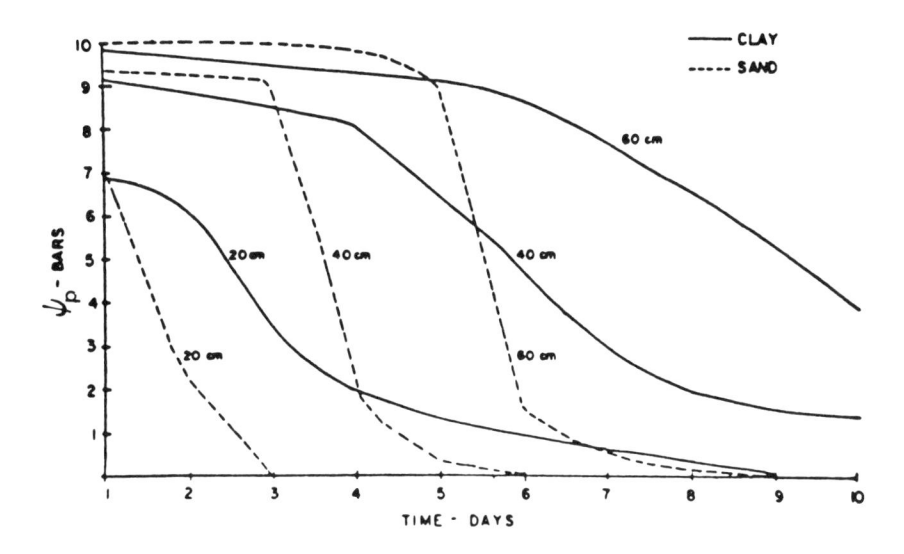

FIGURE 51. Time changes in the midday tugor pressure during the 10-day drying cycle for two soil types and three root zone depths. (From Zur, B. and Jones, J. W., *Water Resourc. Res.*, 17, 311, 1981. Copyright by the American Geophysical Union.)

An irrigation technique that is not necessarily new, but which has been given considerable attention and study in recent years is drip or trickle irrigation. This is a distribution system where water is released to the soil-water supply through very small openings or emitters under low pressure at a rate equal to or near the rate water is transpired by the plants. Properly managed, this technique provides for the plants to be under a minimum of moisture stress at all times and is the most efficient in water utilization in that very little is lost by soil evaporation. This type of management would suggest the highest water-use efficiency.

MODELING OF SOIL-PLANT-WATER RELATIONSHIPS

There are in the recent literature a number of models developed to show the integrated relationship of soil, plant, water, and climate on crop yield. These are helpful in the understanding of the system. There is no attempt here to review all these models; however, some of the results for one are shown graphically below (Figures 51 through 55).

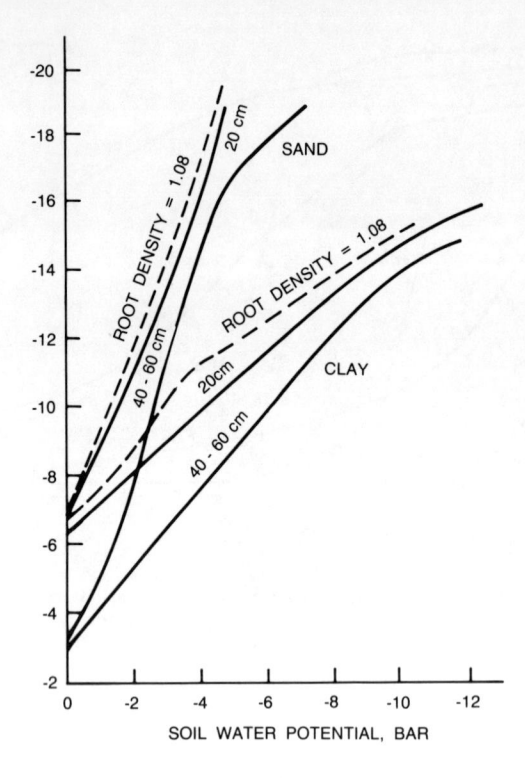

FIGURE 52. Simulated relationship between midday leaf water potential and soil water potential for two soil types and a root density of 2.16 cm/cm^3. Also shown are 1.08 cm/cm^3 root densities for the 40-cm root-zone depth. (From Zur, B. and Jones, J. W., *Water Resourc. Res.*, 17, 311, 1981. Copyright by the American Geophysical Union.)

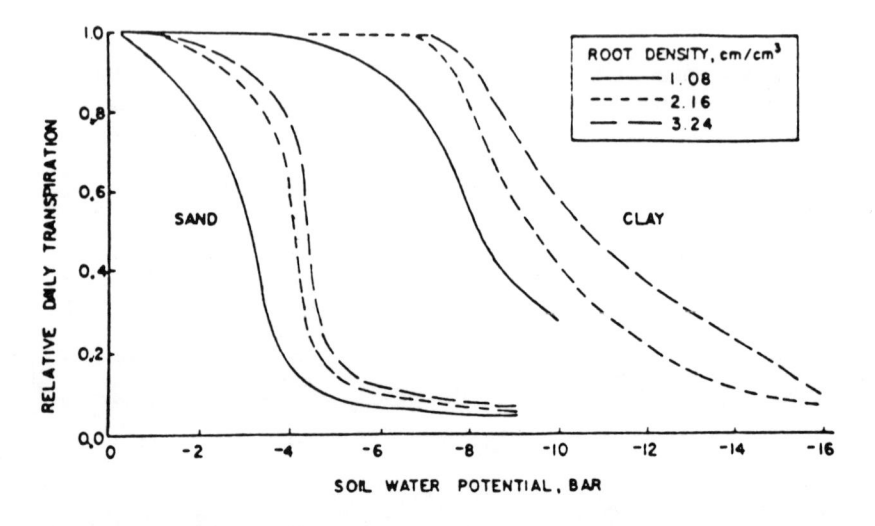

FIGURE 53. Simulated relationship between transpiration and soil water potential for two soil types and three root densities. (From Zur, B. and Jones, J. W., *Water Resourc. Res.*, 17, 311, 1981. Copyright by the American Geophysical Union.)

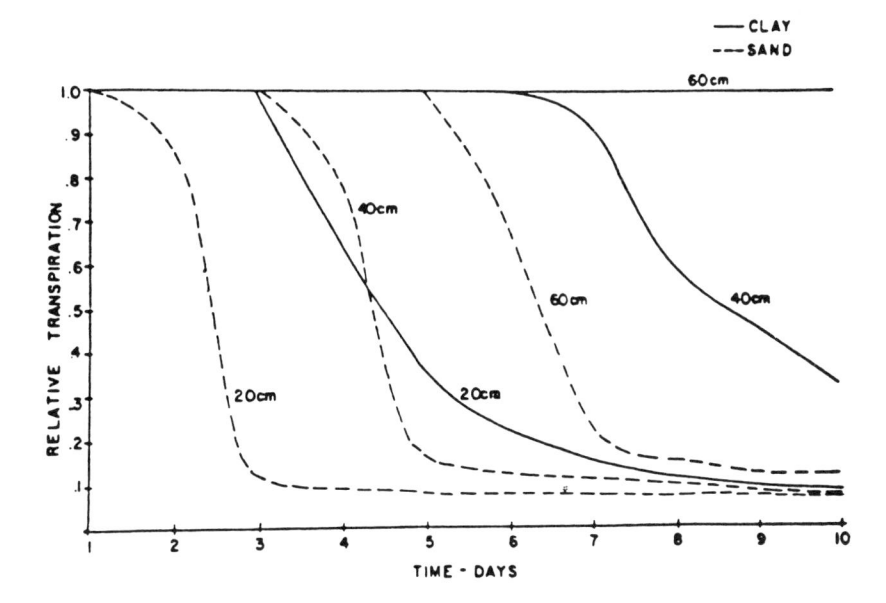

FIGURE 54. Time changes in simulated transpiration as affected by root-zone depth and soil type. (From Zur, B. and Jones, J. W., *Water Resourc. Res.*, 17, 311, 1981. Copyright by the American Geophysical Union.)

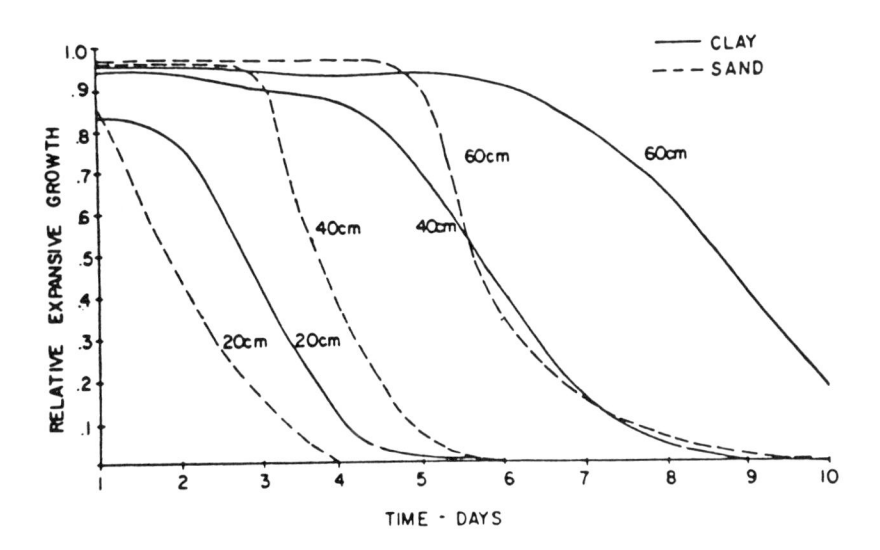

FIGURE 55. Time changes in daily relative expansive growth as affected by root-zone depth and soil types. (From Zur, B. and Jones, J. W., *Water Resourc. Res.*, 17, 311, 1981. Copyright by the American Geophysical Union.)

REFERENCES

1. Moisture and Fertility, American Potash Institute, Atlanta, 1973.
2. **Erie, L. J., French, O. F., and Harris, K.,** Consumptive use of water by Arizona crops, Tech. Bull. 169, Arizona Agricultural Experiment Station, Tucson, 1965.
3. **Begg, J. E. and Turner, N. C.,** Crop water deficits, *Adv. Agron.,* 28, 161, 1976.
4. **Bierhuizen, J. F. and Slatyer, R. O.,** Effect of atmospheric concentration of water vapor and CO_2 in determining transpiration-photosynthesis relationship of cotton leaves, *Agric. Meteorol.,* 2, 259, 1965.
5. **Boyer, J. S.,** Relationship of water potential to growth of leaves, *Plant Physiol.,* 43, 1056, 1968.
6. **Boyer, J. S.,** Water deficits and photosynthesis, in *Water Deficits and Plant Growth,* Vol. 4, Kozlowski, T. T., Ed., Academic Press, New York, 1976.
7. **Boyer, J. S.,** Plant productivity and environment, *Science,* 218, 443, 1982.
8. **Boyer, J. S. and Bowen, B. L.,** Inhibition of oxygen evolution and chloroplasts from leaves with low water potentials, *Plant Physiol.,* 45, 612, 1970.
9. **Brady, N. C.,** *The Nature and Properties of Soils,* 8th ed., Macmillan, New York, 1974.
10. **Briggs, L. J. and Shantz, H. L.,** Relative water requirements of plants, *Agric. Res.,* 3, 1, 1914.
11. **Brix, H.,** The effect of water stress on the rates of photosynthesis and respiration in tomato plants and loblolly pine seedlings, *Physiol. Plant.,* 15, 10, 1962.
12. **Brouwer, R.,** The influence of the suction tension of the nutrient solutions on growth, transpiration and diffusion pressure deficit on bean leaves *(Phaseolus vulgaris), Acta Bot. Neerl.,* 12, 248, 1963.
13. **Cowan, I. R.,** Transport of water in soil-plant-atmosphere, *J. Appl. Ecol.,* 2, 221, 1965.
14. **Curtis, O. F. and Clark, D. G.,** *An Introduction to Plant Physiology,* McGraw Hill, New York, 1950.
15. **Denmead, O. T. and Shaw, R. H.,** Availability of soil water to plants as affected by soil moisture content and meteorological conditions, *Agron. J.,* 54, 385, 1962.
16. **de Roo, H. C.,** Root growth in Connecticut tobacco soils, Bull. 608, Connecticut Agricultural Experiment Station, New Haven, 1957.
17. **Devlin, R. M.,** *Plant Physiology,* Reinhold, New York, 1968.
18. **Doorenbos, J. and Pruitt, W. O.,** Crop Water Requirements, Irrigation Drainage Paper 24, Food and Agriculture Organization, Rome, 1977.
19. **Eagleson, P. S.,** *Dynamic Hydrology,* McGraw-Hill Cook Company, 1970.
20. **Ferry, J. F. and Ward, H. S.,** *Fundamentals of Plant Physiology,* Macmillan, New York, 1959.
21. **Foth, H. D.,** *Fundamentals of Soil Science,* 6th ed., John Wiley & Sons, New York, 1978.
22. **Fuller, H. J. and Tippo, O.,** *College Botany,* rev. ed., Henry Holt, 1954.
23. **Galston, A. W., Davies, P. J., and Satter, R. L.,** *The Life of the Green Plant,* 3rd ed., Prentice-Hall, Englewood Cliffs, N.J., 1980.
24. **Gardner, W. R.,** Relation of root distribution to water uptake and availability, *Agron. J.,* 56, 41, 1964.
25. **Gardner, W. and Widtsoe, J. A.,** The movement of soil moisture, *Soil Sci.,* 11, 215, 1921.
26. **Hagan, R. M., Vaadia, Y., and Russell, M. B.,** Interpretations of plant responses to soil moisture regimes, *Adv. Agron.,* 11, 77, 1959.
27. **Hansen, V. E., Israelsen, O. W., and Stringham, G. E.,** *Irrigation Principles and Practices,* 4th ed., John Wiley & Sons, New York, 1980.
28. **Henderson, D. W., Hagan, R. M., and Mikkelsen, D. S.,** Water and efficiency in irrigation agriculture, in *Moisture and Fertility,* American Potash Institute, Atlanta, 1963.
29. **Hillel, D.,** *Soil and Water,* Academic Press, New York, 1971.
30. **Hillel, D.,** *Applications of Soil Physics,* Academic Press, New York, 1980.
31. **Hofler, K.,** Ein Schema für die osmotische Leistung der Pflanzenzelle, *Ber. Dtsch. Bot. Ges.,* 38, 288, 1920.
32. Irrigation Guide for Indiana, Soil Conservation Service, United States Department of Agriculture, Indianapolis, 1977.
33. **Jamison, V. C. and Beale, O. W.,** Irrigating corn in humid regions, *U.S. Dept. Agric. Farmers' Bull.,* No. 2143, 1959.
34. **Kelly, O. J.,** Requirement and availability to soil water, *Adv. Agron.,* 6, 67, 1954.
35. **Kohnke, H.,** *Soil Physics,* McGraw-Hill, New York, 1968.
36. **Lee, R.,** The hydrologic importance of transpiration control by stomata, *Water Resourc. Res.,* 3, 737, 1967.
37. **Levitt, J.,** *An Introduction to Plant Physiology,* 2nd ed., C. V. Mosby, St. Louis, 1974.
38. **Lundstrom, D. R., Stegman, E. C., and Werner, H. D.,** Irrigation scheduling by the checkbook method, in *Irrigation Scheduling for Water Energy Conservation in the 80s, Proc. Agricultural Engineers Irrigation Scheduling Conf.,* American Society of Agricultural Engineers, St. Joseph, Mich., 1981, 187.
39. **Martin, E. V. and Clements, F. E.,** Studies of the effects of artificial wind on growth and transpiration in *Helianthus annuus, Plant Physiol.,* 10, 613, 1935.

40. **Merva, G. E.,** *Physioengineering Principles,* AVI, Westport, Conn., 1975.
41. **Meyer, B. S., Anderson, D. B., Bohning, R. H., and Fratianne, D. G.,** *Introduction to Plant Physiology,* 2nd ed., D Van Nostrand, New York, 1973.
42. **Miller, D. H.,** *Water at the Surface of the Earth,* Academic Press, New York, 1977.
43. **Pereira, H. C.,** *Land Use and Water Resources in Temperate and Tropical Climates,* Cambridge University Press, London, 1973.
44. **Peters, D. B.,** Use of water by plants, *Plant Food Rev.,* 10(2), 12, 1964.
45. **Philips, J. R.,** Evaporation moisture and heat fields in the soil, *J. Meteorol.,* 14, 354, 1957.
46. **Robins, J. S. and Domingo, C. E.,** Some effects of severe soil moisture deficits on specific growth stages in corn, *Agron. J.,* 45, 618, 1953.
47. **Russell, M. B.,** Interactions of water and soil, *Adv. Agron.,* 11, 35, 1959.
48. **Shaw, R. H.,** Estimation of Soil Moisture Under Corn, Bull. 520, Iowa Agricultural and Home Economics Experiment Station, Ames, 1963.
49. **Shaw, R. H.,** Estimates of yield reductions in corn caused by water in temperature stress, in *Crop Reactions to Water and Temperature Stresses in Humid, Temperate Climates,* Raper, C. D., Jr., and Kramer, P. J., Eds., Westville Press, Boulder, Colo., 1983, 49.
50. **Shaw, R. H. and Laing, D. R.,** Moisture stress and plant response, in *Plant Environment and Efficient Water Use,* Pierre, W. H., Ed., American Society of Agronomy, Madison, 1965.
51. **Singh, S. D.,** Moisture sensitive growth stages of dwarf wheat and optional sequencing of evapotranspiration deficits, *Agron. J.,* 73, 387, 1981.
52. **Slatyer, R. O.,** The influence of progressive increases in total soil moisture stress on transpiration, growth, and internal water relations in plants, *Aust. J. Biol. Sci.,* 10, 320, 1957.
53. **Slatyer, R. O.,** *Plant-Water Relationships,* Academic Press, New York, 1967.
54. **Sutcliffe, J.,** *Plants and Water,* No. 14, The Institute of Biology's Studies in Biology, St. Martin's Press, New York, 1968.
55. **Tanner, C. B.,** Evaporation of water from plants and soil, in *Water Deficits and Plant Growth,* Vol. 1, Kozlowski, T. T., Ed., Academic Press, New York, 1968.
56. **Thorne, D. W. and Peterson, H. B.,** *Irrigated Soils, Their Fertility and Management,* Blakiston, Philadelphia, 1949.
57. **Thornthwaite, C. W.,** The climate of North America according to a new classification, *Geog. Rev.,* 21, 633, 1931.
58. **Trewartha, G. T.,** *An Introduction to Climate,* McGraw-Hill, New York, 1954.
59. Soil Conservation Bull. 199, U.S. Department of Agriculture, Washington, D.C.
60. **Wadleigh, C. H., Raney, W. A., and Herchfield, D. M.,** The moisture problem, in *Plant Environment and Efficient Water Use,* Pierre, W. H., Ed., American Society of Agronomy, Madison, 1966.
61. **Weaver, J. E.,** *The Ecological Relation of Roots,* Publ. 286, Carnegie Institute, Washington, D.C., 1919.
62. **Weaver, J. E. and Clements, F. E.,** *Plant Ecology,* 2nd ed., McGraw-Hill, New York, 1938.
63. **Zur, B. and Jones, J. W.,** A model for the water relations, photosynthesis, and expansive growth of crops, *Water Resourc. Res.,* 17, 311, 1981.

SALT TOLERANCE OF PLANTS

E. V. Maas*

INTRODUCTION

Salt tolerance of plants not only varies considerably among species but also depends heavily upon the cultural conditions under which the crop is grown. Many plant, soil, water, and environmental factors interact to influence the salt tolerance of a plant. Consequently, plant responses to known salt concentrations cannot be predicted on an absolute basis. Nevertheless, plants can be compared on a relative basis to provide general salt-tolerance guidelines. This chapter presents these salt-tolerance data as well as tolerance limits for boron, chloride, and sodium.

SALT-TOLERANCE CRITERIA

Plant tolerance to salinity is usually appraised in one of three ways: (1) the ability of a plant to survive on saline soils, (2) the absolute plant growth or yield, and (3) the relative growth or yield on saline soil compared with that on nonsaline soil. Although plant survival on saline soils is useful to ecologists, this criterion is of little value for growers because it often bears little relation to yield reduction within commercially acceptable limits.

Absolute yields permit direct estimations of economic returns under specified salinity conditions but they also reflect responses to many other environmental parameters; e.g., climate, soil moisture regime, soil fertility, cultural practices, pest and disease control, etc. Furthermore, absolute yield responses do not permit intercrop comparisons because yields for different crops are not expressed in comparable terms. Relative yield, the third criterion, is the yield of a crop grown under saline conditions expressed as a fraction of its yield obtained under nonsaline conditions. On a relative basis, one can compare crops whose yields are expressed in different units. The reliability of relative salt-tolerance data depends upon the degree to which yield reductions are unaffected by the interaction with other essential factors. If reductions in relative yield are independent of differences in absolute yield caused by irrigation, climate, fertility, or other variables, the relative yield-salinity relationship permits a useful expression of plant tolerance.

Salinity affects plants at all stages of development but sensitivity sometimes varies from one growth stage to the next. A comparison of salt tolerance during germination and emergence with later growth stages is difficult because different criteria must be used to evaluate plant response. A comparison of 50% reduction in yield and emergence of several crops** is shown in Table 1. Tolerance at emergence is based on survival while tolerance after emergence is based on decreases in growth or yield. Usually crops are as tolerant, if not more so, at the germination stage as at later stages. One known exception is sugar beet which is more sensitive during germination. Other crops, e.g., barley, corn, cowpea, rice, sorghum, and wheat, are most sensitive during early seedling growth and then become increasingly tolerant during later stages of growth and development.

FACTORS INFLUENCING SALT TOLERANCE

Many environmental and edaphic factors interact with salinity to influence crop salt

---

* The article cited was prepared by a U.S. Government employee as part of his official duties and cannot legally be copyrighted.

** Botanical and common names used throughout this chapter follow the conventions of *Hortus Third*[1] where possible.

Table 1
RELATIVE SALT TOLERANCE OF VARIOUS CROPS AT EMERGENCE AND DURING GROWTH TO MATURITY

		Electrical conductivity of soil saturation extract		
Common name	Botanical name[a]	50% yield[b] (dS/m)	50% emergence[c] (dS/m)	Source of emergence data
Barley	*Hordeum vulgare*	18	16—24	6—9
Cotton	*Gossypium hirsutum*	17	15	9, 10
Sugar beet	*Beta vulgaris*	15	6—12	6, 7
Sorghum	*Sorghum bicolor*	15	13	7, 11
Safflower	*Carthamus tinctorius*	14	12	12
Wheat	*Triticum aestivum*	13	14—16	7, 8, 13
Beet, red	*Beta vulgaris*	9.6	13.8	8
Cowpea	*Vigna unguiculata*	9.1	16	14
Alfalfa	*Medicago sativa*	8.9	8—13	6—8
Tomato	*Lycopersicon lycopersicum*	7.6	7.6	8
Cabbage	*Brassica oleracea capitata*	7.0	13	10
Corn	*Zea mays*	5.9	21—24	6, 7, 9, 13
Lettuce	*Lactuca sativa*	5.2	11	8
Onion	*Allium cepa*	4.3	5.6—7.5	7, 16
Rice	*Oryza sativa*	3.6	18	9—15
Bean	*Phaseolus vulgaris*	3.6	8.0	6

[a] Botanical and common names follow the conventions of *Hortus Third*[1] where possible.
[b] Yield data from References 2, 12, and 14.
[c] Emergence percentage of saline treatments determined when nonsaline control treatments attained maximum emergence.

tolerance. Temperature, relative humidity, and air pollution are important climatic factors that influence plant response to salinity. Most crops are more sensitive to salinity under hot, dry conditions than under cool, humid ones. Air pollution increases the apparent salt tolerance of oxidant-sensitive crops. Ozone, a major air pollutant, decreases yields of some crops more under nonsaline than saline conditions. Soil conditions also influence the apparent salt tolerance of many crops. Plants grown on infertile soils may seem more salt tolerant than those grown with adequate fertility because fertility, not salinity, is the primary factor limiting growth. Proper fertilization would increase yields whether the soil was saline or not, but proportionately more if it were nonsaline. Water stress is another important factor. As plants extract water from the soil, the remaining soil water becomes more concentrated; consequently, plants experience increased salt stress as well as water stress when the soil water is depleted.

Except for controlled experimental conditions, soil salinity is seldom constant with time or uniform in space. In fact salinities often vary from concentrations approximately that of the irrigation water near the soil surface to concentrations many times higher at the bottom of the root zone. Since soil salinity also increases between irrigations as the soil water is evapotranspired, plants must respond to a heterogeneous system that is continually changing. Current evidence indicates that plant growth is most closely related to the soil-water salinity in that part of the root zone where maximum water uptake occurs, i.e., zones with the highest total water potential. Crop yields therefore should be related to time-integrated salinity measured in the zone where roots absorb most of the water. Under high-frequency irrigation, this zone corresponds primarily to the upper part of the root zone where soil salinity is influenced mostly by the salinity of the irrigation water. With infrequent irrigation the zone of maximum water uptake becomes larger as the plant is forced to extract increasingly saline water from increasingly greater depths.

SALT TOLERANCE OF HERBACEOUS CROPS

This assessment of the salt tolerance of agricultural crops was made by updating a previous comprehensive analysis of all available salt-tolerance data.[2] Much of the data were obtained where crops were grown under conditions simulating recommended cultural and management practices for commercial production. Consequently, the data provide relative tolerances of different crops grown under different conditions and not under some standardized set of conditions.

Yield-response curves indicate that most crops tolerate salinity up to a threshold level above which yields decrease approximately linearly as salinity increases. Most deviations from linearity can be disregarded because they occur at salinities that reduce yields to commercially unacceptable levels. Yield-response curves provide two essential parameters sufficient for expressing salt tolerance: threshold, the maximum allowable salinity without yield reduction below that for nonsaline conditions, and slope, the precent yield decrease per unit increase in salinity beyond the threshold. Table 2 lists these salt-tolerance parameters for many field, forage, vegetable, and fruit crops. The data are presented in terms of the electrical conductivity of an extract of a saturated-soil paste,[3] κ_e, and the osmotic potential of the soil water at field capacity, $\psi_{o,fc}$, both at 25°C. The first expresses salt tolerance in terms of the traditional soil salinity measurement with units of decisiemens per meter (dS/m = mmho/cm). The second expresses tolerance in terms of the primary variable influencing plant response, i.e., the osmotic potential of the soil solution bathing the roots. Two significant digits are given for the convenience of separating crops but do *not* necessarily imply that level of accuracy. These data serve only as a guideline to relative tolerances among crops. Absolute tolerances vary depending upon climate, soil conditions, and cultural practices.

From these parameters, relative yield, Y_r, for any given soil salinity exceeding the threshold, can be calculated by either Equation 1 or 2:

$$Y_r = 100 - B(\kappa_e - A) \qquad (1)$$

$$Y_r = 100 - B(\psi_{o,fc} - A) \qquad (2)$$

in which A = the salinity threshold, and B = the percent yield decrease per unit salinity increase above the threshold. For example, barley yields decrease approximately 5% per dS/m when κ_e exceeds 8 dS/m (see Table 2); therefore, at a soil salinity of 13 dS/m, the relative yield, $Y_r = 100 - 5(13 - 8) = 75\%$. Although $\psi_{o,fc}$ is not a linear function of κ_e (see Footnote c, Table 2), relative yields calculated from Equation 2 are within 1 to 2% of those calculated from Equation 1.

The data in Table 2 and subsequent salt-tolerance tables apply to soils where chloride is the predominant anion. Because of the dissolution of gypsum in preparing saturated-soil pastes, κ_e of gypsiferous soils will range from 1 to 3 dS/m higher than that of nongypsiferous soils having the same conductivity in the soil water at field capacity.[4] Therefore, plants grown on gypsiferous soils will tolerate κ_es approximately 2 dS/m higher than those indicated in the tables.

Four qualitative salt-tolerance ratings with limits defined by the division boundaries in Figure 1 are also given for quick relative comparisons among crops. Some crops are listed with only a qualitative rating because experimental data are inadequate to permit a quantitative evaluation.

Since salinity treatments were usually applied after seedlings became established in nonsaline plots, the data indicated salt tolerance of crops after the late seedling stage. Of the crops tested, only a few indicated greater sensitivity during germination and early seedling stages than at later stages. The tolerance levels for those crops having exceptionally sensitive stages are indicated in the footnotes.

Table 2
SALT TOLERANCE OF HERBACEOUS CROPS[a]

		Electrical conductivity of saturated-soil extract		Osmotic potential of soil solution at field capacity[c]			
Common name	Botanical name	Threshold[b] (dS/m)	Slope (% per dS/m)	Threshold (bar)	Slope (%/bar)	Rating[d]	Ref.

Fiber, Grain, and Special Crops

Common name	Botanical name	Threshold (dS/m)	Slope (% per dS/m)	Threshold (bar)	Slope (%/bar)	Rating	Ref.
Barley[e]	*Hordeum vulgare*	8.0	5.0	6.6	5.5	T	2
Bean	*Phaseolus vulgaris*	1.0	19	0.7	23	S	2
Broad bean	*Vicia faba*	1.6	9.6	1.2	11	MS	2
Corn	*Zea mays*	1.7	12	1.3	14	MS	2
Cotton	*Gossypium hirsutum*	7.7	5.2	6.3	5.7	T	2
Cowpea	*Vigna unguiculata*	4.9	12	3.9	14	MT	14
Flax	*Linum usitatissimum*	1.7	12	1.3	14	MS	2
Guar	*Cyamopsis tetragonoloba*	—	—	—	—	MT	22
Millet, foxtail	*Setaria italica*	—	—	—	—	MS	2
Oats	*Avena sativa*	—	—	—	—	MT*	
Peanut	*Arachis hypogaea*	3.2	29	2.5	34	MS	2
Rice, paddy[f]	*Oryza sativa*	3.0[g]	12[g]	2.3	14	S	2
Rye	*Secale cereale*	—	—	—	—	MT*	
Safflower	*Carthamus tinctorius*	—	—	—	—	MT	2
Sesame	*Sesamum indicum*	—	—	—	—	S	17
Sorghum	*Sorghum bicolor*	6.8	16	5.5	18	MT	18
Soybean	*Glycine max*	5.0	20	4.0	23	MT	2
Sugar beet[h]	*Beta vulgaris*	7.0	5.9	5.7	6.5	T	2
Sugarcane	*Saccharum officinarum*	1.7	5.9	1.3	6.7	MS	2
Sunflower	*Helianthus annuus*	—	—	—	—	MS*	
Triticale	× *Triticosecale*	6.1	2.5	4.9	2.7	T	22
Wheat[e]	*Triticum aestivum*	6.0	7.1	4.8	7.9	MT	2
Wheat (semidwarf)[i]	*T. aestivum*	8.6	3.0	7.1	3.2	T	19
Wheat, durum	*T. turgidum*	5.9	3.8	4.8	4.2	T	19

Grasses and Forage Crops

Common name	Botanical name	Threshold (dS/m)	Slope (% per dS/m)	Threshold (bar)	Slope (%/bar)	Rating	Ref.
Alfalfa	*Medicago sativa*	2.0	7.3	1.5	8.4	MS	2
Alkaligrass, Nuttall	*Puccinellia airoides*	—	—	—	—	T*	
Alkali sacaton	*Sporobolus airoides*	—	—	—	—	T*	
Barley (forage)[e]	*Hordeum vulgare*	6.0	7.1	4.8	7.9	MT	2
Bent grass	*Agrostis stolonifera palustris*	—	—	—	—	MS	2
Bermuda grass[j]	*Cynodon dactylon*	6.9	6.4	5.6	7.1	T	2
Bluestem, Angleton	*Dichanthium aristatum*	—	—	—	—	MS*	
Brome, mountain	*Bromus marginatus*	—	—	—	—	MT*	
Brome, smooth	*B. inermis*	—	—	—	—	MS	2
Buffel grass	*Cenchrus ciliaris*	—	—	—	—	MS*	
Burnet	*Poterium sanguisorba*	—	—	—	—	MS*	
Canary grass, reed	*Phalaris arundinacea*	—	—	—	—	MT	2
Clover, alsike	*Trifolium hybridum*	1.5	12	1.1	14	MS	2

Table 2 (continued)
SALT TOLERANCE OF HERBACEOUS CROPS[a]

Common name	Botanical name	Electrical conductivity of saturated-soil extract Threshold[b] (dS/m)	Slope (% per dS/m)	Osmotic potential of soil solution at field capacity[c] Threshold (bar)	Slope (%/bar)	Rating[d]	Ref.
\multicolumn{8}{c}{Grasses and Forage Crops}							
Clover, Berseem	*T. alexandrinum*	1.5	5.7	1.1	6.5	MS	2
Clover, Hubam	*Melilotus alba*	—	—	—	—	MT*	
Clover, ladino	*Trifolium repens*	1.5	12	1.1	14	MS	2
Clover, red	*T. pratense*	1.5	12	1.1	14	MS	2
Clover, strawberry	*T. fragiferum*	1.5	12	1.1	14	MS	2
Clover, sweet	*Melilotus*	—	—	—	—	MT*	
Clover, white Dutch	*Trifolium repens*	—	—	—	—	MS*	
Corn (forage)	*Zea mays*	1.8	7.4	1.4	8.5	MS	2
Cowpea (forage)	*Vigna unguiculata*	2.5	11	1.9	13	MS	14
Dallis grass	*Paspalum dilatatum*	—	—	—	—	MS*	
Fescue, tall	*Festuca elatior*	3.9	5.3	3.1	5.9	MT	2
Fescue, meadow	*F. pratensis*	—	—	—	—	MT*	
Foxtail, meadow	*Alopecurus pratensis*	1.5	9.6	1.1	11	MS	2
Grama, blue	*Bouteloua gracilis*	—	—	—	—	MS*	
Harding grass	*Phalaris tuberosa*	4.6	7.6	3.7	8.6	MT	2
Kallar grass	*Diplachne fusca*	—	—	—	—	T*	
Love grass[k]	*Eragrostis* sp.	2.0	8.4	1.5	9.7	MS	2
Milk vetch, Cicer	*Astragalus cicer*	—	—	—	—	MS*	
Oatgrass, tall	*Arrhenatherum, Danthonia*	—	—	—	—	MS*	
Oats (forage)	*Avena sativa*	—	—	—	—	MS*	
Orchard grass	*Dactylis glomerata*	1.5	6.2	1.1	7.1	MS	2
Panic grass, blue	*Panicum antidotale*	—	—	—	—	MT*	
Rape	*Brassica napus*	—	—	—	—	MT*	
Rescue grass	*Bromus unioloides*	—	—	—	—	MT*	
Rhodes grass	*Chloris gayana*	—	—	—	—	MT	2
Rye (forage)	*Secale cereale*	—	—	—	—	MS*	
Ryegrass, Italian	*Lolium italicum multiflorum*	—	—	—	—	MT*	
Ryegrass, perennial	*L. perenne*	5.6	7.6	4.5	8.5	MT	2
Salt grass, desert	*Distichlis stricta*	—	—	—	—	T*	
Sesbania[f]	*Sesbania exaltata*	2.3	7.0	1.8	8.0	MS	2
Siratro	*Macroptilium atropurpureum*	—	—	—	—	MS	20
Sphaerophysa	*Sphaerophysa salsula*	2.2	7.0	1.7	8.0	MS	21
Sudan grass	*Sorghum sudanense*	2.8	4.3	2.2	4.8	MT	2
Timothy	*Phleum pratense*	—	—	—	—	MS*	
Trefoil, big	*Lotus uliginosus*	2.3	19	1.8	23	MS	2
Trefoil, narrow-leaf birdsfoot	*L. corniculatus tenuifolium*	5.0	10	4.0	11	MT	2
Trefoil, broad-leaf birdsfoot[l]	*L. corniculatus arvensis*	—	—	—	—	MT	2
Vetch, common	*Vicia angustifolia*	3.0	11	2.3	13	MS	2
Wheat (forage)[i]	*Triticum aestivum*	4.5	2.6	3.6	2.8	MT	19
Wheat, durum (forage)	*T. turgidum*	2.1	2.5	1.6	2.7	MT	19

Table 2 (continued)
SALT TOLERANCE OF HERBACEOUS CROPS[a]

Common name	Botanical name	Electrical conductivity of saturated-soil extract Threshold[b] (dS/m)	Slope (% per dS/m)	Osmotic potential of soil solution at field capacity[c] Threshold (bar)	Slope (%/bar)	Rating[d]	Ref.
\multicolumn{8}{c}{Grasses and Forage Crops}							
Wheat grass, standard crested	*Agropyron sibiricum*	3.5	4.0	2.7	4.4	MT	2
Wheat grass, fairway crested	*A. cristatum*	7.5	6.9	6.1	7.7	T	2
Wheat grass, intermediate	*A. intermedium*	—	—	—	—	MT*	
Wheat grass, slender,	*A. trachycaulum*	—	—	—	—	MT	2
Wheat grass, tall	*A. elongatum*	7.5	4.2	6.1	4.6	T	2
Wheat grass, western	*A. smithii*	—	—	—	—	MT*	
Wild rye, Altai	*Elymus angustus*	—	—	—	—	T	2
Wild rye, beardless	*E. triticoides*	2.7	6.0	2.1	6.8	MT	2
Wild rye, Canadian	*E. canadensis*	—	—	—	—	MT*	
Wild rye, Russian	*E. junceus*	—	—	—	—	T	2
\multicolumn{8}{c}{Vegetable and Fruit Crops}							
Artichoke	*Helianthus tuberosus*	—	—	—	—	MT*	
Asparagus	*Asparagus officinalis*	4.1	2.0	3.2	2.1	T	22
Bean	*Phaseolus vulgaris*	1.0	19	0.7	23	S	2
Beet, red[h]	*Beta vulgaris*	4.0	9.0	3.2	10	MT	2
Broccoli	*Brassica oleracea botrytis*	2.8	9.2	2.2	11	MS	2
Brussels sprouts	*B. oleracea gemmifera*	—	—	—	—	MS*	
Cabbage	*B. oleracea capitata*	1.8	9.7	1.4	12	MS	2
Carrot	*Daucus carota*	1.0	14	0.7	17	S	2
Cauliflower	*B. oleracea botrytis*	—	—	—	—	MS*	
Celery	*Apium graveolens*	1.8	6.2	1.4	7.1	MS	25
Corn, sweet	*Zea mays*	1.7	12	1.3	14	MS	2
Cucumber	*Cucumis sativus*	2.5	13	1.9	15	MS	2
Eggplant	*Solanum melongena esculentum*	—	—	—	—	MS*	
Kale	*B. oleracea acephala*	—	—	—	—	MS*	
Kohlrabi	*B. oleracea gongylodes*	—	—	—	—	MS*	
Lettuce	*Lactuca sativa*	1.3	13	1.0	16	MS	2
Muskmelon	*Cucumis melo*	—	—	—	—	MS	23
Okra	*Abelmoschus esculentus*	—	—	—	—	S	2
Onion	*Allium cepa*	1.2	16	0.9	19	S	2
Parsnip	*Pastinaca sativa*	—	—	—	—	S*	
Pea	*Pisum sativum*	—	—	—	—	S*	
Pepper	*Capsicum annuum*	1.5	14	1.1	17	MS	2
Potato	*Solanum tuberosum*	1.7	12	1.3	14	MS	2
Pumpkin	*Cucurbita pepo pepo*	—	—	—	—	MS*	

Table 2 (continued)
SALT TOLERANCE OF HERBACEOUS CROPS[a]

Common name	Botanical name	Electrical conductivity of saturated-soil extract Threshold[b] (dS/m)	Slope (% per dS/m)	Osmotic potential of soil solution at field capacity[c] Threshold (bar)	Slope (%/bar)	Rating[d]	Ref.
Vegetable & Fruit Crops							
Radish	*Raphanus sativus*	1.2	13	0.9	16	MS	2
Spinach	*Spinacia oleracea*	2.0	7.6	1.5	8.8	MS	2
Squash, scallop	*Cucurbita pepo melopepo*	3.2	16	2.5	19	MS	24
Squash, zucchini	*C. pepo melopepo*	4.7	9.4	3.7	11	MT	24
Strawberry	*Fragaria* sp.	1.0	33	0.7	41	S	2
Sweet potato	*Ipomoea batatas*	1.5	11	1.1	13	MS	2
Tomato	*Lycopersicon lycopersicum*	2.5	9.9	1.9	12	MS	2
Turnip	*Brassica rapa*	0.9	9.0	0.6	11	MS	26
Watermelon	*Citrullus lanatus*	—	—	—	—	MS*	

[a] These data serve only as a guideline to relative tolerances among crops. Absolute tolerances vary depending upon climate, soil conditions, and cultural practices.

[b] In gypsiferous soils plants will tolerate κ_es about 2 dS/m higher than indicated.

[c] $\psi_{o,fc}$ was calculated from κ_e by the equation, $\psi_{o,fc} = -0.725 \kappa_e^{1.06}$. This relationship was obtained from Figure 6 of the *USDA Handbook No. 60*,[3] after converting osmotic pressure in atmospheres at 0°C to osmotic potential in bars at 25°C, and is based on the assumption that the soluble salt concentration in the soil water at field capacity is twice that of the saturated-soil extract.

[d] Ratings are defined by the boundaries in Figure 1. Ratings with an asterisk (*) are estimates. For references consult the indexed bibliography by Francois and Maas.[27,28]

[e] Less tolerant during emergence and seedling stage. κ_e at this stage should not exceed 4 or 5 dS/m.

[f] Less tolerant during emergence and seedling stage.

[g] Because paddy rice is grown under flooded conditions, values refer to the electrical conductivity of the soil water while the plants are submerged.

[h] Sensitive during germination. κ_e should not exceed 3 dS/m.

[i] Data from one cultivar, 'Probred'.

[j] Average of several varieties. 'Suwannee' and 'Coastal' are about 20% more tolerant and common and 'Greenfield' are about 20% less tolerant than the average.

[k] Average for 'Boer', 'Wilman', 'Sand', and 'Weeping' cultivars. 'Lehmann' seems about 50% more tolerant.

[l] Broadleaf birdsfoot trefoil seems less tolerant than narrowleaf.

Although differences in salt tolerance are not common among cultivars, significant differences have been observed for Bermuda grass, bromegrass, birdsfoot trefoil, rice, creeping bent grass, barley, wheat, soybean, and berseem clover. It is not always clear, however, whether cultivar differences reflect differences in salt tolerance or differences in adaptation to climatic or nutritional conditions under which the tests were conducted. Consequently, the data in Table 2 are means for all the cultivars studied. Some known cultivar differences in salt tolerance are indicated in footnotes.

SALT TOLERANCE OF WOODY CROPS

The salt tolerance of woody crops is complicated because they are also influenced by specific salt constituents. Hence, the concentration of specific ions, as well as total salinity, must be taken into account. Tree, vine, and other woody crops are susceptible to foliar injury caused by the toxic accumulation of Cl^- and/or Na^+ in the leaves. Because different

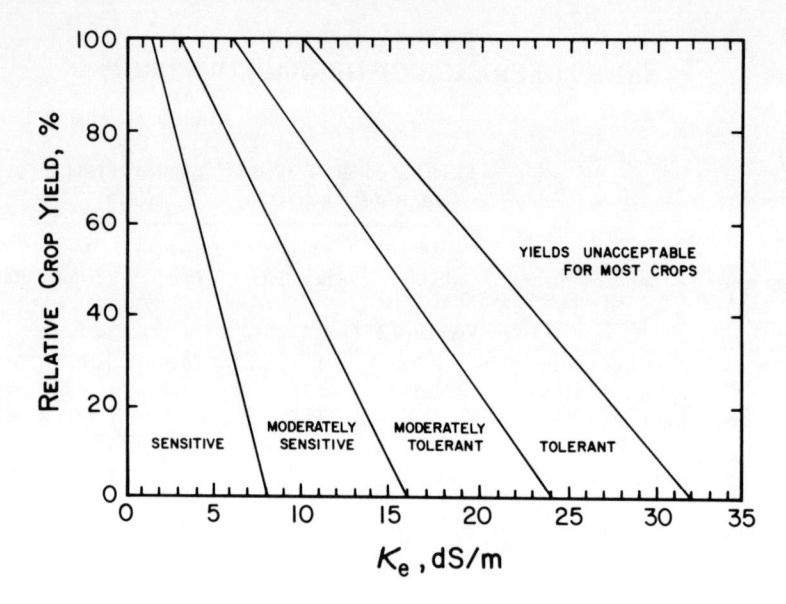

FIGURE 1. Divisions for classifying crop tolerance to salinity.

cultivars or rootstocks absorb Cl⁻ and Na⁺ at markedly different rates, each cultivar or rootstock must be considered individually. Tolerance to specific ions will be discussed later. With cultivars or rootstocks that restrict the uptake of Cl⁻ or Na⁺, osmotic effects predominate and salt tolerance can be expressed as a function of $\psi_{o,fc}$ or κ_e. The data for woody crops generally are not as comprehensive as for herbaceous crops. Nevertheless, the salt-tolerance ratings in Table 3 are believed to be reasonably accurate in the absence of specific-ion toxicities. Some tolerances are based on vegetative growth because of the lack of other yield data (see footnotes). In contrast to other crop groups, most woody fruit and nut crops tend to be salt sensitive, even in the absence of specific-ion toxicity. Only date palm is relatively salt tolerant, while olive and a few others are moderately salt tolerant.

SALT TOLERANCE OF ORNAMENTALS

Aesthetic factors, such as size and appearance, are the important criteria in setting permissible levels of salinity for ornamentals. Salt-tolerance limits for 49 ornamental shrub, tree, and ground-cover species grown in Riverside, Calif., are given in Table 4. The data indicate the maximum permissible κ_e at which an acceptable appearance can be expected. In some cases growth may be reduced as much as 50% but the plants should appear healthy and attractive. Some woody ornamentals are very salt sensitive and exhibit the characteristic leaf symptoms associated with excess Cl⁻ or Na⁺ accumulation, e.g., rose, Burford holly, and cotoneaster.

SALT TOLERANCE OF SPRINKLED CROPS

Crops irrigated by sprinkler systems are subject to additional salt damage when the foliage is directly wetted by saline water. Susceptibility to foliar injury varies considerably among plant species and depends more on leaf characteristics and rate of foliar absorption than on tolerance to soil salinity. Leaves of deciduous fruit trees (almonds, apricots, plums, etc.) absorb Na⁺ and Cl⁻ readily and are severely damaged. Citrus leaves absorb these ions at a somewhat slower rate but still fast enough to cause damage. Avocado and strawberry, which are very sensitive to soil salinity, absorb salt so slowly into the leaves that foliar

Table 3
SALT TOLERANCE OF WOODY CROPS[a]

Common name	Botanical name	Electrical conductivity of saturated-soil extract Threshold[b] (dS/m)	Slope (% per dS/m)	Osmotic potential of soil solution at field capacity[c] Threshold (bar)	Slope (%/bar)	Rating[d]	Ref.
Almond[e]	Prunus dulcis	1.5	19	1.1	23	S	2
Apple	Malus sylvestris	—	—	—	—	S	2
Apricot[e]	P. armeniaca	1.6	24	1.2	29	S	2
Avocado[e]	Persea americana	—	—	—	—	S	2
Blackberry	Rubus sp.	1.5	22	1.1	26	S	2
Boysenberry	Rubus ursinus	1.5	22	1.1	26	S	2
Castor bean	Ricinus communis	—	—	—	—	MS*	
Cherimoya	Annona cherimola	—	—	—	—	S*	
Cherry, sweet	Prunus avium	—	—	—	—	S*	
Cherry, sand	P. besseyi	—	—	—	—	S*	
Currant	Ribes sp.	—	—	—	—	S*	
Date palm	Phoenix dactylifera	4.0	3.6	3.2	4.0	T	2
Fig	Ficus carica	—	—	—	—	MT*	
Gooseberry	Ribes sp.	—	—	—	—	S*	
Grape[e]	Vitis sp.	1.5	9.6	1.1	11	MS	2
Grapefruit[e]	Citrus paradisi	1.8	16	1.4	19	S	2
Guayule	Parthenium argentatum	15	13	13	15	T	29
Jojoba[e]	Simmondsia chinensis	—	—	—	—	T	30
Jujube	Ziziphus jujuba	—	—	—	—	MT*	
Lemon[e]	C. limon	—	—	—	—	S	2
Lime	C. aurantiifolia	—	—	—	—	S*	
Loquat	Eriobotrya japonica	—	—	—	—	S*	
Mango	Mangifera indica	—	—	—	—	S*	
Olive	Olea europaea	—	—	—	—	MT	2
Orange	C. sinensis	1.7	16	1.3	19	S	2
Papaya[e]	Carica papaya	—	—	—	—	MT	31
Passion fruit	Passiflora edulis	—	—	—	—	S*	
Peach	Prunus persica	1.7	21	1.3	25	S	2
Pear	Pyrus communis	—	—	—	—	S*	
Persimmon	Diospyros virginiana	—	—	—	—	S*	
Pineapple	Ananas comosus	—	—	—	—	MT*	
Plum; prune[e]	Prunus domestica	1.5	18	1.1	22	S	2
Pomegranate	Punica granatum	—	—	—	—	MT*	
Pummelo	Citrus maxima	—	—	—	—	S*	
Raspberry	Rubus idaeus	—	—	—	—	S	2
Rose apple	Syzygium jambos	—	—	—	—	S*	
Sapote, white	Casimiroa edulis	—	—	—	—	S*	
Tangerine	Citrus reticulata	—	—	—	—	S*	

[a] These data are applicable when rootstocks are used that do not accumulate Na^+ or Cl^- rapidly or when these ions do not predominate in the soil

[b] In gypsiferous soils plants will tolerate κ_es about 2 dS/m higher than indicated.

[c] $\psi_{o,fc}$ was calculated from κ_e by the equation $\psi_{o,fc} = -0.725\, \kappa_e^{1.06}$. This relationship was obtained from Figure 6 of the *USDA Handbook No. 60*,[3] after converting osmotic pressure in atmospheres at 0°C to osmotic potential in bars at 25°C and is based on the assumption that the soluble salt concentration in the soil water at field capacity is twice that of the saturated-soil extract.

[d] Ratings are defined by the boundaries in Figure 1. Ratings with an asterisk (*) are estimates. For references consult the indexed bibliography by Francois and Maas.[27,28]

[e] Tolerance is based on growth rather than yield.

Table 4
SALT TOLERANCE OF ORNAMENTAL SHRUBS, TREES, AND GROUND COVER[a]

Common name	Botanical name	Maximum permissible κ_e[b] (dS/m)
Very sensitive		
Star jasmine	*Trachelospermum jasminoides*	1—2
Pyrenees cotoneaster	*Cotoneaster congestus*	1—2
Oregon grape	*Mahonia aquifolium*	1—2
Photinia	*Photinia × fraseri*	1—2
Sensitive		
Pineapple guava	*Feijoa sellowiana*	2—3
Chinese holly, cv. Burford	*Ilex cornuta*	2—3
Rose, cv. Grenoble	*Rosa* sp.	2—3
Glossy abelia	*Abelia × grandiflora*	2—3
Southern yew	*Podocarpus macrophyllus*	2—3
Tulip tree	*Liriodendron tulipifera*	2—3
Algerian ivy	*Hedera canariensis*	3—4
Japanese pittosporum	*Pittosporum tobira*	3—4
Heavenly bamboo	*Nandina domestica*	3—4
Chinese hibiscus	*Hibiscus rosa-sinensis*	3—4
Laurustinus, cv. Robustum	*Viburnum tinus*	3—4
Strawberry tree, cv. Compact	*Arbutus unedo*	3—4
Crape myrtle	*Lagerstroemia indica*	3—4
Moderately sensitive		
Glossy privet	*Ligustrum lucidum*	4—6
Yellow sage	*Lantana camara*	4—6
Orchid tree	*Bauhinia purpurea*	4—6
Southern magnolia	*Magnolia grandiflora*	4—6
Japanese boxwood	*Buxus microphylla* var. *japonica*	4—6
Xylosma	*Xylosma congestum*	4—6
Japanese black pine	*Pinus thunbergiana*	4—6
Indian hawthorn	*Raphiolepis indica*	4—6
Dodonaea, cv. Atropurpurea	*Dodonaea viscosa*	4—6
Oriental arborvitae	*Platycladus orientalis*	4—6
Thorny elaeagnus	*Elaeagnus pungens*	4—6
Spreading juniper	*Juniperus chinensis*	4—6
Pyracantha, cv. Graberi	*Pyracantha fortuneana*	4—6
Cherry plum	*Prunus cerasifera*	4—6
Moderately tolerant		
Weeping bottlebrush	*Callistemon viminalis*	6—8
Oleander	*Nerium oleander*	6—8
European fan palm	*Chamaerops humilis*	6—8
Blue dracaena	*Cordyline indivisa*	6—8
Spindle tree, cv. Grandiflora	*Euonymus japonica*	6—8
Rosemary	*Rosmarinus officinalis*	6—8
Aleppo pine	*Pinus halepensis*	6—8
Sweet gum	*Liquidambar styraciflua*	6—8
Tolerant		
Brush cherry	*Syzygium paniculatum*	>8[c]
Ceniza	*Leucophyllum frutescens*	>8[c]
Natal plum	*Carissa grandiflora*	>8[c]
Evergreen pear	*Pyrus kawakamii*	>8[c]
Bougainvillea	*Bougainvillea spectabilis*	>8[c]
Italian stone pine	*Pinus pinea*	>8[c]
Very tolerant		
White ice plant	*Delosperma alba*	>10[c]
Rosea ice plant	*Drosanthemum hispidum*	>10[c]
Purple ice plant	*Lampranthus productus*	>10[c]
Croceum ice plant	*Hymenocyclus croceus*	>10[c]

Table 4 (continued)
SALT TOLERANCE OF ORNAMENTAL SHRUBS, TREES, AND GROUND COVER[a]

[a] Species are listed in order of increasing tolerance based on appearance as well as growth reduction. Data compiled from References 32 through 34.
[b] Salinities exceeding the maximum permissible κ_e may cause leaf burn, loss of leaves and/or excessive stunting.
[c] Maximum permissible κ_e is unknown. No injury symptoms or growth reduction was apparent at 7 dS/m. The growth of all ice-plant species was increased by soil salinity of 7 dS/m.

Table 5
RELATIVE SUSCEPTIBILITY OF CROPS TO FOLIAR INJURY FROM SALINE SPRINKLING WATERS[a]

Na or Cl conc (mol/m³) causing foliar injury[b]

<5	5—10	10—20	>20
Almond	Grape	Alfalfa	Cauliflower
Apricot	Pepper	Barley	Cotton
Citrus	Potato	Corn	Sugar beet
Plum	Tomato	Cucumber	Sunflower
		Safflower	
		Sesame	
		Sorghum	

[a] Susceptibility based on direct accumulation of salts through the leaves. Data compiled from References 35 through 39.
[b] Foliar injury is influenced by cultural and environmental conditions. These data are presented only as general guidelines for daytime sprinkling.

absorption is a negligible factor in sprinkling. Some nonwoody species that do not appear specifically sensitive to Na$^+$ or Cl$^-$ when surface irrigated may be injured by sprinkling with irrigation waters containing over 10 to 20 mol/m³ Na$^+$ or Cl$^-$. Table 5 lists some susceptible crops and gives approximate salt concentrations in sprinkler irrigation waters that can cause foliar injury symptoms. The list is not all inclusive, as foliar absorption and injury of many crops have yet to be determined.

CHLORIDE TOLERANCE

Most nonwoody crops are not specifically sensitive to Cl$^-$. One exception to this generalization involves certain cultivars of soybeans. Salt-sensitive cultivars accumulate excessive amounts of Cl$^-$ which are toxic to the plants. This problem has been avoided, however, by breeding cultivars that exclude Cl$^-$. For most nonwoody crops, tolerance to Cl$^-$ salinity can be calculated from the κ_e values given in the salt-tolerance tables. If one assumes that the soil salinity consists predominately of Cl$^-$ salts, multiplying κ_e by 10 gives the approximate Cl$^-$ concentration in moles per cubic meter in the saturated-soil extract.

Unlike herbaceous crops, woody plant species generally are susceptible to Cl$^-$ toxicity. However, tolerances vary among species and even among varieties or rootstocks within a

Table 6
CHLORIDE-TOLERANCE LIMITS OF SOME FRUIT-CROP CULTIVARS AND ROOTSTOCKS

Crop	Rootstock or cultivar	Maximum permissible Cl⁻ in soil water without leaf injury[a] (mol/m³)
Rootstocks		
Avocado		
(*Persea americana*)	West Indian	15
	Guatemalan	12
	Mexican	10
Citrus		
(*Citrus* spp.)	Sunki mandarin, grapefruit, Cleopatra mandarin, Rangpur lime	50
	Sampson tangelo, rough lemon,[b] sour orange, Ponkan mandarin	30
	Citrumelo 4475, trifoliate orange, Cuban shaddock, Calamondin, sweet orange, Savage citrange, Rusk citrange, Troyer citrange	20
Grape		
(*Vitis* spp.)	Salt Creek, 1613-3	80
	Dog ridge	60
Stone fruit		
(*Prunus* spp.)	Marianna	50
	Lovell, Shalil	20
	Yunnan	15
Cultivars		
Berries[c]		
(*Rubus* spp.)	Boysenberry	20
	Olallie blackberry	20
	Indian Summer raspberry	10
Grape		
(*Vitis* spp.)	Thompson seedless, Perlette	40
	Cardinal, black rose	20
Strawberry		
(*Fragaria* spp.)	Lassen	15
	Shasta	10

[a] For some crops these concentrations may exceed the osmotic threshold and cause some yield reduction. Data compiled from References 40 through 43.
[b] Data from Australia indicate that rough lemon is more sensitive to Cl⁻ than sweet orange.[44]
[c] Data available for one variety of each species only.

species. These differences usually reflect the capability of the plant to prevent or retard Cl⁻ accumulation in the tops. The tolerances of many fruit crops can be significantly improved by selecting varieties or rootstocks that restrict Cl⁻ accumulation. Table 6 lists Cl⁻-tolerance limits for some fruit-crop rootstocks and cultivars. The data indicate the maximum Cl⁻ concentrations permissible in the soil water that do not cause leaf injury. In some cases, however, the osmotic threshold may be exceeded so that yield is decreased without obvious injury. The list is by no means complete. Quantitative data for many cultivars and rootstocks are not available.

SODIUM TOLERANCE

Plant growth and yield may be affected by the Na⁺ status of soils as well as by salinity.

The effects of Na$^+$ can be both direct and indirect. Direct effects are due to the accumulation of toxic levels of Na$^+$ and are generally limited to woody species. Their sensitivity varies considerably because Na$^+$ uptake varies widely among species and rootstocks. Na$^+$ injury in avocado, citrus, and stone-fruit trees is rather widespread and can occur at Na$^+$ concentrations as low as 5 mol/m^3 in soil water. Na$^+$ is normally retained in the roots and lower trunk of stone-fruit trees, but after 3 or 4 years, the conversion of sapwood to heartwood apparently releases the accumulated Na$^+$ which is transported to the leaves causing leaf burn.[5]

Indirect effects include both nutritional imbalance and impairment of soil physical conditions. The nutritional effects of Na$^+$ are not simply related to the exchangeable Na$^+$ percentage of soils but depend upon the concentrations of Na$^+$, Ca^{2+}, and Mg^{2+} in the soil solution. In sodic, nonsaline soils, total soluble salt concentrations are low and, consequently, Ca^{2+} and/or Mg^{2+} concentrations are often nutritionally inadequate. These deficiencies and not Na$^+$ toxicity per se are usually the primary cause of poor plant growth among nonwoody species. Furthermore, since Na$^+$ uptake by plants is strongly regulated by Ca^{2+} in the soil solution, the presence of sufficient Ca^{2+} is essential to prevent the accumulation of toxic levels of Na$^+$. This is particularly important with Na$^+$-sensitive woody crops. As a general guide, Ca^{2+} and Mg^{2+} concentrations in the soil solution above 1 mol/m^3 each, with Ca^{2+} $\geq$ Mg^{2+}, are nutritionally adequate in nonsaline, sodic soils. As the total salt concentration increases into the saline range, Ca^{2+} concentrations become adequate for most plants and osmotic effects begin to predominate.

Sodic soil conditions affect almost all crops because of the deterioration of soil physical conditions. Dispersion of soil aggregates in sodic soils decreases soil permeability to water and air, thereby reducing plant growth. Therefore, yield reductions may occur in crops that are not specifically sensitive to Na$^+$ because of the combined effects of nutritional problems and impaired soil physical condition.

BORON TOLERANCE

Boron (B) is an essential plant element but it can become toxic to some plants when soil-water concentrations exceed only slightly that required for optimum plant growth. Generally, toxic concentrations in the soil are found only in arid regions of the world. Most surface irrigation waters contain acceptable B levels, but there are some areas where well waters contain toxic levels of B. Since plants vary in their tolerance to B, a water that is marginal for sensitive crops may be used for the more tolerant plants. The relative tolerance of various agronomic and horticultural crops to B in the soil water is indicated in Table 7. These data, with few exceptions, are based on crop tolerance to B levels in sand cultures. They indicate maximum permissible concentrations in the soil water that do not cause yield reductions. Some crops, however, may exhibit leaf injury at lower concentrations without decreasing yield. The crops have been classified in six groups ranging from very sensitive to very tolerant. Like salt tolerance, B tolerance varies with climate, soil conditions, and crop cultivars; consequently a concentration range rather than a specific concentration is given. Yield reductions as a function of B concentrations generally can not be predicted from available data.

Different rootstocks of citrus and stone fruits absorb B at different rates so that tolerance may be improved by using rootstocks that restrict boron uptake. A number of these rootstocks are listed in order of increasing B accumulation in Table 8.

The tolerance of ornamental plants to B is listed in Table 9. Since appearance is often more important than the growth rate of ornamentals, leaf injury as well as growth reduction were taken into account in setting maximum permissible B levels.

Symptoms of excess B often include characteristic chlorotic and necrotic patterns of leaves

Table 7
BORON TOLERANCE LIMITS FOR AGRICULTURAL CROPS

Common name	Botanical name	Threshold[a] (g/m³)	Ref.
Very sensitive			
Lemon[b]	*Citrus limon*	<0.5	45,46
Blackberry[b]	*Rubus* sp.	<0.5	45
Sensitive			
Avocado[b]	*Persea americana*	0.5—0.75	46
Grapefruit[b]	*C. × paradisi*	0.5—0.75	46
Orange[b]	*C. sinensis*	0.5—0.75	46
Apricot[b]	*Prunus armeniaca*	0.5—0.75	47
Peach[b]	*P. persica*	0.5—0.75	45,46
Cherry[b]	*P. avium*	0.5—0.75	45
Plum[b]	*P. domestica*	0.5—0.75	47
Persimmon[b]	*Diospyros kaki*	0.5—0.75	45
Fig, kadota[b]	*Ficus carica*	0.5—0.75	45
Grape[b]	*Vitis vinifera*	0.5—0.75	45
Walnut[b]	*Juglans regia*	0.5—0.75	46
Pecan[b]	*Carya illinoinensis*	0.5—0.75	46
Cowpea[b]	*Vigna unguiculata*	0.5—0.75	45
Onion	*Allium cepa*	0.5—0.75	45
Garlic	*A. sativum*	0.75—1.0	48
Sweet potato	*Ipomea batatas*	0.75—1.0	45
Wheat	*Triticum aestivum*	0.75—1.0	49,50
Barley	*Hordeum vulgare*	0.75—1.0	45,49
Sunflower	*Helianthus annuus*	0.75—1.0	51
Bean, mung[b]	*Vigna radiata*	0.75—1.0	49
Sesame[b]	*Sesamum indicum*	0.75—1.0	49
Lupine[b]	*Lupinus hartwegii*	0.75—1.0	45
Strawberry[b]	*Fragaria* sp.	0.75—1.0	45
Artichoke, Jerusalem[b]	*Helianthus tuberosus*	0.75—1.0	45
Bean, kidney[b]	*Phaseolus vulgaris*	0.75—1.0	45
Bean, lima[b]	*P. lunatus*	0.75—1.0	45
Peanut	*Arachis hypogaea*	0.75—1.0	52
Moderately sensitive			
Broccoli	*Brassica oleracea botrytis*	1.0—2.0	65
Pepper, red	*Capsicum annuum*	1.0—2.0	45
Pea[b]	*Pisum sativa*	1.0—2.0	45
Carrot	*Daucus carota*	1.0—2.0	45
Radish	*Raphanus sativus*	1.0—2.0	45,65
Potato	*Solanum tuberosum*	1.0—2.0	45
Cucumber	*Cucumis sativus*	1.0—2.0	53
Moderately tolerant			
Lettuce[b]	*Lactuca sativa*	2.0—4.0	45
Cabbage[b]	*Brassica oleracea capitata*	2.0—4.0	45
Celery[b]	*Apium graveolens*	2.0—4.0	45
Turnip	*B. rapa*	2.0—4.0	45
Bluegrass, Kentucky[b]	*Poa pratensis*	2.0—4.0	45
Oats	*Avena sativa*	2.0—4.0	45
Corn	*Zea mays*	2.0—4.0	45,53
Artichoke[b]	*Cynara scolymus*	2.0—4.0	45
Tobacco[b]	*Nicotiana tabacum*	2.0—4.0	45
Mustard[b]	*Brassica juncea*	2.0—4.0	45
Clover, sweet[b]	*Melilotus indica*	2.0—4.0	45
Squash	*Cucurbita pepo*	2.0—4.0	53
Muskmelon[b]	*Cucumis melo*	2.0—4.0	45,53
Cauliflower	*B. oleracea botrytis*	2.0—4.0	65
Tolerant			
Sorghum	*Sorghum bicolor*	4.0—6.0	50

Table 7 (continued)
BORON TOLERANCE LIMITS FOR AGRICULTURAL CROPS

Common name	Botanical name	Threshold[a] (g/m³)	Ref.
Tomato	Lycopersicon lycopersicum	4.0—6.0	64,45
Alfalfa[b]	Medicago sativa	4.0—6.0	45
Vetch, purple[b]	Vicia benghalensis	4.0—6.0	45
Parsley[b]	Petroselinum crispum	4.0—6.0	45
Beet, red	Beta vulgaris	4.0—6.0	45
Sugar beet	B. vulgaris	4.0—6.0	45,54
Very tolerant			
Cotton	Gossypium hirsutum	6.0—10.0	45
Asparagus[b]	Asparagus officinalis	10.0—15.0	45

[a] Maximum permissible concentration in soil water without yield reduction. Boron tolerances may vary depending upon climate, soil conditions, and crop varieties.

[b] Tolerance based on reductions in vegetative growth.

although some sensitive fruit crops such as stone and pome fruits may be damaged without exhibiting visible leaf-injury symptoms. Crop leaves normally contain about 40 to 100 mg of B per kilogram of dry weight but often contain more than 250 mg/kg when soil B approaches toxic levels. Boron concentrations in leaves may exceed 700 to 1000 mg/kg under extreme conditions of B toxicity. Pome and stone fruits do not accumulate high levels of B in their leaves but develop twig die-back and gummosis as a result of B toxicity.

Table 8
CITRUS AND STONE-FRUIT ROOTSTOCKS RANKED IN ORDER OF INCREASING BORON ACCUMULATION AND TRANSPORT TO SCIONS[a]

Common name	Botanical name
Citrus	
Alemow	*Citrus macrophylla*
Gajanimma	*C. pennivesiculata* or *C. moi*
Chinese box orange	*Severina buxifolia*
Sour orange	*C. aurantium*
Calamondin	× *Citrofortunella mitis*
Sweet orange	*C. sinensis*
Yuzu	*C. junos*
Rough lemon	*C. limon*
Grapefruit	*C.* × *paradisi*
Rangpur lime	*C.* × *limonia*
Troyer citrange	× *Citroncirus webberi*
Savage citrange	× *Citroncirus webberi*
Cleopatra mandarin	*C. reticulata*
Rusk citrange	× *Citroncirus webberi*
Sunki mandarin	*C. reticulata*
Sweet lemon	*C. limon*
Trifoliate orange	*Poncirus trifoliata*
Citrumelo 4475	*Poncirus trifoliata* × *C. paradisi*
Ponkan mandarin	*C. reticulata*
Sampson tangelo	*C.* × *Tangelo*
Cuban shaddock	*C. maxima*
Sweet lime	*C. aurantiifolia*
Stone fruit	
Almond	*Prunus dulcis*
Myrobalan plum	*P. cerasifera*
Apricot	*P. armeniaca*
Marianna plum	*P. domestica*
Shalil peach	*P. persica*

[a] Data compiled from References 43, 55, and 56.

Table 9
BORON TOLERANCE LIMITS FOR ORNAMENTALS[a]

Common name	Botanical name	Threshold (g/m³)[b]	Ref.
Very sensitive			
Oregon grape	*Mahonia aquifolium*	<0.5	57
Photinia	*Photinia* × *fraseri*	<0.5	57
Xylosma	*Xylosma congestum*	<0.5	57
Thorny elaeagnus	*Elaeagnus pungens*	<0.5	57
Laurustinus	*Viburnum tinus*	<0.5	57
Wax-leaf privet	*Ligustrum japonicum*	<0.5	57
Pineapple guava	*Feijoa sellowiana*	<0.5	57
Spindle tree	*Euonymus japonica*	<0.5	57
Japanese pittosporum	*Pittosporum tobira*	<0.5	57
Chinese holly	*Ilex cornuta*	<0.5	57
Juniper	*Juniperus chinensis*	<0.5	57
Yellow sage	*Lantana camara*	<0.5	57
American elm	*Ulmus americana*	<0.5	45

Table 9 (continued)
BORON TOLERANCE LIMITS FOR ORNAMENTALS[a]

Common name	Botanical name	Threshold (g/m³)[b]	Ref.
Sensitive			
Zinnia	*Zinnia elegans*	0.5—1.0	45
Pansy	*Viola tricolor*	0.5—1.0	45
Violet	*V. odorata*	0.5—1.0	45
Larkspur	*Delphinum* sp.	0.5—1.0	45
Glossy abelia	*Abelia × grandiflora*	0.5—1.0	57
Rosemary	*Rosmarinus officinalis*	0.5—1.0	57
Oriental arborvitae	*Platycladus orientalis*	0.5—1.0	57
Geranium	*Pelargonium × hortorum*	0.5—1.0	58
Moderately sensitive			
Gladiolus	*Gladiolus* sp.	1.0—2.0	59
Marigold	*Calendula officinalis*	1.0—2.0	57
Poinsettia	*Euphorbia pulcherrima*	1.0—2.0	60
China aster	*Callistephus chinensis*	1.0—2.0	61
Gardenia	*Gardenia* sp.	1.0—2.0	62
Southern yew	*Podocarpus macrophyllus*	1.0—2.0	57
Brush cherry	*Syzygium paniculatum*	1.0—2.0	57
Blue dracaena	*Cordyline indivisa*	1.0—2.0	57
Ceniza	*Leucophyllum frutescens*	1.0—2.0	57
Moderately tolerant			
Bottlebrush	*Callistemon citrinus*	2.0—4.0	57
California poppy	*Eschscholzia californica*	2.0—4.0	45
Japanese boxwood	*Buxus microphylla*	2.0—4.0	57
Oleander	*Nerium oleander*	2.0—4.0	57
Chinese hibiscus	*Hibiscus rosa-sinensis*	2.0—4.0	57
Sweet pea	*Lathyrus odoratus*	2.0—4.0	45
Carnation	*Dianthus caryophyllus*	2.0—4.0	63
Tolerant			
Indian hawthorn	*Raphiolepis indica*	6.0—8.0	57
Natal plum	*Carissa grandiflora*	6.0—8.0	57
Oxalis	*Oxalis bowiei*	6.0—8.0	45

[a] Species listed in order of increasing tolerance based on appearance as well as growth reduction.
[b] Boron concentrations exceeding the threshold may cause leaf burn and loss of leaves.

REFERENCES

1. Liberty Hyde Bailey Hortorium Staff, *Hortus Third. A Concise Dictionary of Plants Cultivated in the United States and Canada*, Macmillan, New York, 1976.
2. **Maas, E. V. and Hoffman, G. J.,** Crop salt tolerance — current assessment, *J. Irrig. and Drainage Div.*, ASCE 103(IR2), 115, 1977.
3. United States Salinity Laboratory Staff, Diagnosis and Improvement of Saline and Alkali Soils, *U.S. Dep. Agric. Agric. Handbook 60*, 1954.
4. Bernstein, Leon, Salt-affected soils and plants, in The Problems of the Arid Zones, Proc. UNESCO Symp. (Paris, France) 18th, United Nations Educational, Scientific and Cultural Organization, Geneva, 1962, 139.
5. **Bernstein, L., Brown, J. W. and Hayward, H. E.,** The influence of rootstock on growth and salt accumulation in stone-fruit trees and almonds, *Proc. Am. Soc. Hortic. Sci.*, 68, 86, 1956.
6. Ayers, A. D. and Hayward, H. E., A method for measuring the effects of soil salinity on seed germination with observations on several crop plants, *Soil Sci. Soc. Am. Proc.*, 13, 224, 1948.
7. **Harris, F. S. and Pittman, D. W.,** Relative resistance of various crops to alkali, *Utah Agric. Exp. Stn. Bull.*, No. 168, 1919.

8. **Lopez, G.,** Germination capacity of seeds in saline soil, in *Saline Irrigation for Agriculture and Forestry,* Boyko, H., Ed., W. Junk, The Hague, 1968, 11.
9. **Wahhab, A.,** Salt tolerance of various varieties of agricultural crops at the germination stage, in *Salinity Problems in the Arid Zone, Proc. Teheran Symp. Arid Zone Res.,* UNESCO, Paris, 14, 185, 1961.
10. **Mehta, B. V. and Desai, R. S.,** Effect of soil salinity on germination of some seeds, *J. Soil Water Conserv. India,* 6, 168, 1958.
11. **Lyles, L. and Fanning, C. D.,** Effects of presoaking, moisture tension, and soil salinity on the emergence of grain sorghum, *Agron. J.,* 56, 518, 1964.
12. **Francois, L. E. and Bernstein, L.,** Salt tolerance of safflower, *Agron. J.,* 56, 38, 1964.
13. **Mehrotra, C. L. and Gangwar, B. R.,** Studies on salt and alkali tolerance of some important agricultural crops of Uttar Pradesh, *J. Indian Soc. Soil Sci.,* 12, 75, 1964.
14. **West, D. W. and Francois, L. E.,** Effects of salinity on germination, growth and yield of cowpea, *Irrig. Sci.,* 3, 169, 1982.
15. **Pearson, G. A., Ayers, A. D., and Eberhard, D. L.,** Relative salt tolerance of rice during germination and early seedling development, *Soil Sci.,* 102, 151, 1966.
16. **Aziz, M. A., Fattah, A., Salam, A. S. A., Elmofty, I. A., and Abdel-Gawwad, M. M.,** Salt tolerance of onion during germination and early seedling growth, *Desert Inst. Bull. ARE,* 22, 157, 1972.
17. **Yousif, Y. H., Bingham, F. T., and Yermanos, D. M.,** Growth, mineral composition, and seed oil of sesame (*Sesamum indicum* L.) as affected by NaCl, *Soil Sci. Soc. Am. Proc.,* 36, 450, 1972.
18. **Francois, L. E., Donovan, T., and Maas, E. V.** Salinity effects on seed yield, growth and germination of grain sorghum, *Agron. J.,* 76, 741, 1984.
19. **Francois, L. E., Maas, E. V., Donovan, T. J., and Youngs, V. L.,** Effect of salinity on grain yield and quality, vegetative growth, and germination of semi-dwarf and durum wheat, *Agron. J.,* 78, 1053, 1986.
20. **Maas, E. V.,** U.S. Salinity Laboratory, Riverside, Calif., unpublished data.
21. **Francois, L. E. and Bernstein, L.,** Salt tolerance of *Sphaerophysa salsula,* Report To Collaborators, U.S. Salinity Laboratory, Riverside, Calif., 1964.
22. **Francois, L. E.,** U.S. Salinity Laboratory, Riverside, Calif., unpublished data.
23. **Shannon, M. C. and Francois, L. E.,** Salt tolerance of three muskmelon cultivars, *J. Am. Soc. Hortic. Sci.,* 103, 127, 1978.
24. **Francois, L. E.,** Salinity effects on germination, growth, and yield of two squash cultivars, *HortScience,* 20, 1102, 1985.
25. **Francois, L. E. and West, D. W.** Reduction in yield and market quality of celery caused by soil salinity, *J. Am. Soc. Hortic. Sci.,* 107, 952, 1982.
26. **Francois, L. E.,** Salinity effects on germination, growth, and yield of turnips, *HortScience,* 19, 82, 1984.
27. **Francois, L. E. and Maas, E. V.,** Plant responses to salinity: an indexed bibliography, ARM-W-6, Agricultural Reviews and Manuals, Western Series, Science and Education Administration, U.S. Department of Agriculture, Washington, D.C., 1978.
28. **Francois, L. E. and Maas, E. V.,** Plant responses to salinity: a supplement to an indexed bibliography, ARS-24, Agricultural Research Service, U.S. Department of Agriculture, 1985.
29. **Maas, E. V., Donovan, T. J., Francois, L. E., and Hamerstrand, G. E.,** Salt tolerance of Guayule, *Proc. 4th Int. Conf. Guayule Res. and Development,* Tucson, Arizona, Oct. 16-19, 1985, in press.
30. **Yermanos, D. M., Francois, L. E., and Tammadoni, T.,** Effects of soil salinity on the development of jojoba, *Econ. Bot.,* 21, 69, 1967.
31. **Siegel, S. M.,** University of Hawaii, Honolulu, personal communication, 1984.
32. **Bernstein, L., Francois, L. E., and Clark, R. A.,** Salt tolerance of ornamental shrubs and ground covers, *J. Am. Soc. Hortic. Sci.,* 97, 550, 1972.
33. **Francois, L. E.,** Salt tolerance of eight ornamental tree species, *J. Am. Soc. Hortic. Sci.,* 107, 66, 1982.
34. **Francois, L. E. and Clark, R. A.,** Salt tolerance of ornamental shrubs, trees, and iceplant, *J. Am. Soc. Hortic. Sci.,* 103, 280, 1978.
35. **Ehlig, C. F. and Bernstein, L.,** Foliar absorption of sodium and chloride as a factor in sprinkler irrigation, *Proc. Am. Soc. Hortic. Sci.,* 74, 661, 1959.
36. **Francois, L. E. and Clark, R. A.,** Accumulation of sodium and chloride in leaves of sprinkler-irrigated grapes, *J. Am. Soc. Hortic. Sci.,* 104, 11, 1979.
37. **Gornat, B., Goldberg, D., Rimon, D., and Ben-Asher, J.,** The physiological effect of water quality and method of application on tomato, cucumber, and pepper, *J. Am. Soc. Hortic. Sci.,* 98, 202, 1973.
38. **Maas, E. V., Clark, R. A., and Francois, L. E.,** Sprinkling-induced foliar injury to pepper plants: effects of irrigation frequency, duration and water composition, *Irrig. Sci.,* 3, 101, 1982.
39. **Maas, E. V., Grattan, S. R., and Ogata, G.,** Foliar salt accumulation and injury in crops sprinkled with saline water, *Irrig. Sci.,* 3, 157, 1982.
40. **Bernstein, L.,** Salt tolerance of fruit crops, *U.S. Dept. Agric. Inf. Bull.,* 292, 1965.

41. **Bernstein, L., Ehlig, C. F., and Clark, R. A.**, Effect of grape rootstocks on chloride accumulation in leaves, *J. Am. Soc. Hortic. Sci.*, 94, 584, 1969.
42. **Embleton, T. W., Matsumura, M., Storey, W. B., and Garber, M. J.**, Chlorine and other elements in avocado leaves as influenced by rootstock, *Proc. Am. Soc. Hortic. Sci.*, 80, 230, 1962.
43. **Embleton, T. W., Jones, W. W., Labanauskas, C. K., and Reuther, W.**, Leaf analysis as a diagnostic tool and guide to fertilization, *The Citrus Industry*, Vol. 3, Reuther, W., Ed., University of California, 1973, chapt. 6.
44. **Grieve, A. M. and Walker, R. R.**, Uptake and distribution of chloride, sodium and potassium ions in salt-treated citrus plants, *Aust. J. Agric. Res.*, 34, 133, 1983.
45. **Eaton, F. M.**, Deficiency, toxicity, and accumulation of boron in plants, *J. Agric. Res.*, 69, 237, 1944.
46. **Haas, A. R. C.**, Toxic effects of boron on fruit trees, *Bot. Gaz.*, 88, 113, 1929.
47. **Woodbridge, C. G.**, The boron requirements of stone fruit trees, *Can. J. Agric. Sci.*, 35, 282, 1955.
48. **Singh, R. N. and Singh, J. R.**, Studies on the influence of boron nutrition on the growth characteristics of garlic (*Allium sativum* L.), *Indian J. Hortic.*, 31, 255, 1974.
49. **Khudairi, A. K.**, Boron toxicity and plant growth, in *Salinity Problems in the Arid Zones, Proc. Teheran Symp., Arid Zone Res., UNESCO*, Paris, 14, 175, 1961.
50. **Bingham, F. T., Strong, J. E., Rhoades, J. D., and Keren, R.**, An application of the Maas-Hoffman salinity response model for boron toxicity, *Soil Sci. Soc. Am. J.*, 49, 672, 1985.
51. **Pathak, A. M., Singh, R. K., and Singh, R. S.**, Effect of different concentrations of boron in irrigation water on sunflower, *J. Indian Soc. Soil Sci.*, 23, 388, 1975.
52. **Gopal, N. H.**, Influence of boron on growth and yield in groundnut, *Turrialba*, 21, 435, 1971
53. **El-Sheikh, A. M., Ulrich, A., Awad, S. K., and Mawardy, A. E.**, Boron tolerance of squash, melon, cucumber, and corn, *J. Am. Soc. Hortic. Sci.*, 96, 536, 1971.
54. **Vlamis, J. and Ulrich, A.**, Boron tolerance of sugar beets in relation to the growth and boron content of tissues, *J. Am. Soc. Sugar Beet Technol.*, 17, 280, 1973.
55. **Hansen, C. J.**, Influence of the rootstock on injury from excess boron in French (agen) prune and president plum, *Proc. Am. Soc. Hortic. Sci.*, 51, 239, 1948.
56. **Hansen, C. J.**, Influence of the rootstock on injury from excess boron in nonpareil almond and elberta peach, *Proc. Am. Soc. Hortic. Sci.*, 65, 128, 1955.
57. **Francois, L. E. and Clark, R. A.**, Boron tolerance of twenty-five ornamental shrub species, *J. Am. Soc. Hortic. Sci.*, 104, 319, 1979.
58. **Kofranek, A. M., Kohl, H. C., and Lunt, O. R.**, Effects of excess salinity and boron of geraniums, *Proc. Am. Soc. Hortic. Sci.*, 71, 516, 1958.
59. **Kofranek, A. M., Lunt, O. R., and Kohl, H. C.**, Tolerance of gladioli to salinity and boron, *Proc. Am. Soc. Hortic. Sci.*, 69, 556, 1957.
60. **Kofranek, A. M., Lunt, O. R., and Kohl, H. C.**, Tolerance of poinsettias to saline conditions and high boron concentrations, *Proc. Am. Soc. Hortic. Sci.*, 68, 551, 1956.
61. **Kohl, H. C., Kofranek, A. M., and Lunt, O. R.**, Response of china asters to high salt and boron concentrations, *Proc. Am. Soc. Hortic. Sci.*, 70, 437, 1957.
62. **Lunt, O. R., Kohl, H. C., and Kofranek, A. M.**, Tolerance of azaleas and gardenias to salinity conditions and boron, *Proc. Am. Soc. Hortic. Sci.*, 69, 543, 1957.
63. **Lunt, O. R., Kohl, H. C., and Kofranek, A. M.**, Tolerance of carnations to saline conditions and boron, *Carnation Craft*, 36, 5, 1956.
64. **Francois, L. E.**, Effect of excess boron on tomato yield, fruit size and vegetative growth, *J. Am. Soc. Hortic. Sci.*, 109, 322, 1984.
65. **Francois, L. E.**, Effect of excess boron on broccoli, cauliflower and radish, *J. Am. Soc. Hortic. Sci.*, 111, 494, 1986.

WORLD CROP PRODUCTION AND DISTRIBUTION

B. E. Coulman*

The world is dependent upon relatively few plant species as sources of food, oil, and fiber. The following tables summarize production statistics for the major crop categories and species of the world. These statistics are given for the major regions of the world.

The information in Tables 1 through 13 comes from the 1983 Production Yearbook of the Food and Agriculture Organization (FAO) of the United Nations. For some crop categories and species, data on area harvested, yield per hectare, and total production are given. For other categories and species only total production is given as the other statistics were not available from FAO. It should be pointed out that some of the figures in these tables are FAO estimates, as no official figures were available from certain countries.

Cereals are the most important crop group produced in the world, with an annual production of over one and one half billion metric tons. Wheat, maize, and rice are the world's most important food crops. All areas of the world produce large quantities of cereals (Table 2), with Asia being the largest producer. Yields of most cereal crops are highest in Europe and North America and lowest in Africa.

Root and tuber crops also represent an important source of food. Europe and Asia are the largest producers of these crops, with potatoes being the major crop in this category.

Vegetable and fruit production is substantial in all regions of the world (Tables 4, 5, and 6). Tomatoes and cabbage are the world's leading vegetable crops and grapes and oranges the most important fruits. It should be noted that the figures for orange production also include the production of tangerines, mandarins, clementines, and satsumas.

The world production of sugarcane is approximately three times as great as that of sugar beets. Asia is the major cane-producing region, while Europe is the most important producer of beets (Table 7).

The figures for oil crops (Tables 8 and 9) do not relate to the actual production of oil but rather to the potential production. In other words, the total production of some crops, such as groundnuts and sunflowers, is not processed into oil. Soybeans are by far the most important oil crop, with the majority of the production coming from the U.S.

Pulse crops are important sources of protein in many countries. Asia is the largest producer of pulse crops, with dry beans and peas being the most important (Table 10).

Cotton is the world's major fiber crop, followed by jute (Table 11). Asia is the region that produces the greatest quantities of these two crops.

Almost one half of the coffee in the world is produced in South America, while over one half of the cocoa comes from Africa (Table 12). Asia is the major producer of tea and tobacco.

Almonds and walnuts are the most important treenut crops (Table 13). Asia is the leading producer of most treenut species.

Data on forage crop production are difficult to find. The use of several species in a forage seeding, the indeterminate life of a forage stand, and the use of some stands for both pasture and hay in the same year make accurate data collection of forage crop production almost impossible. In the U.S., for example, the only individual forage crop on which data are available now is alfalfa. As can be seen from Table 14, alfalfa is present in approximately 40% of the hay-producing area. Some world data are available on alfalfa production (Table 15), but most of these date to the mid-to-late 1960s.

---

* Tables follow text.

Table 1
MEAN ANNUAL PRODUCTION OF THE PRINCIPAL CROP GROUPS AND MISCELLANEOUS CROPS (1981—1983)

Crop or crop group	Production (1000 MT)
Cereals	1,666,568
Root crops	555,936
Vegetables and melons	367,215
Fruits	294,077
Sugar (centrifugal, raw)	97,458
Oil crops (oil equiv.)	54,815
Pulses	43,317
Fibers, vegetable	21,383
Tobacco	6,321
Coffee, green	5,502
Nuts	3,814
Natural rubber	3,805
Tea	1,939
Cocoa beans	1,624

Note: MT, metric ton.

Table 2
MEAN ANNUAL AREA, YIELD, AND PRODUCTION (1981—1983) BY REGION OF THE MAJOR CEREAL CROPS OF THE WORLD

Crop	Production statistic[a]	Africa	North America	South America	Asia	Europe	Oceania	U.S.S.R.	World
Cereals, total	Harvested area	72,046	106,436	39,136	304,497	69,931	17,596	117,991	727,633
	Yield	996	3,483	1,916	2,279	3,687	1,336	1,448	2,291
	Production	71,835	373,283	74,990	694,129	257,884	23,814	170,634	1,666,568
Wheat	Harvested area	8,112	43,767	10,091	80,541	26,168	12,118	55,789	236,586
	Yield	1,169	2,356	1,501	1,910	3,776	1305	1,494	2,029
	Production	9,489	102,908	15,247	153,982	98,867	15,982	83,000	479,475
Rice, paddy	Harvested area	4,907	2,005	7,085	128,495	349	116	644	143,602
	Yield	1,759	4,440	1,928	3,052	5,072	6,268	3,831	2,983
	Production	8,630	8,947	13,647	392,173	1,772	732	2,467	428,368
Maize	Harvested area	20,084	37,374	17,190	36,303	11,336	84	4,298	126,670
	Yield	1,393	5,219	2,026	2,374	5,057	4,128	2,613	3,276
	Production	27,989	197,794	34,844	86,131	57,299	349	11,333	415,738
Barley	Harvested area	4,863	9,097	618	11,218	19,638	2,871	31,106	79,410
	Yield	816	2,675	1,169	1,493	3,417	1,268	1,434	2,031
	Production	3,953	24,338	718	16,731	67,059	3,729	44,667	161,194
Sorghum	Harvested area	15,528	7,197	3,053	20,811	164	670	141	47,565
	Yield	673	3,443	3,046	1,039	3,524	1,756	915	1,433
	Production	10,468	24,975	9,276	21,616	580	1,172	127	68,214
Oats	Harvested area	594	5,530	593	742	5,196	1,546	12,158	26,360
	Yield	452	1,994	1,284	1,553	2,728	1,043	1,207	1661
	Production	265	11,063	766	1,152	14,200	1,663	14,667	43,776
Millet	Harvested area	16,102		160	22,900	16	32	2,798	42,008
	Yield	589		1,184	763	1,791	1,002	677	694
	Production	9,498		190	17,485	29	32	1,900	29,134
Rye	Harvested area	39	757	185	1,215	5,545	34	9,179	16,955
	Yield	232	1,937	896	1,341	2,507	373	1,242	1,689
	Production	9	1,465	166	1,631	13,919	13	11,500	28,702

[a] Harvested area in 1000 ha; yield in kg/ha; production in 1000 MT.

Table 3
MEAN ANNUAL AREA, YIELD, AND PRODUCTION (1981—1983) BY REGION OF THE MAJOR ROOT AND TUBER CROPS OF THE WORLD

Crop	Production statistic	Africa	North America	South America	Asia	Europe	Oceania	U.S.S.R.	World
Roots and tubers, total	Harvested area (1000 ha)	13,483	1,195	3,797	16,768	5,388	260	6,866	47,758
	Yield (kg/ha)	6,537	18,558	10,832	13,212	18,993	10,543	11,326	11,641
	Production (1000 MT)	88,151	22,169	41,141	221,531	102,430	2,740	77,775	555,936
Potatoes	Harvested area (1000 ha)	603	717	945	5,731	5,374	45	6,866	20,281
	Yield (kg/ha)	8,786	26,935	10,462	12,289	19,016	24,111	11,326	14,106
	Production (1000 MT)	5,297	19,313	9,898	70,433	102,277	1,090	77,775	286,082
Cassava	Harvested area (1000 ha)	8,139	158	2,567	4,028		22		14,914
	Yield (kg/ha)	6,100	5,519	11,350	11,804		10,885		8,542
	Production (1000 MT)	49,655	871	29,134	47,472		235		127,367
Sweet potatoes	Harvested area (100 ha)	799	208	160	6,711	13	114		8,006
	Yield (kg/ha)	6,088	6,393	8,686	15,058	10,679	5,435		13,661
	Production (1000 MT)	4,865	1,131	1,393	100,954	136	622		109,301

Table 4
MEAN ANNUAL AREA, YIELD, AND PRODUCTION (1981—1983) BY REGION OF THE MAJOR VEGETABLE CROPS OF THE WORLD

Crop	Production statistic[a]	Africa	North America	South America	Asia	Europe	Oceania	U.S.S.R.	World
Vegetables, total (includes melons)	Production	24,954	33,559	11,598	197,521	65,076	1,759	32,747	367,215
Tomatoes	Harvested area	375	306	124	737	469	11	395	2,416
	Yield	14,469	31,477	23,502	17,473	31,688	25,683	18,389	22,032
	Production	5,424	9,620	2,911	12,873	14,862	277	7,258	53,225
Cabbage	Harvested area	29	99	22	779	313	4	412	1,657
	Yield	24,779	19,167	27,339	19,235	24,847	27,148	23,954	21,689
	Production	733	1,891	594	14,975	7,768	109	9,873	35,943
Watermelons	Harvested area	130	115	111	885	149	4	395	1,789
	Yield	16,383	14,362	8,934	16,167	20,371	11,142	10,266	14,652
	Production	2,131	1,657	993	14,302	3,035	49	4,056	26,224
Onions, dry	Harvested area	132	65	120	912	235	6	172	1,642
	Yield	12,484	29,860	13,688	11,038	18,113	30,889	11,701	13,247
	Production	1,651	1,932	1,642	10,072	4,251	197	2,009	21,754
Cucumbers, gherkins	Harvested area	21	81	3	289	128	1	177	800
	Yield	15,671	13,460	17,730	14,420	21,369	12,703	8,601	14,189
	Production	335	1,092	45	5604	2,742	14	1,525	11,357
Carrots	Harvested area	33	46	27	147	128	5	137	522
	Yield	12,096	30,170	19,554	19,463	28,674	29,150	15,287	21,203
	Production	395	1,395	528	2,857	3,659	144	2,099	11,076
Chillies, peppers	Harvested area	163	84	21	556	147			970
	Yield	6,871	9,547	8,282	5,396	15,398	13,000		7,570
	Production	1,118	806	169	2,997	2,254	1		7,344

[a] Harvested areas in 1000 ha; yield in kg/ha; production in 1000 MT.

Table 5
MEAN ANNUAL PRODUCTION (1000 MT) BY REGION OF OTHER IMPORTANT VEGETABLE CROPS

Crop	Africa	North America	South America	Asia	Europe	Oceania	U.S.S.R.	World
Cantaloupe and other melons	697	1,201	325	3,766	1,621	1		7,610
Pumpkins, squash, gourds	978	239	702	2,259	1,111	86		5,374
Eggplant	418	64	8	3,785	576			4,850
Cauliflower	148	301	72	1,879	2,162	109	15	4,687
Green peas	125	1,168	106	616	2,145	166	237	4,564
Green beans	185	197	107	717	1,166	41		2,414
Garlic	164	127	138	1,405	432	1	30	2,296
Artichokes	92	51	84	14	945			1,185

Table 6
MEAN ANNUAL PRODUCTION (1000 MT) BY REGION OF THE MAJOR FRUIT CROPS OF THE WORLD (1981—1983)

Crop	Africa	North America	South America	Asia	Europe	Oceania	U.S.S.R.	World
Total fruits	32,212	44,115	40,958	80,423	74,885	3,766	17,716	294,077
Grapes	2,322	5,563	5,209	7,494	36,923	833	7,415	65,758
Oranges	3,779	11,785	12,786	10,381	5,532	443	307	45,012
Bananas	4,444	7,352	11,511	15,562	497	1,129		40,494
Apples	526	4,397	1,398	8,287	14,893	531	6,978	37,011
Plantains	12,331	1,588	4,253	1,706		2		19,881
Mangoes	836	1,333	821	10,773		11		13,774
Pears	238	823	266	3,084	3,973	146	584	9,114
Pineapples	1,230	1,263	1,165	4,849	2	134		8,640
Peaches, nectarines	242	1,426	541	1,211	3,451	93	440	7,405
Plums	66	692	92	926	3,193	24	1,027	6,020
Lemons and limes	247	1,783	712	1,288	1,422	42		5,495
Grapefruit and pomelo	285	2,821	319	886	25	36		4,372
Dates	1,022	21		1,607	11			2,662
Papayas	221	313	613	791		18		1,956
Strawberries	3	483	15	304	929	8	107	1,849
Apricots	175	100	33	444	716	37	234	1,739
Avocados	115	962	317	142	2	3		1,541
Currants					426	2	79	507
Raspberries		25			136	2	101	264

Table 7
MEAN AREA, YIELD, AND PRODUCTION (1981—1983) BY REGION OF SUGARCANE AND SUGAR BEETS AND MEAN ANNUAL PRODUCTION OF NATURAL RUBBER AND HOPS

Crop	Production statistic	Africa	North America	South America	Asia	Europe	Oceania	U.S.S.R.	World
Sugarcane	Harvested area (1000 ha)	1,000	2,871	4,093	6,336	5	376		14,682
	Yield (kg/ha)	66,200	59,300	61,238	53,490	63,569	73,716		58,172
	Production (1000 MT)	66,160	170,296	251,112	338,513	330	27,719		854,129
Sugar beets	Harvested area (1000 ha)	70	476	48	1,055	3,815		3558	9,022
	Yield (kg/ha)	37,327	46,378	35,139	25,236	40,963		20,115	31,160
	Production (1000 MT)	2,608	22,164	1,662	26,637	156,607		71,437	281,116
Natural rubber	Production (1000 MT)	189		41	3,571		5		3,805
Hops	Production (1000 MT)	35			3	79	2	8	128

Table 8
MEAN ANNUAL AREA, YIELD, AND PRODUCTION (1981—1983) BY REGION OF THE MAJOR OIL CROPS OF THE WORLD

Crop	Production statistic	Africa	North America	South America	Asia	Europe	Oceania	U.S.S.R.	World
Soybeans	Harvested area (1000 ha)	366	27,504	10,834	10,529	527	42	875	50,677
	Yield (kg/ha)	1,031	1,968	1,751	1,117	1,259	1,522	680	1,710
	Production (1000 MT)	378	54,330	18,970	11,751	666	63	596	86,754
Groundnuts (in shell)	Harvested area (1000 ha)	6,285	715	474	11,459	10	41	1	18,986
	Yield (kg/ha)	753	2,487	1,445	1,100	2,140	1,258	1,900	1,047
	Production (1000 MT)	4,736	1,781	684	12,609	22	51	2	19,885
Sunflower	Harvested area (1000 ha)	520	1,678	1,719	1,831	2,440	182	4,271	12,640
	Yield (kg/ha)	903	1,246	1,103	1,214	1,419	634	1,195	1,219
	Production (1000 MT)	476	2,104	1,919	2,227	3,465	116	5,106	15,412
Rapeseed	Harvested area (1000 ha)	54	1,838	41	8,798	1,863	17	110	12,721
	Yield (kg/ha)	415	1,245	919	834	2,185	807	486	1,085
	Production (1000 MT)	22	2,262	36	7,358	4,077	14	54	13,823

Table 8 (continued)
MEAN ANNUAL AREA, YIELD, AND PRODUCTION (1981—1983) BY REGION OF THE MAJOR OIL CROPS OF THE WORLD

Crop	Production statistic	Africa	North America	South America	Asia	Europe	Oceania	U.S.S.R.	World
Linseed	Harvested area (1000 ha)	90	798	913	1,956	230	8	1,108	5,102
	Yield (kg/ha)	640	980	788	300	533	1,005	203	491
	Production (1000 MT)	57	790	720	588	123	8	226	2,512
Sesame seed	Harvested area (1000 ha)	1,451	200	118	4,769	2			6,540
	Yield (kg/ha)	319	591	560	289	501			309
	Production (1000 MT)	462	119	65	1,373	1			2,020
Safflower	Harvested area (1000 ha)	66	454	1	742	18	37	7	1,324
	Yield (kg/ha)	482	1,099	850	508	617	577	398	711
	Production (1000 MT)	32	494	1	377	11	24	3	943
Castor beans	Harvested area (1000 ha)	72	11	418	825	15		156	1498
	Yield (kg/ha)	576	629	593	648	349		391	592
	Production (1000 MT)	42	7	244	537	5		50	885

Table 9
MEAN ANNUAL PRODUCTION (1000 MT) BY REGION OF OTHER IMPORTANT OIL CROPS (1981—1983)

Crop	Africa	North America	South America	Asia	Europe	Oceania	U.S.S.R.	World
Cottonseed	2,042	5,026	1,813	13,126	359	181	5718	28,265
Palm oil	1,372	40	165	4,184		89		5,848
Coconut (copra)	177	190	38	4,003		320		4,729
Palm kernels	738	18	302	984		31		2,075
Olive (oil)	154	3	16	195	1,418			1,785

Table 10
MEAN ANNUAL AREA, YIELD, AND PRODUCTION (1981—1983) BY REGION OF THE MAJOR PULSE CROPS OF THE WORLD

Crop	Production statistic	Africa	No. America	So. America	Asia	Europe	Oceania	U.S.S.R.	World
Pulses, total	Harvested area (1000 ha)	12,112	3,876	6,192	33,998	2,557	305	5,687	64,726
	Yield (kg/ha)	465	927	512	637	1,047	1,076	1,104	669
	Production (1000 MT)	5,636	3,598	3,192	21,640	2,680	327	6,245	43,318
Beans, Dry	Harvested area (1000 ha)	2,233	3,247	5,698	12,123	1,238	4	52	24,596
	Yield (kg/ha)	703	878	504	475	587	803	1,490	565
	Production (1000 MT)	1,569	2,860	2,900	5,756	727	4	77	13,892
Peas, dry	Harvested area (1000 ha)	433	137	132	1,909	310	99	4,668	7,689
	Yield (kg/ha)	745	1,988	710	1,263	2,579	1,644	1,078	1,182
	Production (1000 MT)	323	273	93	2,411	804	163	5,000	9,066
Broad beans, dry	Harvested area (1000 ha)	756	75	225	2,059	323	14		3,452
	Yield (kg/ha)	1,325	1,134	519	1,177	1,384	538		1,184
	Production (1000 MT)	1,006	85	117	2,420	447	8		4,083
Chickpeas	Harvested area (1000 ha)	370	157	42	8,787	140			9,496
	Yield (kg/ha)	679	1,076	498	645	584			652
	Production (1000 MT)	248	169	21	5,660	82			6,179
Lentils	Harvested area (1000 ha)	132	133	74	1,642	100		12	2,092
	Yield (kg/ha)	628	1,078	525	676	639		704	691
	Production (1000 MT)	82	144	38	1,114	64		8	1,449

Table 11
MEAN ANNUAL PRODUCTION (1000 MT) BY REGION OF THE MAJOR FIBER CROPS OF THE WORLD (1981—1983)

Crop	Africa	North America	South America	Asia	Europe	Oceania	U.S.S.R.	World
Cotton lint	1,161	3,025	959	6,698	193	111	2,813	14,962
Jute and jute-like fibers	21	11	87	3,822			50	3,990
Flax fiber and tow	25	3	3	109	140		366	644
Sisal	169	12	229	4		2		414
Hemp fiber and tow			4	122	46		48	220

Table 12
MEAN ANNUAL AREA, YIELD, AND PRODUCTION (1981—1983) BY REGION OF MAJOR NONALCOHOLIC BEVERAGE CROPS AND TOBACCO

Crop	Production statistic	Africa	North America	South America	Asia	Europe	Oceania	U.S.S.R.	World
Coffee, green	Harvested area (1000 ha)	3,394	1,508	4,146	989		44		10,081
	Yield (kg/ha)	356	649	629	642		1,244		545
	Production (1000 MT)	1,207	978	2,627	635		55		5,502
Tea	Harvested area (1000 ha)	181		46	2,232		4	79	2,543
	Yield (kg/ha)	1,161		1,056	685		2,021	1,802	762
	Production (1000 MT)	210		49	1,529		8	142	1,939
Cocoa beans	Harvested area (1000 ha)	3,288	260	952	135		86		4,721
	Yield (kg/ha)	283	391	513	510		395		344
	Production (1000 MT)	931	102	488	69		34		1,624
Tobacco leaves	Harvested area (1000 ha)	305	555	425	2,431	500	8	177	4,400
	Yield (kg/ha)	948	1,354	1,287	1,369	1,490	1,958	1,734	1,438
	Production (1000 MT)	289	1,094	547	3,323	744	16	308	6,321

Table 13
MEAN ANNUAL PRODUCTION (1000 MT) BY REGION OF THE MAJOR TREENUT CROPS OF THE WORLD (1981—1983)

Crop	Africa	North America	South America	Asia	Europe	Oceania	U.S.S.R.	World
Treenuts, total	309	648	142	1532	1101	8	76	3814
Almonds	50	246		223	518	2	8	1046
Walnuts	6	201		258	250		57	780
Chestnuts		12	11	380	151		6	549
Hazelnuts	2	3		320	153		4	492
Cashew nuts	179		81	205				468
Pistachios		13		66	4			82

Table 14
PRODUCTION STATISTICS FOR VARIOUS CLASSES OF HAY PRODUCED IN THE U.S. IN 1970 AND FROM 1976—1978

	1970 Area (1000 ha)	1970 Production (1000 MT)	1976—1978 Area (1000 ha)	1976—1978 Yield (MT/ha)	1976—1978 Production (1000 MT)
Alfalfa and alfalfa mixtures	11,038	76,782	10,985	7.30	80,256
Clover and timothy	5,226	23,777			
Lespedeza	444	1,640			
Grain crop for hay	1,081	3,983			
Wild hay	3,460	8,292			
Other hay	3,503	14,415			
All hay	24,871	129,000	24,613	5.40	133,293

[a] Statistics on classes of hay other than alfalfa are not available after 1970.

Table 15
AREA UNDER ALFALFA PRODUCTION BY CONTINENT

Continent	Area (1000 ha)[a]
Africa	174
North America	13142
South America	7800
Asia	1323
Europe	9363
Oceania	1213
World	33105

[a] Figures for continents are based on summed estimates from individual countries. Most of these estimates were made during the years 1967—1969, although some are from earlier years.

REFERENCES

1. FAO Production Yearbook, FAO Statistics Series No. 34, Food and Agricultural Organization of the United Nations, Rome, 1983.
2. Agricultural Statistics, United States Department of Agriculture, Washington, D.C., 1979.
3. **Bolton, J. L., Goplen, B. P., and Baenziger, H.,** World distribution and historical development, in *Alfalfa Science and Technology,* Hanson, C. H., Ed., Agronomy Monograph No. 15, American Society of Agronomy, Madison, 1972.

DRY-MATTER PRODUCTION FROM CROPS

M. Tollenaar

Whereas agricultural productivity is based on the conversion of radiant solar energy into the chemical energy which is stored in the carbon-to-carbon bonds of organic matter, environmental and managerial factors strongly influence its biological and economic potential. Vegetative productivity may be restricted by temperature, moisture, and photoperiod (through their effects on duration of the growing season and rate of crop development), by the physical and chemical conditions of the soil, or by diseases and pests. Management systems may alleviate several of the limiting factors restricting vegetative productivity through the use of adapted genotypes, the application of fertilizer and irrigation, and the use of crop-protection practices. Agricultural practices may sometimes reduce potential productivity by increasing soil erosion or deteriorating soil structure and may increase the vulnerability to disease epidemics by growing crop varieties with little genotypic variability over large areas (e.g., maize in the North American continent). Hence, an examination of agricultural productivities on a global basis reveals that agricultural yields are highly variable and commercial production in any region seldom provides a clear indication of the biological and commercial potentials.

An estimate of potential vegetative productivities on a global basis can be obtained by examining incident solar radiation, efficiency of conversion of solar radiant energy into plant dry matter, and the duration of periods of partial and full light interception by crop canopies. Incident photosynthetic active radiation under a clear sky depends on solar height (β) and can be expressed as follows:

$$S = 640 \times \sin(\beta) \times \exp(K/\sin(\beta)) \qquad (1)$$

where 640 represents the solar constant for photosynthetic active radiation and K ranges between 0.1 for a clear atmosphere and 0.18 for a rather humid and dusty one.[1] Photosynthetic active radiation is approximately 50% of global irradiation under a clear sky. Actual incident radiation can be calculated from the ratio of the durations of clear to overcast skies. In general, crop photosynthesis,[2] short-term crop growth rate,[3] and total dry-matter production[4-6] are highly correlated with intercepted incident radiation, in the absence of any apparent environmental stresses. Consequently, potential vegetative productivity can be estimated for various conversion efficiencies when the total of intercepted irradiance by a crop canopy is known.[7] Computer simulation models which describe the environment-physiological interactions in considerable detail have been developed to calculate potential production. Goudriaan and van Laar[8] calculated daily gross CO_2 assimilation of crop canopies with a leaf area index of 5 for clear and overcast skies (i.e., overcast sky radiation was assumed to be 20% of clear sky radiation) using a computer simulation program developed by de Wit and co-workers.[9] In their model, leaf photosynthetic efficiency was 14×10^{-9} kg CO_2 J^{-1} and variations in leaf photosynthetic capacity of crops were expressed by a range of values for the rate of leaf photosynthesis at light saturation (PMAX). For C_4 plants (without photorespiration) a typical value of PMAX of 60 kg CO_2 ha^{-1} hr^{-1} can be used, and for C_3 plants (with photorespiration) a typical PMAX value is 30 kg CO_2 ha^{-1} hr^{-1}. Values in Table 2 are rates of gross CO_2 fixation; growth and maintenance respiration have to be subtracted and rates have to be adjusted for the conversion of CO_2 into plant dry matter to find dry-matter production (e.g., if total respiration is 30% of gross photosynthesis and all CO_2 is converted into CH_2O, then a gross CO_2 fixation of 1000 kg/ha^{-1} corresponds to a dry matter production of 477 kg ha^{-1}). An indication of potential productivity can also be obtained from tabulations of short-term crop-growth rates and maximum yields which have been

Table 1
DAILY TOTAL INCIDENT PHOTOSYNTHETIC ACTIVE RADIATION (400—700 nm)[a] IN MJ m^{-2} FOR A STANDARD CLEAR DAY

North latitude (°)	Jan. 1	Jan. 15	Feb. 1	Feb. 15	March 1	March 15	April 1	April 15	May 1	May 15	June 1	June 15
0	13.81	14.00	14.38	14.71	15.00	15.52	15.14	14.95	14.61	14.23	13.93	13.77
5	12.88	13.14	13.64	14.13	14.61	14.98	15.23	15.26	15.12	14.93	14.71	14.61
10	11.87	12.17	12.80	13.44	14.09	14.67	15.19	15.43	15.52	15.48	15.39	15.34
15	10.77	11.13	11.86	12.63	13.45	14.23	15.01	15.47	15.78	15.91	15.96	15.96
20	9.61	10.00	10.84	11.73	12.70	13.67	14.71	15.38	15.91	16.21	16.40	16.46
25	9.39	8.82	9.73	10.73	11.85	12.99	14.27	15.15	15.93	16.39	16.73	16.85
30	7.14	7.59	8.57	9.65	10.89	12.20	13.71	14.80	15.80	16.44	16.93	17.11
35	5.86	6.33	7.35	8.50	9.85	11.30	13.04	14.33	15.55	16.36	17.01	17.25
40	4.60	5.06	6.10	7.30	8.73	10.31	12.24	13.73	15.18	16.16	16.96	17.27
45	3.36	3.81	4.84	6.05	7.54	9.22	11.34	13.01	14.68	15.85	16.81	17.18
50	2.20	2.61	3.50	4.79	6.31	8.06	10.34	12.18	14.07	15.42	16.55	17.00
55	1.16	1.52	2.40	3.54	5.04	6.84	9.24	11.25	13.36	14.89	16.20	16.72
60	0.37	0.62	1.32	2.33	3.77	5.57	8.08	10.23	12.55	14.29	15.79	16.40
65	0.00	0.07	0.46	1.24	2.52	4.26	6.83	9.13	11.68	13.64	15.38	16.10
70	0.00	0.00	0.02	0.38	1.37	2.96	5.54	7.97	10.78	13.02	15.13	16.05
75	0.00	0.00	0.00	0.00	0.43	1.71	4.21	6.77	9.95	12.72	15.32	16.37
80	0.00	0.00	0.00	0.00	0.00	0.62	2.87	5.63	9.56	12.82	15.58	16.68
85	0.00	0.00	0.00	0.00	0.00	0.02	1.60	4.94	9.53	12.93	15.76	16.88
90	0.00	0.00	0.00	0.00	0.00	0.00	0.69	4.82	9.54	12.97	15.83	16.95

North latitude (°)	July 1	July 15	Aug. 1	Aug. 15	Sept. 1	Sept. 15	Oct. 1	Oct. 15	Nov. 1	Nov. 15	Dec. 1	Dec. 15
0	13.79	13.98	14.34	14.68	15.02	15.17	15.14	14.95	14.57	14.23	13.92	13.77
5	14.62	14.75	14.90	15.16	15.26	15.19	14.91	14.51	13.92	13.45	13.02	12.84
10	15.35	15.41	15.50	15.51	15.38	15.09	14.55	13.95	13.16	12.55	12.04	11.81
15	15.96	15.95	15.89	15.74	15.36	14.85	14.07	13.28	12.28	11.57	10.97	10.71
20	16.46	16.38	16.16	15.83	15.20	14.48	13.47	12.49	11.33	10.51	9.83	9.54
25	16.83	16.68	16.31	15.80	14.92	13.99	12.75	11.60	10.28	9.37	8.63	8.32
30	17.09	16.86	16.32	15.64	14.51	13.37	11.91	10.62	9.16	8.17	7.39	7.06
35	17.22	16.92	16.21	15.35	13.97	12.64	10.99	9.55	7.98	6.94	6.12	5.79
40	17.23	16.86	15.98	14.93	13.32	11.79	9.96	8.41	6.75	5.68	4.86	4.52
45	17.14	16.68	15.62	14.39	12.55	10.85	8.85	7.21	5.49	4.42	3.61	3.28
50	16.94	16.39	15.16	13.74	11.67	9.80	7.67	5.96	4.24	3.19	2.43	2.13
55	16.66	16.02	14.39	12.99	10.69	8.67	6.43	4.69	3.01	2.03	1.36	1.11
60	16.32	15.58	13.95	12.14	9.62	7.47	5.15	3.43	1.85	1.01	0.50	0.30
65	16.01	15.14	13.25	11.22	8.47	6.20	3.86	2.21	0.85	0.26	0.30	0.00
70	15.93	14.82	12.57	10.27	7.26	4.88	2.58	1.10	.15	0.00	0.00	0.00
75	16.23	14.96	12.15	9.35	6.01	3.55	1.37	0.26	0.00	0.00	0.00	0.00
80	16.54	15.21	12.18	8.79	4.77	2.22	0.39	0.00	0.00	0.00	0.00	0.00
85	16.74	15.38	12.27	8.71	3.81	0.98	0.00	0.00	0.00	0.00	0.00	0.00
90	16.80	15.44	12.31	8.70	3.00	0.19	0.00	0.00	0.00	0.00	0.00	0.00

[a] Photosynthetic active radiation constitutes 0.5 of total short-wave irradiance.

reported in the literature.[10-12] Figures for maximum short-term crop growth rates (i.e., vegetative productivity over a 2- to 3-week period) indicate that maximum crop growth rates of C_4 crops are 40 to 60% higher than those of C_3 crops (Table 3). Although part of the difference may be environmental (C_4 crops growing at lower latitudes receive in general higher insolation than C_3 crops growing in more temperate climates), high crop-growth rates of C_4 crops at moderately high insolation (i.e., 20 MJ m^{-2} day^{-1}) are associated with a

Table 2
CALCULATED DAILY GROSS CO$_2$ ASSIMILATION IN KILOGRAMS OF CO$_2$ ha^{-1} OF A CLOSED CANOPY WITH A SPHERICAL LEAF ANGLE DISTRIBUTION FOR STANDARD CLEAR (PC) AND OVERCAST (PO) DAYS

PMAX = 30 kg of CO$_2$ ha^{-1} hr^{-1}

North. Latitude (°)		15 Jan.	15 Feb.	15 Mar.	15 Apr.	15 May	15 June	15 July	15 Aug.	15 Sept.	15 Oct.	15 Nov.	15 Dec.
0	PC	623	642	654	648	630	616	622	641	654	648	629	616
	PO	293	305	312	309	297	289	292	304	312	309	297	289
10	PC	560	600	638	664	670	669	670	669	652	616	572	549
	PO	259	282	304	318	320	318	319	320	311	291	266	252
20	PC	486	545	610	668	699	711	707	684	637	570	503	469
	PO	217	250	286	318	334	340	338	327	301	264	227	208
30	PC	396	475	566	657	716	742	732	686	607	510	419	375
	PO	169	211	260	309	341	353	349	325	282	230	181	159
40	PC	294	389	507	633	721	763	747	676	562	433	321	270
	PO	117	164	225	292	339	360	352	315	254	187	130	105
50	PC	183	288	429	593	716	776	753	652	499	339	211	158
	PO	63	112	181	265	329	359	348	296	217	137	76	51
60	PC	66	175	333	536	704	790	756	615	417	230	98	38
	PO	15	57	130	229	312	354	338	268	170	81	25	8
70	PC	0	45	220	467	699	846	784	572	318	109	0	0
	PO	0	10	72	184	293	357	331	234	116	27	0	0

PMAX = 60 kg of CO$_2$ ha^{-1} hr^{-1}

		15 Jan.	15 Feb.	15 Mar.	15 Apr.	15 May	15 June	15 July	15 Aug.	15 Sept.	15 Oct.	15 Nov.	15 Dec.
0	PC	894	926	946	937	906	883	892	925	947	937	904	883
	PO	321	336	345	341	327	316	321	335	345	341	326	316
10	PC	796	859	920	960	967	964	966	966	941	884	815	777
	PO	282	309	335	351	353	350	352	353	344	320	290	274
20	PC	680	773	873	963	1010	1027	1021	988	915	812	707	654
	PO	234	272	314	351	369	375	373	361	332	289	245	224
30	PC	543	663	803	942	1032	1070	1056	987	865	716	576	511
	PO	180	227	283	340	376	390	385	358	309	248	194	168
40	PC	389	529	707	898	1033	1095	1071	964	790	595	427	354
	PO	122	174	242	318	372	396	387	344	275	199	137	109
50	PC	227	377	584	829	1014	1104	1069	918	688	451	266	192
	PO	64	116	193	286	358	393	379	320	232	144	78	52
60	PC	71	212	437	733	980	1107	1057	850	558	289	107	40
	PO	15	58	135	244	336	383	365	287	180	84	25	8
70	PC	0	47	268	615	948	1151	1066	766	403	119	0	0
	PO	0	10	74	193	311	381	353	247	120	28	0	0

From Goudriaan, J. and van Laar, H. H., *Neth. J. Agric. Sci.*, 26, 373, 1978. With permission.

higher efficiency of conversion of solar radiation into plant dry matter. The higher annual productivity of C$_4$ crops (Table 4) can probably be attributed both to higher crop-growth rates and to longer duration of the growing season.

Actual dry-matter production is usually substantially lower than aforementioned estimates of potential and maximum productivity. Rates of dry-matter accumulation may be reduced by soil moisture deficit, physical and chemical conditions of the soil, or by plant diseases and pests. However, even under optimal conditions for plant growth, rates of crop dry matter accumulation during the period of full light interception by crop canopies frequently do not reflect their genotypic potential. For instance, we recorded maximum crop-growth rates in

Table 3
SOME MAXIMUM SHORT-TERM ABOVE-GROUND GROWTH RATES (CGR) FOR SELECTED PLANT SPECIES

Group	Species	Location	CGR[a] (kg ha^{-1} day^{-1})	Irradiance[b] (MJ m^{-2} day^{-1})	Conversion efficiency[c]	Ref.
C$_4$	Sorghum	Texas	690	—	—	16
	Napier grass	N.T., Australia	540	21.3	8.9	17
	Maize	California	520	30.8	5.9	18
	Maize	Kikyogahara, Japan	520	20.3	8.6	19
	Sorghum	California	510	28.8	6.2	20
	Sugarcane	Hawaii	380	—	—	21
C$_3$	Corn cockle	England	390	—	—	22
	Potato	California	370	—	—	23
	Rice	Hainuzuka, Japan	360	20.5	6.3	19
	Cattail	Oklahoma	340	—	—	24
	Rye	Ontario, Canada	320	22.9	4.9	7
	Sugar beet	Netherlands	310	—	—	24
	Soybeans	Morioka, Japan	270	12.1	8.7	19

[a] CGR includes underground parts.
[b] Total short-wave radiation.
[c] Conversion of photosynthetic active radiation (400—700 nm) into plant dry matter. Energy content of 1 g dry matter (all parts) is assumed to be 17.5 KJ for all species except soybeans. Energy content soybeans is assumed 23.1 KJ for seeds and 16.1 J for other parts.

the range 400 to 500 kg ha^{-1} day^{-1} for maize grown hydroponically in the field at Guelph, Ontario,[13] whereas the mean rate for field-grown maize at nearby locations was 230 kg ha^{-1} during periods of full light interception. In addition, the crop growth rates of field-grown maize were not significantly correlated with incident short-wave radiation, which ranged from 15 to 22 MJ m^{-2} day^{-1}, although symptoms of stress were not apparent at any time. Indeed, crop growth rates are frequently fairly constant, which may be the result of a combination of various above- and below-ground factors which cause an "invisible" stress, and rates of both C$_3$ and C$_4$ species tend to be in the range from 200 to 250 kg ha^{-1} day^{-1}. As a rule of thumb, dry matter accumulation at the end of the growing season (DWF) can be estimated as follows:

$$DWF - DWI + CGR \times FLID \qquad (2)$$

where DWI is crop dry weight at the onset of the period of full light interception by the canopy, FLID is the duration of the period of full light interception, and values for crop growth rate (CGR) range from 200 to 250 kg ha^{-1} day^{-1}. A crop canopy attains full light interception at a leaf area index of 3.0 to 6.0 (depending on various canopy characteristics) and under optimal conditions for plant growth, the duration of full light interception by a crop canopy is likely the most important determinant of dry matter production. Duration of the period of full light interception varies with length of the growing season, accumulated thermal units or degree days, crop species, crop cultivar, and management practices.

Whereas dry matter accumulation of various crops (e.g., maize, wheat, potato) may be estimated using Equation 2, several exceptions apply. For instance, when environmental conditions deviate substantially from optimal conditions for crop growth (e.g., temperature extremes, low irradiance, "visible" moisture or nutrient stress), rates of dry matter accumulation will be lower. In addition, lower rates of dry-matter accumulation of oil and protein crops during seed filling can be attributed to high respiratory demands associated with the formation of a high-energy product. Also, lower crop growth rates of perennial rye

Table 4
ESTIMATES OF MAXIMUM VEGETATIVE PRODUCTIVITY FOR SELECTED CROP SPECIES[a]

Group	Species	Location	Production (t ha^{-1})	Notes	Ref.
C$_4$	Sugarcane	Hawaii	111.8	365 days, 250 t ha^{-1} canestalks; 0.17 taken as sucrose content; sucrose taken as 0.38 of total crop.	26
	Napier grass	El Salvador	85.4	365 days	27
	Sugarcane	Hawaii	78.0	365 days	28
	Sorghum	California	46.6	210 days	29
	Maize	Michigan	39.4	157 days; 352.6 bu A^{-1} at 15.5% moisture; taken as 0.5 of total crop	30
	Bermuda grass	Texas	32.6	365 days	31
C$_3$	Sugar beet	California	44.7	240 days; 115 t ha^{-1} fresh beets at 16.5% sucrose; 26.7 t dry matter in beets and 18 t dry matter in tops	32
	Wheat	Washington	29.8	209 bu A^{-1} grain at 14% moisture; taken as 0.4 of total crop	33
	Alfalfa	California	29.7		12
	Potato	Netherlands	25.0	177 days; 19 t ha^{-1} tuber dry weight; taken as 0.76 of total crop	34, 35
	Rice	Takada, Japan	19.7	161 days	18
	Soybeans	Illinois	13.0	177 days; 83.5 bu A^{-1} at 14% moisture; taken as 0.37 of total crop	36

[a] Above-ground parts, except where tubers or storage roots are considered.

grass during the fall in England have been attributed to the low irradiance, due to mutual shading in the canopy, under which new leaves have been developed,[14] and crop-growth rates of several crop species may also be reduced by an unbalanced source-sink ratio.[15] Indeed, numerous factors can modify rate of dry matter accumulation, and the relative constancy in crop-growth rates among crop species and environments is surprising. A major limitation to crop productivity of field-grown crops may be root dry-matter accumulation and development rather than above-ground factors, but very little is known about the effects of root growth on dry-matter accumulation of field-grown crops.

Economic yields can be derived from estimates of vegetative productivity by taking into account the appropriate harvest index (Table 5). Harvest index is defined as the fraction of vegetative dry matter that is economically useful (e.g., grain in wheat). It should be noted that factors such as management practices, crop cultivar, moisture conditions, and climate cause the value of the harvest index to fluctuate considerably and that under less favorable conditions actual values may be 10 to 40% lower than indicated in Table 5. Average commercial productivity for some selected agricultural crops grown in various countries (Table 6) can be compared with estimates of maximum vegetative productivity described by taking into account the appropriate harvest index and percent moisture of the commercial product (Table 5).

Table 5
APPROXIMATE HARVEST INDEX, MOISTURE CONTENT AT WHICH COMMERCIAL PRODUCTS ARE TRADED, AND RATIO OF AVERAGE WORLD PRODUCTIVITY (TABLE 6) OVER MAXIMUM REPORTED ECONOMIC PRODUCTIVITY FOR SELECTED AGRICULTURAL CROP SPECIES

Species	Economic dry-matter yield)/(total above-ground dry matter)[a]	Moisture content commercial product[b] (%)	(Av world productivity)/ (max. reported economic productivity)
Wheat	0.40	14.5	0.14
Barley	0.45	14.8	—
Rice	0.40	14.0	0.25
Maize	0.50	15.5	0.13
Sunflower	0.45	9.5	—
Rapeseed	0.30	10.0	—
Soybeans	0.35	14.0	0.29
Sugar beet	0.60	77.0	0.27
Sugarcane	2.24[c]	—	0.23
Potato	0.76	80.0	0.15

[a] Total vegetative dry matter (i.e., top + roots) where tuber or storage roots are considered.
[b] From *Official Grain Grading Guide*, Canadian Grain Commission, 1985. Agriculture Canada; figures for rice, sugar beet, sugarcane, and potato estimated from data in Evans, L. T., *Crop Physiology: Some Case Histories*, Cambridge University Press, Cambridge, 1975.
[c] Fresh weight of cane stalks/total above-ground dry matter.

Table 6
AREA OF PRODUCTION AND AVERAGE ECONOMIC YIELD OF NINE CROP SPECIES FOR THE WORLD AND SELECTED COUNTRIES (MEANS FOR THE PERIOD 1978—1980)

	Wheat		Barley		Rice		Maize		Sunflower	
	Area × 1,000 ha	Av prod'n. (t/ha)	Area × 1,000 ha	Av prod'n. (t/ha)	Area × 1,000 ha	Av prod'n. (t/ha)	Area × 1,000 ha	Av prod'n. (t/ha)	Area × 1,000 ha	Av prod'n. (t/ha)
World	231,202	1.91	84,818	1.98	142,232	2.72	128,130	3.13	12,006	1.15
Africa	8,388	1.04	5,154	.80	4,743	1.73	21,472	1.23	561	0.92
Egypt	575	3.24	45	2.69	426	5.64	797	3.88	5	1.82
Ghana	—	—	—	—	80	0.74	337	1.03	—	—
South Africa	1,771	0.98	91	1.32	1	2.31	5,667	1.67	348	1.07
Asia	79,102	1.65	14,016	1.25	127,332	2.77	36,565	2.20	988	1.09
India	22,020	1.50	1,859	1.08	39,485	1.92	5,771	1.05	—	—
Japan	151	3.32	111	3.35	2,474	5.84	2	2.84	—	—
Turkey	9,397	1.84	2,759	1.87	76	4.57	572	2.21	458	1.25
Europe	25,700	3.58	20,412	3.46	384	4.86	11,872	4.45	2,010	1.41
Italy	3,441	2.69	308	2.80	183	5.29	930	6.80	33	1.81
Hungary	1,245	3.95	244	3.12	20	1.87	1,338	5.24	218	1.59
Netherlands	135	6.24	62	4.80	—	—	1	6.83	—	—
North/Central America	37,096	2.12	7,793	2.45	1,973	4.16	38,795	5.23	1,768	1.38
Canada	10,722	1.79	4,273	2.33	—	—	880	5.47	129	1.30
Mexico	715	3.65	281	1.64	144	3.17	6,647	1.51	26	0.76
U.S.	25,604	2.22	3,238	2.68	1,232	5.04	29,325	6.31	1,613	1.39
South America	9,105	1.35	794	1.26	7,059	1.81	16,450	1.74	1,953	0.86
Argentina	4,751	1.69	262	1.41	93	3.18	2,624	3.12	1,819	0.86
Brazil	3,262	0.85	82	1.35	5,757	1.42	11,292	1.48	10	1.00
Columbia	33	1.24	67	1.60	434	4.25	637	5.23	—	—
Australia	10,969	1.37	2,624	1.35	105	5.66	46	3.07	235	0.71
U.S.S.R.	60,754	1.69	33,949	1.53	618	3.92	2,901	3.15	4,491	1.15

Table 6 (continued)
AREA OF PRODUCTION AND AVERAGE ECONOMIC YIELD OF NINE CROP SPECIES FOR THE WORLD AND SELECTED COUNTRIES (MEANS FOR THE PERIOD 1978—1980)

	Rapeseed		Soybeans		Sugar beet		Sugarcane		Potato	
	Area × 1,000 ha	Av prod'n. (t/ha)	Area × 1,000 ha	Av prod'n. (t/ha)	Area × 1,000 ha	Av prod'n. (t/ha)	Area × 1,000 ha	Av prod'n. (t/ha)	Area × 1,000 ha	Av prod'n. (t/ha)
World	11,915	0.89	51,000	1.64	8,738	31.07	13,283	56.53	18,285	14.30
Africa	53	0.41	320	0.90	68	35.56	868	68.26	564	8.19
Egypt	—	—	37	2.48	—	—	104	82.68	64	15.30
Ghana	—	—	—	—	—	—	9	21.37	—	—
South Africa	—	—	26	1.31	—	—	216	79.08	50	52.63
Asia	7,621	0.57	11,166	1.05	713	29.00	5,630	50.57	3,124	10.97
India	3,539	0.47	297	0.92	—	—	2,979	51.54	716	12.34
Japan	3	1.75	129	1.48	59	55.05	35	68.67	125	26.76
Turkey	16	1.48	3	1.00	275	32.02	—	—	174	17.07
Europe	1,361	2.17	458	1.30	3,683	38.57	6	64.14	5,834	19.10
Italy	0.33	1.97	—	2.62	279	45.71	—	—	169	17.62
Hungary	51	1.43	20	1.66	110	36.94	—	—	104	15.41
Netherlands	8	2.88	—	—	126	47.95	—	—	167	37.54
North/Central America	2,775	1.15	27,807	1.98	507	44.09	2,897	60.15	719	26.48
Canada	2,771	1.15	284	2.23	25	36.07	—	—	2	12.24
Mexico	4	1.00	261	1.67	—	—	538	64.74	63	13.11
U.S.	1	1.33	27,260	1.98	482	44.51	299	81.86	513	29.77
South America	62	1.22	10,353	1.50	33	33.16	3,547	58.02	1,018	9.37
Argentina	16	0.61	1,550	2.11	—	—	321	46.87	110	14.72
Brazil	—	—	8,293	1.39	—	—	2,525	55.00	198	10.28
Columbia	—	—	73	1.97	—	—	282	86.85	153	13.18
Australia	30	1.02	53	1.65	—	—	271	82.16	36	22.86
U.S.S.R.	12	0.92	843	6.24	3,735	22.24	—	—	6,982	11.64

From *FAO Production Yearbook*, Vol. 34, Food and Agriculture Organization, Rome 1980. With permission of the Food and Agriculture Organization of the United Nations.

REFERENCES

1. **Goudriaan, J.,** Potential production processes, in *Simulation of Plant Growth and Production,* Penning de Vries, F. W. T. and van Laar, H. H., Eds., Centre for Agricultural Publishing and Documentation, Wageningen, The Netherlands, 1982, 98
2. **Hesketh, J. and Baker, D.,** Light and carbon assimilation by plant communities, *Crop Sci.,* 7, 285, 1967.
3. **Gallagher, J. N. and Biscoe, P. V.,** Radiation absorption, growth and yield of cereals, *J. Agric. Sci. Camb.,* 91, 47, 1978.
4. **Monteith, J. L.,** Climate and efficiency of crop production in Britain, *Phil. Trans. R. Soc. London Ser. B,* 281, 277, 1977.
5. **Jong, S. K., Brewbaker, J. L., and Lee, C. H.,** Effects of solar radiation on the performance of maize in 41 successive monthly plantings in Hawaii, *Crop Sci.,* 22, 13, 1982.
6. **Bonhomme, R., Ruget, R., Derieux, M., and Vincourt, P.,** Relations entre production de matiere seche aerienne et energie interceptee chex differents genotypes de mais, *C. R. Acad. Sci. Paris,* 294, 393, 1982.
7. **Tollenaar, M.,** Potential vegetative productivity in Canada, *Can. J. Plant Sci.,* 63, 1, 1983.
8. **Goudriaan, J. and van Laar, H. H.,** Calculation of dialy totals of gross CO_2 assimilation of leaf canopies, *Neth. J. Agric. Sci.,* 26, 373, 1978.
9. **de Wit, C. T.,** *Simulation of assimilation, respiration and transpiration of crops,* Centre for Agricultural Publishing and Documentation, Wageningen, The Netherlands, 1978, 141.
10. **Westlake, D. F.,** Comparisons of plant productivity, *Biol. Rev.,* 38, 385, 1963.
11. **Cooper, J. P.,** Potential production and energy conversion in temperate and tropical grasses, *Herb. Abstr.,* 40, 1, 1970.
12. **Loomis, R. S. and Gerakis, P. A.,** Productivity of agricultural ecosystems, in *Photosynthesis and Productivity in Different Environments,* Cooper, J. P., Ed., Cambridge University Press, London, 1975, 145.
13. **Tollenaar, M. and Migus, W.,** Dry matter accumulation of maize grown hydroponically under controlled-environment and field conditions, *Can. J. Plant Sci.,* 64, 475, 1984.
14. **Parsons, A. J. and Robson, M. J.,** Seasonal changes in the physiology of S24 perennial ryegrass (*Lolium perenne* L.). II. Potential leaf and canopy photosynthesis during the transition from vegetative to reproductive growth, *Ann. Bot.,* 47, 249, 1981.
15. **Tollenaar, M. and Daynard, T. B.,** Effect of source-sink ratio on dry matter accumulation and leaf senescence of maize, *Can. J. Plant Sci.,* 62, 855, 1982.
16. **Reeves, S. A., Hipp, B. W., and Smith, B. A.,** Sweet sorghum biomass, *Sugar Azucar,* 74, 23, 1979.
17. **Begg, J. E.,** High photosynthetic activity in a low latitude environment, *Nature (London),* 205, 1025, 1965.
18. **Williams, W. A., Loomis, R. S., and Lepley, C. R.,** Vegetative growth of corn as affected by population density. I. Productivity in relation to interception of solar radiation, *Crop Sci.,* 5, 211, 1965.
19. **Murata, Y. and Togari, Y.,** Summary of data, in *Crop Productivity and Solar Energy,* Murata, Y., Ed., University of Tokyo Press, Tokyo, 1975, 9.
20. **Loomis, R. S. and Williams, W. A.,** Maximum crop productivity: an estimate, *Crop Sci.,* 3, 67, 1963.
21. **Borden, R. J.,** A search for guidance in the nitrogen fertilization of the sugar cane crop. I. The plant crop, *Hawaii. Plant Rec.,* 46, 191, 1942.
22. **Newton, J.,** Some Factors Influencing the Productivity of Vegetation, Ph.D. thesis, University of Oxford, Oxford, England, 1968.
23. **Lorenz, O. A.,** Studies on potato nutrition. II. Nutrient uptake at various stages of growth by Kern County potatoes, *Am. Soc. Hortic. Sci. Proc.,* 44, 389, 1944.
24. **Penfound, W. T.,** Primary productivity of vascular aquatic plants, *Limnol. Oceanogr.,* 1, 192, 1956.
25. **Sibma, L.,** Maximization of arable crop yield in the Netherlands, *Neth. J. Agric. Sci.,* 25, 278, 1977.
26. **Bull, T. A. and Glasziou, K. T.,** Sugar cane in *Crop Physiology, Some Case Histories,* Evans, L. T., Ed., Cambridge University Press, New York, 1976, 51.
27. **Watkins, J. M. and Lewy-van Severen, M.,** Effect of frequency and height of cutting on the yield, stand and protein content of some forages in El Salvador, *Agron. J.,* 43, 291, 1951.
28. **Borden, R. J.,** The effect of nitrogen fertilization upon the yield and composition of sugar cane. I. The plant crop, *Hawaii. Plant Rec.,* 49, 259, 1945.
29. **Worker, G. F., Jr. and Marble, V. L.,** Comparison of sorghum forage types as to yield and chemical composition, *Agron. J.,* 60, 669, 1968.
30. **Anon.,** Unpublished leaflet, DeKalb Seed Co., DeKalb, Ill., 1977.
31. **Fisher, F. L. and Caldwell, A. G.,** The effects of continued use of heavy rates of fertilizers on forage production and quality of coastal Bermudagrass, *Agron. J.,* 51, 99, 1959.
32. **Fick, G. W., Loomis, R. S., and Williams, W. A.,** Sugar beet, in *Crop Physiology, Some Case Histories,* Evans, L. T., Ed., Cambridge University Press, New York, 1976, 259.
33. **Reitz, L. P.,** Wheat distribution and importance of wheat, in *Wheat and Wheat Improvement,* Quisenberry, K. S. and Reitz, L. P., Eds., American Society of Agronomy, 1967, 1.

34. **Moorby, J. and Milthorpe, F. L.,** Potato, in *Crop Physiology, Some Case Histories,* Evans, L. T., Ed., Cambridge University Press, New York, 1976, 225.
35. **Bodlaender, K. B. A. and Algra, S.,** Influence of the growth retardent B 995 in growth and yield of potatoes, *Eur. Potato J.,* 9, 242, 1966.
36. **Anon.,** Illinois is a national soybean yield champion, *Soybean Dig.,* 32, 5, 1972.

WEED SCIENCE AS IT RELATES TO CROP PRODUCTION

W. Powell Anderson

INTRODUCTION

Weed science is concerned with understanding and preventing the adverse effects of weeds on crop production. An understanding of these adverse effects is achieved through studies to determine the growth and propagation characteristics of weed species infesting croplands and the resulting crop-weed interactions in these lands. Prevention of these adverse effects is accomplished through development and use of practices that control weeds without adversely affecting crop production.

Weeds have been a constant companion of crop plants in croplands since man replaced native vegetation with more nutritious, productive, and economical plants to meet the needs of an ever-increasing human population.[1,2] When the natural ecology of an area was altered to favor production of selected plant species, man began a constant battle with unwanted species that thrived in his croplands under the very conditions provided for his crop plants. This is when man discovered *weeds,* and when the embryo of *weed science* was conceived.

Considering the great diversity in crop plants and cultural practices, this perspective of weed science as it relates to crop production is directed more toward cultivated crops than noncultivated crops, although much of the material presented here applies to either kind of crop.

WEED AND CROP PLANTS DEFINED

A weed may be defined as any plant growing where it is not wanted.[3,4] King has provided a history of the term "weed".[5] A crop plant is any plant grown for its value to man at a given time. Thus, both weeds and crops are plants. Plants, in general, have similar growth and propagation characteristics, and they require essentially the same environmental ingredients to support these characteristics.

ECONOMIC CROP LOSSES DUE TO WEEDS

Weeds cause economic losses in crop production. If this were not so, man would not expend the tremendous amount of time, energy, and money on weed control. In spite of modern weed-control technology, weeds continue to cause annual losses of about 10% in agricultural production. The annual monetary loss caused by weeds in agricultural production is estimated at more than $18.2 billion, with about $12 billion of this amount attributed to the production losses caused by weeds. Another $3.6 billion is spent on chemical weed control, and $2.6 billion is spent on cultural, ecological, and biological weed-control methods.[6,7]

From 1972 to 1976, the estimated monetary loss from weeds in 13 major agronomic crops was about $5 billion, with losses in seven of these crops accounting for 93% of this total.[8] The seven agronomic crops involved and their respective percentage of this loss were: soybeans, 32; corn, 30; wheat, 16; grain sorghum, 5; cotton, 4; rice, 4; and peanuts, 2. During this same 5-year period, the estimated monetary loss from weeds in 22 vegetable crops was about $313 million, with losses in seven of these crops accounting for 73% of this loss.[8] The seven vegetable crops involved and their respective percentage of this total were dry beans, 19; tomatoes (fresh and processed), 17; potatoes, 12; lettuce, 9; sweet corn, 7; onions, 5; and green snapbeans, 4.

ADVERSE EFFECTS OF WEEDS IN CROPLANDS

Weeds growing among crop plants adversely affect yield and quality of the harvested commodity and increase production costs, resulting in economic loss. Weeds reduce crop yields by decreasing the amount of the desired commodity produced. For example, sugar beet (*Beta vulgaris* L.) root yields were reduced 48% and recoverable sucrose decreased 46% as the result of competition with common lambsquarters (*Chenopodium album* L.).[9] Bearded sprangletop (*Leptochloa fascicularis* [Lam.] Gray)[10] reduced grain yields 36% in rice (*Oryza sativa* L.), and the grain yields of corn (*Zea mays* L.) and soybeans (*Glycine max* Merr.) were reduced 25 and 28%, respectively, by competition with giant foxtail (*Setaria faberii* Herrm.).[11] Weeds lower crop quality by contaminating the harvested product with their seeds and vegetative parts, impeding vegetative, fruit and seed formation of the crop, and in some cases (such as cotton fiber), staining the harvested commodity with plant pigments. Weeds also harbor insects and plant pathogens. Weeds increase production costs by reducing the productivity of the land, interfering with crop harvesting, requiring the cleaning of the harvested commodity to remove unwanted weed parts, and impeding water flow in irrigation and drainage channels.

CLASSIFICATION AND PROPAGATION OF WEEDS

Weed species infesting croplands are primarily herbaceous grass and broadleaf (dicotyledonous) plants. Based on their respective life cycles, they may be classified as annual, biennial, or perennial plants.[12] Annual weeds complete their life cycles, from seed germination through seed formation and death of the parent plant, in one growing season (1 year or less). Annuals that complete their life cycles during the warm growth period from spring to fall are called *summer annuals*. Those that complete their life cycles during the colder period from fall to spring are called *winter annuals*. Those that complete their life cycles at any time throughout the year, independent of season, are referred to as *indeterminate annuals*. Annual weeds propagate by seeds.

Biennials need two growing seasons in which to complete their life cycles, producing seed during the second growing season. Biennial weeds propagate by seeds.

Perennials are relatively long-lived plants, living for 3 or more years. Most perennial weeds propagate by seeds, but many species also propagate asexually. Perennial weed species may be subgrouped as *simple perennials,* producing new plants each year from a root crown supported by a fleshy taproot, and *creeping perennials,* propagating asexually from horizontal roots, rhizomes, and/or stolons. Other asexual propagules of perennial weed species include bulbs and tubers. Perennial weed species may produce seeds and asexual propagules the first year of establishment and each year thereafter.

INTRODUCTION OF WEEDS INTO CROPLANDS

Weeds commonly arise and infest croplands from seeds and asexual propagules already present in or on the soil when a crop is planted. If weed propagules are not present when the crop is planted, and none are subsequently introduced, weeds will not infest the crop. Similarly, plants of a specific weed species will not be found among the weed population of a field if none of its propagules are present or introduced to the land.

The seeds and asexual propagules of weed species are introduced to croplands by the intentional and unintentional actions of man, and by natural means of plant dissemination. The most common intentional ways man introduces weeds to croplands are by planting a crop new to the area that subsequently takes on the characteristics of a weed, or by planting an ornamental plant species that escapes confinement as a weed. The most common unintentional way man introduces weeds to croplands is by planting crop seed contaminated with weed seeds.[13] A survey of planter drill boxes in Utah revealed that more than half of the 1232 seed grain samples collected and analyzed contained weed seeds. The worst sample contained seeds of five noxious weed species, with 167,000 such seeds per 45.5 kg of sample. In this case, an average of five noxious weed seeds were planted, intermixed with the crop seed, on each 0.09 m² of cropland.[14]

Weeds are commonly introduced to irrigated croplands by seed-laden irrigation water moving in open channels. The embankments of such channels are often covered with weeds and, as the weeds mature, their seeds drop into the flowing water and are deposited on cropland irrigated with this water. A study of the Columbia River and two irrigation laterals in Washington involving weekly and biweekly screenings of water in each of these irrigation systems during the irrigation season yielded seeds of 77, 84, and 137 plant species, respectively, and the total number of seeds per 254 kℓ of water averaged 292, 682, and 2220 for the season, respectively.[15] Assuming an even seed distribution in the irrigation water for the season, the number of seeds disseminated would average 14,100, 10,400 and 94,500/ha. Weed control measures were practiced along the irrigation lateral having the fewer kinds of weed species and numbers of weed seeds. In a similar study of the North Platte River and two irrigation canals in Nebraska, two to five times more weed seeds were found in the canals than in the river, and the weed seed content of the water increased dramatically as the water moved through the canals. Most seeds were found floating on the water surface. Among seeds collected from the irrigation water, 75 kinds of weed species were identified, with an average germination of 26%. It was calculated that 48,400 weed seeds per hectare were deposited on cropland irrigated by this water during the irrigation season.[16]

Man also introduces weed propagules to croplands as contaminants in manure spread on the land and in soil surrounding the roots of transplanted crop plants.

Wind is the principal natural means by which weeds are introduced to croplands. The seed-dispersal units of many weed species are well adapted to wind dispersal, having such appendages as "parachutes", wings, or tufts of hair. Wind also tumbles seed-laden plant parts across the landscape, introducing weed seeds from other crop and noncrop areas. Weed seeds may also be introduced to cropland by floodwater, birds, animals, and even the farmer's trousers' cuffs. Creeping perennial weed species can enter a field from its fringes through natural extension of asexual parts (roots, rhizomes, stolons).

PRESENCE OF WEEDS WITHIN CROPLANDS

It is not unusual for croplands to be infested with many species of weeds. Each weed species present in a given field depends on the introducton of one or more of its sexual or asexual propagules into the field, and to the subsequent survival, propagation, and adaptation to the local environment of the resulting plant or plants. The local environment to which weed species must adapt includes all the natural factors such as climate, photoperiod (daylength), soil, insect and plant pathogen pressures, and plant competition and artificial factors such as the cultural practices to which the land is subjected encountered by the weed plants.

Certain weed species are often common to a particular crop because the weed and crop plants have similar growth habits and thrive under the same cultural practices. Some weeds perpetuate themselves in croplands when the land is lying fallow.

Winter annual weeds plague fall-seeded crops that are harvested the following year. Summer annual weeds infest crops that are spring planted and harvested during or after the summer of the same year. Many weed species thrive in row crops, surviving tillage practices by growing among the crop plants in the crop row. Weed species that tolerate continued mowing or grazing thrive in pasture and forage crops. Perennial weeds appear in croplands whenever conditions favor their particular life cycles.

A particular field or region may be free of a given weed species because the species (1) has not been introduced to the area, (2) was introduced and successfully eradicated, or (3) is not adapted to the environment.[17]

DISPERAL OF WEEDS WITHIN CROPLANDS

Weeds already present in croplands produce seeds and/or asexual propagules, and these propagules are dispersed within a given field by man and by natural means of dissemination.[18-20]

The principal way man distributes weed seeds within croplands is by working the soil with tillage and land-leveling equipment. Tillage equipment, as it works the soil, changes the seed location in the soil profile, burying some seeds deeply while moving others nearer the surface. Tillage and land-leveling equipment move soil from one field location to another, redistributing weed seeds mixed with this soil. Weed seeds on the soil surface are also dispersed within a field when mass flow of rain and irrigation waters carry them across the soil surface.

Creeping perennial weed species are distributed in croplands by tillage equipment that cuts and drags asexual reproductive parts from one location to another. Examples of perennial species spread this way within a field are Bermuda grass (*Cynodon dactylon* [L.] Pers.), field binweed (*Convolvulus arvensis* L.), quackgrass (*Agropyron repens* [L.] Beauv.), Johnson grass (*Sorghum halepense* [L.] Pers.), purple nutsedge (*Cyperus rotundus* L.), and yellow nutsedge (*Cyperus esculentus* L.).

PRODUCTION AND LONGEVITY OF WEED SEEDS IN CROPLANDS

Annual weed plants in cultivated lands grow from seeds buried in the soil, deposited there as a legacy from preceding plants. Certain perennial weeds are also perpetuated by seeds in croplands, whereas others arise more freely from viable asexual reproductive parts buried in the soil.

The capacity of an individual weed plant to produce seed is prodigious.[21,22] For example, biennial wormwood (*Artemisia biennis* Willd.) can produce 1 million seeds per plant, redroot pigweed (*Amaranthus retroflexus* L.) 117,000 seeds per plant, kochia (*Kochia scoparia* [L.] Roth.) 14,600 per plant, and foxtail barley (*Hordeum jubatum* L.) 2400 per plant. One study to determine the potential of certain weeds as a high-protein source reported average seed yields of kochia and Russian thistle (*Salsola kali* var. *tenuifolia* Tausch.) as 2170 and 1640 kg/ha (dry basis), respectively.[23]

The longevity of weed seeds buried in cropland soil is a major survival mechanism, allowing many weed species to perpetuate themselves under varied systems of crop production.[18] Many weed seeds disintegrate in soil within days or months of burial, but others remain viable but dormant for years, depending on the weed species involved. Buried in soil, seeds of tall larkspur (*Delphinium barbeyi* Huth.) and western false hellebore (*Veratrum californicum* Durand) either germinate or disintegrate during the first year in the soil, whereas seeds of velvetleaf (*Abutilon theophrasti* Medic.) may remain viable for 40 or more years and seeds of field bindweed (*Convolvulus arvensis* L.) for 20 or more years. Seed longevity, in part, determines the persistence of weed species in cropland.[24,25] Weed species with seeds that remain viable while buried for long periods of time become problem weeds in croplands.

The viability of weed seeds transported by irrigation waters onto irrigated lands is generally unaffected during the relatively short period of time that the weed seeds are in the water. Even seeds that disintegrate comparatively rapidly in water, such as barnyard grass (*Echinochloa crusgalli* [L.] Beauv.), remain viable long enough to survive the journey from their source along the banks of the water channels onto irrigated fields.[26,27]

WEED-CROP COMPETITION

Members of plant communities growing close to one another, where roots intermingle and shoots crowd together, compete for the available essentials of growth and development. Therefore, it is not surprising that competition occurs in intermixed communities of weed and crop plants.[28] The unfortunate consequence of weed-crop competition is that crop yields, the objective of crop production, are usually reduced.

Competition occurs between two or more plants when the supply of one or more factors essential to growth and development falls below the combined demands of the plants.[29] Plants compete for water, nutrients, light, atmospheric carbon dioxide, soil oxygen, and space. The term *interference* is, unfortunately, appearing in the scientific literature to denote the phenomenon that over the years has been referred to as *competition*. Donald[29] weighed the pros and cons of these two terms and chose competition as the preferred term because it is simple and effective, and vividly conveys the active interaction between organisms (plants) for available but limited factors essential to growth, development, and reproduction.

The more aggressive species usually dominate an intermixed community of weed and crop plants.[23] This aggressiveness is associated with differences between members of the plant community in growth habit and rate of root and shoot growth and development. Aggressiveness is favored by greater root elongation and branching, resulting in a vigorous, rapidly spreading, root system that absorbs water, nutrients, and oxygen from the soil at the expense of adjacent plants.[31-33] Aggressiveness is also favored by taller plant species that grow more quickly than adjacent plants or by plants that climb their neighbors as vines, producing foliar canopies that shade slower growing or shorter plants in the community.[34,35]

Some weed species supplement aggressiveness by production of phytotoxic or plant-growth-inhibiting substances that adversely affect growth and development of other plants. These chemicals are released into the soil as root exudates or as leachates of the living or dead plants. This biochemical interaction between plants is called *allelopathy*.[30,36]

Reduction in crop yields is an accepted parameter for determining weed-crop competition.[37] From a literature review by Burnside,[38] who condensed the facts concerning weed-crop competition in soybeans, the following are key factors in crop yield reduction resulting from weed-crop competition: weed species present; weed-crop emergence; competition duration; weed life cycle and growth habit; density of weed and crop plants; crop species and cultivars; crop life cycle and growth habit; crop planting date, depth, and row spacing; and climatic, edaphic, and other biotic factors.

Weed-crop competition in the first 6 weeks or so after crop planting tends to have the greatest adverse effect on crop yields. Soybeans kept weed-free for the first 4 weeks after planting showed little loss in yield from competition with later-emerging weeds, whereas season-long weed competition reduced yields by 50 to 75%, depending on weed species involved and severity of weed density.[38] Weeds that emerge after soybeans are well established are less competitive because established soybean plants compete well with weeds. Soybeans planted in narrow row spacings (51 cm or less) compete better with weeds at an earlier growth stage than those in wide rows because of better root distribution and earlier and more complete shading of the soil surface. When the soil is shaded, weed emergence is essentially stopped.[36]

Different weed species vary in their competitiveness with crop plants. Soybean yields were reduced 7, 8, and 23% by competition from prickly sida (*Sida spinosa* L.), Venice mallow (*Hibiscus trionum* L.), and velvetleaf (*Abutilon theophrasti* Medic.), respectively, emerging with or during the first 10 days after soybean emergence. Soybean yields were not affected by plants of these weed species that emerged 20 or more days after soybean emergence and competed for the remainder of the season.[39]

Individual crops vary in length of time they should be kept weed-free following crop emergence to avoid yield reductions.[40] The following is taken from a literature review by Buchanan and McLaughlin[41] to document this point: sorghum (*Sorghum bicolor* [L.] Moench) and field beans (*Phaseolus vulgaris* L.) kept weed-free 4 to 5 weeks when competing with a mixed annual weed population suffered no yield reduction, similar to sugar beets kept weed-free 4 weeks when competing with kochia, and 12 weeks for common lambsquarters and barnyard grass (*Echinochloa crusgalli* [L.] Beauv.); corn kept weed-free 4 weeks when competing with giant foxtail; and peanuts (*Arachis hypogaea* L.) kept weed-free 6 weeks when competing with smooth pigweed (*Amaranthus hybridus* L.) and large crabgrass (*Digitaria sanguinalis* [L.] Scop.).

First-year production of alfalfa seed was reduced 95% by season-long competition from weeds as compared to weed-free alfalfa.[34] The weeds outgrew the alfalfa seedlings, suppressed them, and some alfalfa plants died. Common lambsquarters, the first weed species to emerge, was the dominant competitor. It grew to a height of 2.4 m, forming a dense canopy over the alfalfa and other weeds. One weed species present, hairy nightshade (*Solanum sarachoides* Sendt.), was eliminated from the weed population as a result of the aggressiveness of common lambsquarters.

In general, crop-yield reduction occurs linearly with an increase in the number of weed plants occupying a given crop row or field, with greater yield reductions occurring as the weed density increases. The 3-year averages of soybean yields were reduced 10, 28, 43, and 52% by full-season competition with common cocklebur (*Xanthium pensylvanicum* Wallr.) at densities of 3,300, 6,600, 13,000, and 26,000 plants per hectare, respectively.[37] In the year of seeding, weed-free alfalfa produced 820 kg/ha of seed. Competition from a dense population (55 plants per meter of row) of mixed broadleaved weeds, a light population (4 plants per meter of row) of mixed broadleaved weeds, or a heavy population (75 culms per meter of row) of barnyard grass reduced seed yields by 90, 62, and 80%, respectively.[34] Common cocklebur at a density of 1 plant per 3 m of row reduced soybean seed yields by 3 to 13%.[42] Sicklepod (*Cassia obtusifolia* L.) at densities of 2, 4, 8, 16, and 32 plants per 15 m of row reduced cotton (*Gossypium hirsutum* L.) seed yields by 11, 15, 23, 42, and 55%, respectively.[43] Tall morning glory (*Ipomoea purpurea* L. Roth) at densities of 4, 8, 16, and 32 plants per 15 m of row reduced cottonseed yields by 19, 41, 64, and 88%, respectively.[44] Broadleaf signalgrass (*Brachiaria platyphylla* [Griseb.] Nash) at densities of 8, 16, and 1,050 plants per 10 m of row reduced peanut (*Arachis hypogaea* L.) seed yields by 14, 28, and 69%, respectively.[45] Bearded sprangletop (*Leptochloa fascicularis* [Lam.] Gray) at densities of 11, 22, 54, and 108 plants per square meter reduced rice (*Oryza sativa* L.) seed yields by 9, 18, 20, and 36%, respectively.[46] Wild oat (*Avena fatua* L.) at densities of 84 and 191 plants per square meter reduced wheat (*Triticum aestivum* L.) seed yields by 22 and 39% and barley (*Hordeum vulgare* L.) seed yields by 7 and 26%, respectively.[47] Corn cockle (*Agrostemma githago* L.) at a density of 340 plants per square meter reduced wheat seed yields by 60%.[48] One kochia plant (*Kochia scoparia* [L.] Roth) per 7.6 m of row reduced the average yield of sugar beet (*Beta vulgaris* L.) roots by 5.9 metric tons (t) and sucrose yield by 1.07 t/ha.[49] The following weed species, at densities of 6, 12, 18, and 24 plants per 30 m of row and competing season-long with sugar beets, reduced root yields by the following respective percentages: sunflower (*Helianthus annuus* L.), 40, 52, 67, and 73; velvetleaf (*Abutilon theophrasti* Medic.), 14, 17, 25, and 30;[50] common lambsquarters, 13, 29, 38, and 48.[51]

WEED CONTROL PRACTICES IN CROPLANDS

It is a common practice in the U.S. and Canada to grow a single crop species on hundreds to thousands of hectares of land in a given crop-production region. This practice subjects the crop to infestation by weed species that thrive under the cultural conditions provided for the crop. The means by which weeds are controlled in croplands is determined largely by the growth and reproductive characteristics of the weeds and by the growth and cultural practices of the crop. Weed control practices may be grouped under five general headings: *preventive, cultural, mechanical* (physical), *biological,* and *chemical.*[52]

Weed populations in croplands are usually not adequately controlled by only one weed-control practice, so the use of several such practices during a single crop season is common. The use of two or more weed-control practices is referred to as *integrated weed control*[52] or *integrated weed management.*[53]

Weeds serve as hosts for insects and plant pathogens that subsequently move to crop plants in the area, causing adverse effects on the crop. Practices directed toward simultaneous control of more than one kind of plant pest is called *integrated pest management.*[6,62,54] Removal of weeds that serve as hosts for insects or plant pathogens removes two plant pests in one operation.

Preventive weed control attempts to block the introduction, establishment, or spread of specific weed species in areas not infested by these species. Preventive weed control is instrumented through laws at national, state, and local levels.[55] The Federal Seed Act of 1939, Public Law 90-583 of 1968, and the Federal Noxious Weed Act of 1974 are preventive measures at the national level.[56] The Federal Seed Act regulates interstate and foreign commerce in seeds. Its purpose is to protect buyers from mislabeled or contaminated crop seed, and it supplements state seed laws in interstate commerce. Public Law 90-583 provides for the control of noxious plants on land under the jurisdiction of the federal government. The Federal Noxious Weed Act regulates importation of foreign weed species into the U.S. and provides for control and eradication of such noxious weeds not widely dispersed within the U.S. In addition to these federal laws, each state has its own seed law to regulate interstate movement of seeds across its borders. These state seed laws identify specific weed species as *prohibited* (primary) noxious, and *restricted* (secondary) *noxious.*[52] Crop seed containing any seeds of weed species designated prohibited noxious cannot be legally shipped into a state that has declared the species prohibited noxious. Each state determines the number of seeds of weed species categorized as restricted noxious that may be present in crop seed entering the state. The Federal Seed Act and state seed laws permit relatively high tolerances for weed-seed contaminants in crop seed. In addition to federal and state laws, some states have established weed districts in an attempt to control specified noxious weeds in their areas. Weed laws are effective as a preventive means of weed control only when the laws are adequately restrictive and enforced.

Preventive weed control is also practiced by planting weed-free crop seed, by use of manure and hay free of weed propagules, by cleaning of harvesting equipment before moving it from a weedy to a weed-free area, by screening of irrigation water to remove weed propagules before the water moves onto cropland, and by the proper education of responsible people.[52,57] In essence, people are the key to preventive weed control.

Cultural weed-control practices include all practices common to good land, crop, and water management, such as smother crops, crop rotation, row spacing, seeding rate, planting date, fertilization, tillage, irrigation management, weed-free crop seed, field sanitation, and use of adapted crop varieties/cultivars.[53,58] Of these practices, only the first two will be discussed here. Any crop that is highly competitive with weeds may be used as a smother crop. Alfalfa is a superior competitive crop. Other examples include barley, buckwheat, clovers, cowpeas, ensilage corn, millet, rye, reed canarygrass, soybeans, sunflower, and forage sorghums. A smother crop is an effective means of weed control when used in conjunction with crop rotation and other weed control practices.

Any crop, cultivated or not, is plagued by specific weed species adapted to the cultural practices and growth habits of the crop. The particular weed species in a given crop may vary from area to area but, in general, they are common to that crop. A planned crop rotation is an effective means of weed control. Crops selected for use in rotation should have sharply contrasting growth habits and cultural requirements, necessitating the use of varied weed-control practices.[52] Crop rotation is the growing of different crops in recurring succession on the same land, in distinction from a one-crop system or a haphazard change of crops determined by opportunism or lack of a definite plan.[58] Weed control is one of the principal reasons for crop rotation, and no one rotation system can be considered typical of the varied crop-production regions of the U.S.[58,60]

Mechanical (physical) weed control is a traditional and well established practice for controlling weeds. Mechanical weed control includes such practices as handpulling, hoeing, machine tillage, smothering with nonliving material, burning, and water management.[52,58,59] Handpulling and hoeing are used primarily to control weeds growing among or near crop plants, weeds not easily controlled by machine-tillage without injury to the crop plants. Machine-tillage allows more rapid and economical weed control over large areas and permits choice from a wide selection of tillage tools. Weed-control by tillage is achieved primarily by (1) burial of small seedling weeds by soil thrown over them through the tillage tool action, and (2) disruption of the intimate soil/root relationship of weed plants by soil disturbance or severing the shoot from the roots. Care must be taken during tillage not to injure roots or shoots of the crop plants.

Weed control by smothering with nonliving material uses mulching material or black plastic film to prevent light from reaching weed seedlings, thereby blocking photosynthesis. Burning sanitizes a field or waste area by combustion of dry vegetation. Burning is also used to a limited extent in cotton and sugar cane to control seedling weeds. Burning or flaming as a means of selective weed control in croplands must provide sufficient heat and exposure time so the cell sap expands, causing the cell walls to rupture. Cotton plants tolerate flaming after their stems are about 5 mm in diameter at ground level. Weed control by water management is practiced primarily in rice and on irrigated, arid-land areas.

Biological weed control uses natural predators and parasites as the agents of weed control.[60-64] The most effective biotic agents used for weed control have been phytophagous (plant-eating) organisms, such as insects, fish, the manatee, and snails.[65,66] More recently, the use of plant pathogens as biotic weed-control agents has proved effective, with more research directed toward the use of fungal pathogens than bacteria, nematodes, viruses, or mycoplasmas.[67-71] Biological weed control reduces weed population densities below the level of economic importance, but does not eradicate the weed plants. Biological weed control reduces the competitive ability of the weed and dissipates its energy reserves, while preventing or curtailing weed reproduction.[66]

It is necessary to evaluate the economic impact of each targeted weed species before beginning a biological-control program.[72] A biological weed-control program requires a lengthy period of research, and the introduction of such a program is undertaken for permanent control. Designation of a specific plant species as an undesired plant is often controversial. To some it is a noxious weed, but to others it is a forage plant for livestock, food or shelter for birds, source of honey for bees, or an ornamental.[73] Obtaining comprehensive economic data on individual weed species selected as targets for biological control is exceedingly difficult. In 1978, calculations indicated that a complete biological weed-control program in Canada would likely require 18.8 to 23.7 scientist years and cost $1.2 to 1.5 million. Such a program is only economical against major weed species.[74]

Biotic agents released to date appear to be more suited for weed control in situations other than cultivated crops, with the exception of certain plant pathogens.[70] Selective livestock grazing as a means of vegetative management is more applicable to grazing lands than cultivated crops.[70] However, geese have been used to selectively control johnsongrass in cotton.[76]

Inevitably, as with any new technique, terminology designating the use of fungal pathogens as biotic agents of weed control is varied, with the terms *mycoherbicide*,[77] *bioherbicide*,[67] and *biological herbicide*[69] so used.

The status of biological weed-control projects with insects and plant pathogens up to 1974 in the U.S. and Canada is available.[78] A summary of biological weed control in the southern U.S. (Texas to North Carolina) is also available.[79] Klingman and Coulson have developed guidelines for introducing foreign organisms into the U.S. for biological weed control.[80]

Chemical weed control in croplands is achieved by approved chemicals that effectively control weeds without significant harm to crop plants or the environment. These phytotoxic chemicals are called *herbicides*. A herbicide is any chemical that kills plants or greatly inhibits plant growth. Herbicides that kill or greatly inhibit growth of most green plants are referred to as *nonselective*. Those that kill or greatly inhibit growth of certain plant species without significant injury to other plant species are called *selective herbicides*. To be of value in crop production, a herbicide must be capable of selectively killing the weed species without significant injury to the crop plants grown on the land.[81]

Herbicides available today can control grass weeds in either grass or broadleaved crops, broadleaved weeds in either grass or broadleaved crops, and grass and broadleaved weeds in either grass or broadleaved crops. Herbicides may be applied to the soil before emergence of the weeds or directly to the foliage of emerged weeds. For selective weed control, herbicides may be applied to the soil before planting the crop or after planting but before crop emergence. Herbicides may be applied after crop emergence as a broadcast over-the-top treatment or as a basally directed application under the foliar canopy of the crop. However, a particular herbicide has limitations with regard to tolerant crop species and susceptible weed species. A single herbicide is not the panacea to the control of all weed species in a given crop or in all croplands. In recognition of this, it is common to rotate herbicides in a given crop, or to use two or more herbicides in combination or as sequential treatments to control a greater weed spectrum than is possible by use of only one herbicide.[81]

Herbicide selectivity is basic to chemical weed control in croplands and it is achieved through *true selectivity* and *placement selectivity*. True selectivity depends on the inherent tolerance of a given plant species to the applied herbicide and the inherent susceptibility of other plant species to the herbicide. Placement selectivity relies on application of the herbicide in such a way or at such a time, that the herbicide does not make contact with the susceptible crop plant, but the chemical does make contact with the susceptible weed plants.[81]

A comprehensive summary of all the aspects of chemical weed control is beyond the scope of this chapter. It is sufficient to mention that chemical weed control in the U.S. is well established, with 76.5 million hectares of cropland treated with herbicides in 1976.[82] Herbicides constitute the most widely used and fastest growing category of pesticides used in croplands. Herbicides accounted for 60% of all pesticides applied to crops in 1976.[82] Properly used, herbicides increase crop yields, improve crop quality, and lower crop production costs.[83-85]

HERBICIDE APPLICATION EQUIPMENT

Herbicides are commonly applied in aqueous sprays or applied in dry granular or pelleted formulations. For dispersal in aqueous mixtures, the herbicidal chemical is formulated with appropriate solvents, emulsifiers, surfactants, or diluents.[52] Targets for applied herbicides are the weed plants or their propagules. Herbicides may be applied directly to the soil for subsequent weed/herbicide interception or directly to shoots of emerged plants.

The most common equipment for applying herbicides on croplands in aqueous spray mixtures is the pressure sprayer mounted on or pulled by tractors or similar powered vehicles, and by aircraft.[86-88] Innovations of this basic applicator include the recirculating sprayer,[89] rope wick and roller applicators,[80,81] controlled-droplet applicators,[82,83] and ICI's Electrodyn that produces electrically charged droplets atomized by electrostatics. Certain herbicides can also be applied through sprinkler irrigation systems (referred to as herbigation)[84] or metered into irrigation water as it flows from distribution channels onto cropland.

CONTRATOXICANTS

A contratoxicant may be defined as any chemical that relieves, prevents, or counteracts the toxicity of one or more chemicals in a biological system. Terms used in the literature to denote herbicide/contratoxicants include herbicide antidotes, seed safeners, chemical plant-protectants, and chemical herbicide-protectants. Contratoxicant chemicals have been developed that (1) protect certain crop plants from the phytotoxicity of a particular herbicide or chemically related herbicides without adversely affecting their herbicidal properties toward weeds growing among the crop plants, and (2) protect certain soil-applied herbicides from degradation by microbial or enzymatic attack in the soil environment, resulting in the herbicides being herbicidally active in the soil for a longer period of time.[95-100]

HERBICIDE INTERACTIONS

Simultaneous or sequential applications of two or more herbicides to control a broader weed spectrum than possible with only one herbicide is common today. When two or more herbicides are introduced into the cellular environment of a plant, conditions for an interaction are created.[100] Herbicide interactions within a plant may be expressed in any of the various plant responses induced by chemicals in general, ranging from a change in foliar pigmentation or carbon dioxide evolution to a change in internode length or death of the plant. Numerous instances of such interactions have been reported, but their systematic analytical evaluation has been hampered by a lack of accepted, unambiguous, terminology.[101] In spite of recognized terminology difficulties, the general effect resulting from herbicide interactions within a plant or plant community may be categorized as (1) *additive*, (2) *synergistic*, (3) *antagonistic*, and (4) *enhancement*.[102] Interactions have also occurred within plants when herbicide-insecticide combinations have been applied.[103-106]

EVOLUTION OF HERBICIDE-RESISTANT WEEDS IN CROPLANDS

Evolution of herbicide-resistant weed biotypes following repeated use of a herbicide year after year in the same area is not unexpected.[107,108] Chemical control of a pest is almost invariably followed by development of a strain (biotype) resistant to the toxic action of the chemical. Documented examples of this phenomenon are numerous in the literature pertaining to applied entomology, bacteriology, and medicine.[109]

Herbicide-resistant biotypes of weed species can evolve through (1) *induced mutation*, (2) *hybridization* (cross-breeding between resistant and susceptible species), and (3) *cross-breeding within a species of surviving plants*, resulting in a greater proportion of the populations possessing the genetic resistance trait.[111,112] The rate at which a weed species responds by natural selection to a given herbicide depends on (1) *the intensity of selection* and (2) *the inheritable variability of the species*.[111] Intensity of selection is favored by a species population reduced to a small resistant residue capable of rapid reproduction. The inheritable variability of a species lies with its (1) *inherent genes*, (2) *system of breeding*, and (3) *extent of cross-breeding*.[112] Rapid evolution of new herbicide-resistant weed strains as a result of selection is not likely to occur as frequently within species that predominantly reproduce (1) *by selfing* (inbreeding), (2) *by apomixis* (involving specialized generative tissues but not dependent on fertilization), or (3) *asexually*.[111] The evolution of resistant biotypes would be expected to be most rapid within weed species that cross-breed. Herbicide resistance is genetically stable once it has evolved.

Pollen does not carry the resistance trait when this trait is maternally (cytoplasmically) inherited, limiting dissemination of this genetic trait. In general, a weed species that has developed resistance to a given herbicide will also have resistance to chemically related herbicides. Known weed species resistant to two or more chemically unrelated herbicides are as yet rare, but it would not be surprising if such resistance occurs, especially in cases where the phytotoxicity of the different herbicides is based upon the exact same mode of action. It will likely require 5 or more years or continuous use of the same herbicide in a given field before a resistant biotype will appear.[112]

Herbicide-resistant biotypes of a given weed species are not morphologically different from susceptible plants of that species. Therefore, resistant biotypes are not readily identifiable from susceptible plants of the same species growing in a mixed resistant-susceptible population. Unless a resistant biotype is recognized and eradicated in the first or second year after its development, subsequent eradication is not likely because of the multitude of seed it will have produced and which have become buried in the soil.[112] A simple, technique for detecting photosynthesis-inhibiting chemicals[113] has been modified and used to identify plant biotypes resistant and susceptible to triazine and urea herbicides.[114] These herbicides inhibit photosynthesis and, more specifically, their site of action is a blockage of electron transport between Q (primary electron acceptor for PSII) and PQ (plastoquinone).[115] Interestingly, this technique showed that triazine-resistant biotypes were susceptible to urea herbicides.[114]

The occurrence of herbicide-resistant biotypes among weed species has been well documented;[116,117] for example, barnyard grass to dalapon,[118] common groundsel (*Senecio vulgaris* L.) to simazine and atrazine,[119] and foxtail barley to siduron.[120] Evolved herbicide resistance to triazine herbicides was first reported in 1970,[109] and has been confirmed in at least 19 genera and 31 weed species.[112] The trait for triazine resistance is inherited cytoplasmically (maternally).[121]

Once recognized or suspected, control or containment of herbicide-resistant weeds within acceptable limits should not be difficult when integrated weed control is practiced.[6,122,123] Rotation of suitable herbicides or use of mixtures or sequential applications of two more herbicides having contrasting modes of action, rotation of crops requiring contrasting cultural practices, soil tillage, and handpulling are all effective in combating herbicide-resistant weeds. It must be remembered that a herbicide-resistant biotype has resistance to a particular herbicide or group of herbicides, but not to all herbicides.

Inheritance of herbicide resistance in crop plants appears feasible via conventional breeding methods[124,125] and genetic-engineering techniques.[126,127]

REFERENCES

1. **Shaw, W. C. and Loustalot, A. J.**, Revolution in weed science, *Agric. Sci. Rev.*, 1(4), 38, 1963.
2. **Danielson, L. L.**, Looking ahead in weed science, *Weed Sci.*, 19, 483, 1971.
3. **Buchholtz, K. P.**, Report of WSA terminology committee, *Weeds*, 15, 389, 1967.
4. **Dayton, W. A.**, Weeds are plants out of place, in Weeds, Yearbook of Agriculture, United States Department of Agriculture, Washington, D.C., 1948, 727.
5. **King, L. J.**, *Weeds of the World*, Interscience, New York, 1966, 1.
6. **Shaw, W. C.**, Integrated weed management systems technology for pest management, *Weed Sci.*, Vol. 30 (Suppl) 2, 1982.
7. **Jansen, J. L.**, Extent and Cost of Weed Control with Herbicides and an Evaluation of Important Weeds, 1968, ARS-H-1, Agriculture Research Service, United States Department of Agriculture, Washington, D.C., 1972.
8. **Chandler, J. J.**, Estimated losses of crops to weeds, in *CRC Hanbook of Pest Management in Agriculture*, Vol. 1, Pimentel, D., Ed., CRC Press, Boca Raton, Fla., 1981, 95.

9. **Schweizer, E. E.,** Common lambsquarters *(Chenopodium album)* interference in sugarbeets, *(Beta vulgaris), Weed Sci.,* 31, 5, 1983.
10. **Smith, R. J., Jr.,** Competition of bearded sprangletop *(Leptochloa fascicularis)* with rice *(Oryza sativa), Weed Sci.,* 31, 120, 1983.
11. **Knake, E. L. and Slife, F. W.,** Competition of *Setaria faberii* with corn and soybeans, *Weeds,* 10, 26, 1962.
12. **Martin, J. H. and Leonard, W. H.,** *Principles of Field Crop Production,* 3rd ed., Macmillan, New York, 1967, 272.
13. **Clark, E. R. and Porter, C. R.,** The seeds in your drill box, in Seeds, Yearbook of Agriculture, United States Department of Agriculture, Washington, D.C., 1961, 474.
14. **Jensen, L. A.,** Are we planting weeds?, *Proc. Western Weed Control Conf.,* 19, 52, 1962.
15. **Kelley, A. D. and Bruns, V. F.,** Dissemination of weed seeds by irrigation water, *Weed Sci.,* 23, 486, 1975.
16. **Wilson, R. G., Jr.,** Dissemination of weed seeds by surface irrigation water in western Nebraska, *Weed Sci.,* 28, 87, 1980.
17. **Isely, D.,** *Weed Identification and Control,* The Iowa State College Press, Ames, 1958.
18. **Anderson, W. P.,** *Weed Science: Principles,* 2nd ed., West Publishing, St. Paul, 1983, chap. 1.
19. **Platt, R.,** *Our Flowering World,* Dodd, Mead & Co., New York, 1947.
20. **Koller, D.,** Germination, *Sci. Am.,* 200(4), 75, 1959.
21. **Salisbury, E.,** *The Reproductive Capacity of Plants,* G. Bell & Sons, London, 1942.
22. **Stevens, O. A.,** The number and weight of seeds produced by weeds, *Am. J. Bot.,* 19, 784, 1932.
23. **Coxworth, E. C. M., Bell, J. M., and Ashford, R.,** Preliminary evaluation of Russian thistle, kochia, and garden atriplex as potential high protein content seed crops for semiarid areas, *Can. J. Plant. Sci.,* 49, 427, 1969.
24. **Burnside, O. C., Fenster, C. R., Evetts, L. L., and Mumm, R. F.,** Germination of exhumed weed seed in Nebraska, *Weed Sci.,* 29, 577, 1981.
25. **Kivilaan, A. and Bandurski, R. S.,** The 90-year period for Dr. Beal's seed viability experiment, *Am. J. Bot.,* 60, 140, 1973.
26. **Burns, V. F. and Rasmussen, L. W.,** The effects of fresh water storage on the germination of certain weed seeds, *Weeds,* 10, 42, 1958.
27. **Shull, H.,** Weed seed in irrigation water, *Weeds,* 10, 248, 1962.
28. **Zimdahl, R. L.,** *Weed-Crop Competition,* International Plant Protection Center, Oregon State University, Corvallis, 1980.
29. **Donald, C. M.,** Competition among crop and pasture plants, *Adv. Agron.,* 15, 1, 1963.
30. **Anderson, W. P.,** *Weed Science: Principles,* 2nd ed., West Publishing, St. Paul, 1983, 33.
31. **Wiese, A. F.,** Rate of weed root elongation, *Weed Sci.,* 16, 11, 1968.
32. **Davis, R. G., Wiese, A. F., and Pafford, J. L.,** Root moisture extraction profiles of various weeds, *Weeds,* 13, 98, 1965.
33. **Scott, H. D. and Oliver, L. R.,** Field competition between tall morningglory and soybean. II. Development and distribution of root systems, *Weed Sci.,* 24, 454, 1976.
34. **Dawson, J. H. and Rincker, C. M.,** Weeds in new seedings of alfalfa *(Medicago sativa)* for seed production, *Weed Sci.,* 30, 20, 1982.
35. **Oliver, L. R., Frans, R. E., and Talbert, R. E.,** Field competition between tall morningglory and soybean. I. Growth analysis, *Weed Sci.,* 24, 482, 1976.
36. **Rice, E. L.,** *Allelopathy,* Academic Press, New York, 1974.
37. **Snipes, C. E., Buchanan, G. A., Street, J. E., and McGuire, J. A.,** Competition of common cocklebur *(Xanthium pensylvanicum)* with cotton *(Gossypium hirsutum), Weed Sci.,* 30, 553, 1982.
38. **Burnside, O. C.,** Soybean, *(Glycine max)* growth as affected by weed removal, cultivar, and row spacing, *Weed Sci.,* 27, 562, 1979.
39. **Eaton, B. J., Russ, O. G., and Feltner, K. C.,** Competition of velvet-leaf, prickly sida, and Venice mallow in soybeans, *Weed Sci.,* 24, 224, 1976.
40. **Dawson, J. H.,** Time and duration of weed infestations in relation to weed-crop competition, *Proc. South. Weed Sci. Soc.,* 23, 13, 1970.
41. **Buchanan, G. A. and McLaughlin, R. D.,** Influence of nitrogen on weed competition in cotton, *Weed Sci.,* 23, 324, 1975.
42. **Bloomberg, J. R., Kirkpatrick, B. L., and Wax, L. M.,** Competition of common cocklebur *(Xanthium pensylvanicum)* with soybean *(Glycine max), Weed Sci.,* 30, 507, 1982.
43. **Buchanan, G. A., Crowley, R. H., Street, J. E., and McGuire, J. A.,** Competition of sicklepod *(Cassia obtusifolia)* and redroot pigweed *(Amaranthus retroflexus)* with cotton *(Gossypium hirsutum), Weed Sci.,* 28, 258, 1980.
44. **Crowley, R. H. and Buchanan, G. A.,** Competition of four morningglory *(Ipomoea* spp.) species with cotton, *(Gossypium hirsutum), Weed Sci.,* 26, 484, 1978.

45. **Chamblee, R. W., Thompson, L., Jr., and Coble, H. D.**, Interference of broadleaf signalgrass *(Brachiaria platyphylla)* in peanuts, *(Arachis hypogaea), Weed Sci.,* 30, 45, 1982.
46. **Smith, R. J., Jr.**, Competition of bearded sprangletop *(Leptochloa fascicularis)* with rice *(Oryza sativa), Weed Sci.,* 31, 120, 1983.
47. **Bell, A. R. and Nalewaja, J. D.**, Competition of wild oat in wheat and barley, *Weed Sci.,* 16, 505, 1968.
48. **Rydrych, D. J.**, Corn cockle *(Agrostemma githago* L.) competition in winter wheat *(Triticum aestivum), Weed Sci.,* 29, 360, 1981.
49. **Weatherspoon, D. M. and Schweizer, E. E.**, Competition between sugar beets and five densities of kochia, *Weed Sci.,* 19, 125, 1971.
50. **Schweizer, E. E. and Bridge, L. D.**, Sunflower *(Helianthus annuus)* and velvetleaf *(Abutilon theophrasti)* interference in sugarbeets *(Beta vulgaris), Weed Sci.,* 30, 514, 1982.
51. **Schweizer, E. E.**, Common lambsquarters *(Chenopodium album)* interference in sugarbeets *(Beta vulgaris), Weed Sci.,* 31, 5, 1983.
52. **Anderson, W. P.**, *Weed Science: Principles,* 2nd ed., West Publishing, St. Paul, 1983, chap. 2.
53. **Walker, R. H. and Buchanan, G. A.**, Crop manipulation in integrated weed management systems, *Weed Sci.,* (Suppl.) 30, 17, 1982.
54. **Bottrell, D. G. and Smith, R. F.**, Integrated pest management, *Environ. Sci. Technol.,* 16, 282A, 1982.
55. **Rollin, S. F. and Johnston, F. A.**, Our laws that pertain to seeds, in Seeds, Yearbook of Agriculture, United States Department of Agriculture, Washington, D.C., 1961, 482.
56. **Shaw, W. C.**, Herbicides: the cost/benefit ratio — the public view, *Proc. South. Weed Sci. Soc.,* 31, 28, 1978.
57. **Slife, F. W.**, Environmental control of weeds, in *Handbook of Pest Management in Agriculture,* Vol. 1, Pimentel, D., Ed., CRC Press, Boca Raton, Fla., 1981, 485.
58. **Crafts, A. S.**, *Modern Weed Control,* University of California Press, Berkeley, 1975, 89.
59. **Zimdahl, R. L.**, Extent of mechanical, cultural, and other nonchemical methods of weed control, in *Handbook of Pest Management in Agriculture,* Vol. 2, Pimentel, D., Ed., CRC Press, Boca Raton, Fla., 1981, 79.
60. **Anderson, W. P.**, *Weed Science: Principles,* 2nd ed., West Publishing, St. Paul, 1983, 83.
61. **Batra, S. W. T.**, Biological control of weeds: principles and prospects, in *Biological Control in Crop Production,* Papavizas, G. C., Ed., Allanheld, Osmun, Toronto, 1981, 45.
62. **Wilson, F.**, The biological control of weeds, *Annu. Rev. Entomol.,* 9, 225, 1964.
63. **Huffaker, C. B.**, Fundamentals of biological control, in *Biological Control of Insect Pests and Weeds,* pp. 631-469. DeBach, P., Ed., Reinhold, New York, 1964, 631.
64. **Erhlich, P. R. and Raven, P. H.**, Butterflies and plants, *Sci. Am.,* 216(6), 104, 1970.
65. **Andres, L. A. and Goeden, R. D.**, The biological control of weeds by introduced natural enemies, in *Biological Control,* Huffaker, C. B., Ed., Plenum Press, New York, 1971, 143.
66. **Blackburn, R. D. and Andres, L. A.**, The snail, the mermaid, and the flea beatle, in Science for Better Living, Yearbook of Agriculture, United States Department of Agriculture, Washington, D.C., 1968, 229.
67. **Emge, R. G. and Templeton, G. E.**, Biological control of weeds with plant pathogens, in *Biological Control in Crop Production,* pp. 219-226. Papavizas, G. C., Ed., Allanheld, Osmun, Toronto, 1981, 219.
68. **Templeton, G. E.**, Biological herbicides: discovery, development, deployment, *Weed Sci.,* 30, 430, 1982.
69. **Quimby, P. C., Jr. and Walker, H. L.**, Pathogens as mechanisms for integrated weed management, *Weed Sci.,* Suppl. 30, 30, 1982.
70. **Daniel, J. T., Templeton, G. E., Smith, R. J., Jr., and Fox, W. T.**, Biological control of northern joint-vetch in rice with an endemic fungal disease, *Weed Sci.,* 21, 303, 1973.
71. **Phatak, S. C., Sumner, D. R., Wells, H. D., Bell, D. K., and Glaze, N. C.**, Biological control of yellow nutsedge with the indigenous rust fungus *Puccinia canaliculata, Science,* 219, 1446, 1983.
72. **Batra, S. W. T.**, Biological control in agroecosystems, *Science,* 215, 134-139.
73. **Andres, L. A.**, Biological control of naturalized and native plants: conflicting interests, in *Biological Control in Crop Production,* pp. 341-349. Papavizas, G. C., Ed., Allanheld, Osmun, Toronto, 1981, 341.
74. **Harris, P.**, Cost of biological control of weeds by insects in Canada, *Weed Sci.,* 27, 242, 1979.
75. **Scifres, C. J.**, Selective grazing as a weed control method, in *Handbook of Pest Management in Agriculture,* Vol.2, Pimentel, D., Ed., CRC Press, Boca Raton, Fla., 1981, 369.
76. **Rodgers, N.**, Fowl tactics get the grass out of cotton, *Farm J.,* June/July, 1974, 28.
77. **Templeton, G. E., TeBeest, D. O., and Smith, R. F., Jr.**, Biological weed control with mycoherbicides, *Annu. Rev. Phytopathol.,* 17, 301, 1979.
78. **Goeden, R. D., Andres, L. A., Freeman, T. E., Harris, P., Pienkowski, R. L., and Walker, C. R.**, Present status of projects on the biological control of weeds with insects and plant pathogens in the United States and Canada, *Weed Sci.,* 22, 490, 1974.
79. **DeLoach, C. J.**, Biological control of weeds in the USDA Southern Region, *Weeds Today,* 11(2), 11, 1980.

80. **Klingman, D. L. and Coulson, J. R.**, Guidelines for introducing foreign organisms into the United States for biological control of weeds, *Weed Sci.*, 30, 661, 1982.
81. **Anderson, W. P.**, *Weed Science: Principles*, 2nd ed., West Publishing, St. Paul, 1983, 101 and chap. 6.
82. **Eichers, T. R., et al.**, Farmers' Use of Pesticices in 1976, A-ER-418, Economic Statistic Cooperative Service, United States Department of Agriculture, Washington, D.C., (December), 1978.
83. **Anon.**, Extent and cost of weed control with herbicides and an evaluation of important weeds, 1968, ARS-H-1, Economic Research Service, Extension Service, and Agricultural Research Service, United States Department of Agriculture, Washington, D.C., November, 1972.
84. **Ennis, W. B., Jr.**, Benefits of agricultural chemicals, *Weed Sci.*, 19, 631, 1971.
85. **McWhorter, C. G. and Holstun, J. T., Jr.**, Science against weeds, in Protecting Our Food, Yearbook of Agriculture, United States Department of Agriculture, Washington, D.C., 1966, 74.
86. **Gebhardt, M. R.**, Methods of pesticide application. in *CRC Handbook of Pest Management in Agriculture*, Vol. 2, Pimentel, D., Ed., CRC Press, Boca Raton, Fla., 1981, 87.
87. **Carleton, W. M., et al.**, The development of equipment in the application of agricultural chemicals in The Nature and Fate of Chemicals Applied to Soils, Plants, and Animals, ARS 20-9, United States Department of Agriculture, Beltsville, Md., 1969, 70.
88. **Weber, D. R.**, Nozzles and spray systems, *Proc. Calif. Weed Conf.*, 23, 26, 1971.
89. **McWhorter, C. G.**, The recirculating sprayer, *Weeds Today*, 6(1), 11, 1975.
90. **Sheldon, E. B.**, Selective equipment to apply glyphosate, *Proc. South. Weed Sci. Soc.*, 34, 300, 1981.
91. **Hite, D.**, Here's how to make a rope wick applicator, *Progr. Farmer*, 94(12), 34, 1979.
92. **Bretches, K.**, Precision spraying, *Progr. Farmer*, 97(12), 16, 1982.
93. **Bretches, K.**, Rotary atomizers, *Progr. Farmer*, 97(12), 29, 1982.
94. **Anon.**, Herbigation: a growing concept in herbicide application, *Solutions* 20(1), 48, 1975.
95. **Pallos, F. M. and Casida, J. E.**, *Chemistry and Action of Herbicide Antidotes*, Academic Press, New York, 1978.
96. **Anderson, W. P.**, *Weed Science: Principles*, 2nd ed., West Publishing, St. Paul, 1983, 398.
97. **Simkins, G. S., Moshier, L. J., and Russ, O. G.**, Influence of acetamide herbicide applications on efficacy of the protectant CGA-43089 in grain sorghum *(Sorghum bicolor)*, *Weed Sci.*, 28, 646, 1980.
98. **Hatzios, K. K. and Penner, D.**, Potential antidotes against buthidazole injury to corn *(Zea mays)*, *Weed, Sci.*, 28, 273, 1980.
99. **Martin, A. R. and Burnside, O. C.**, Protecting corn (*Zea* mays) from herbicide injury with R-25788, *Weed Sci.*, 30, 269, 1982.
100. **Putnam, A. R. and Penner, D.**, Pesticide interactions in higher plants, *Residue Rev.*, 50, 73, 1974.
101. **Morse, P. M.**, Some comments on the assessment of joint action of herbicide mixtures, *Weed Sci.*, 26, 58, 1978.
102. **Akobundu, I. O., Sweet, R. D., and Duke, W. B.**, A method of evaluating herbicide combinations and determining herbicide synergism, *Weed Sci.*, 23, 20, 1975.
103. **El-Refai, A. R. and Mowafy, M.**, Interaction of propanil with insecticides absorbed from soil and translocated into rice plants, *Weed Sci.*, 21, 246, 1973.
104. **Hamill, A. S. and Penner, D.**, Interaction of alachlor and carbofuran, *Weed Sci.*, 21, 330, 1973.
105. **Hamill, A. S. and Penner, D.**, Butylate and carbofuran interaction in barley and corn, *Weed Sci.*, 21, 339, 1973.
106. **Hayes, R. M., Yeargan, K. V., Witt, W. W., and Raney, H. G.**, Interaction of selected insecticide-herbicide combinations on soybeans *(Glycine max)*, *Weed Sci.*, 27, 51, 1979.
107. **LeBaron, H. M. and Gressel, J., Eds.**, *Herbicide Resistance in Plants*, John Wiley & Sons, New York, 1982.
108. **Pfeiffer, R. and Zeller, A.**, Practical aspects of weed control by plant growth-regulating substances in cereals, *Proc. British Weed Control Conf.*, 1, 70, 1953.
109. **Harper, J. L.**, The evolution of weeds in relation to resistance to herbicides, Proc. British Weed Control Conf., Vol. I, 3, 179, 1956.
110. **Anderson, W. P.**, *Weed Science: Principles*, West Publishing, St. Paul, 1983, 601.
111. **Harper, J. L.**, Ecological aspects of weed control, *Outlook Agric.*, 1, 197, 1957.
112. **LeBaron, H. M.**, Resistance of plants to herbicides, *Proc. Wash. State Weed Assoc.*, 32, 83, 1982.
113. **Truelove, B., Davis, D. E., and Jones, L. R.**, A new method for detecting photosynthesis inhibitors, *Weed Sci.*, 22, 15, 1974.
114. **Hensley, J. R.**, A method for identification of triazine resistant and susceptible biotypes of several weeds, *Weed Sci.*, 29, 70, 1981.
115. **Ashton, F. M. and Crafts, A. S.**, *Mode of Action of Herbicides*, 2nd ed., John Wiley & Sons, New York, 1981, 359.
116. **Bandeen, J. D., Stephenson, G. R., and Cowett, E. R.**, Discovery and distribution of herbicide-resistant weeds in North America, in *Herbicide Resistance in Plants*, LeBaron, H. M. and Gressel, J., Eds., John Wiley & Sons, New York, 1982, 9.

117. **Jensen, K. I. N. and Bandeen, J. D.**, Triazine resistance in annual weeds, in *Maize*, Haflinger, E., Ed., CIBA-Geigy Technical Monograph, 1979, 55.
118. **Roche, B. F. and Muzik, T. J.**, Ecological and physiological study of *Echinochloa crus-galli* (L.) Beauv. and response of its biotypes to sodium 2,2-dichloropropionate (dalapon), *Agron. J.*, 56, 155, 1964.
119. **Ryan, G. F.**, Resistance of common groundsel to simazine and atrazine, *Weed Sci.*, 18, 614, 1970.
120. **Schooler, A. B., Bell, A. R., and Nalewaja, J. D.**, Inheritance of siduron tolerance in foxtail barley, *Weed Sci.*, 20, 167, 1972.
121. **Souza Machado, V. and Bandeen, J. D.**, Genetic analysis of chloroplast atrazine resistance in *Brassica campestris* — Cytoplasmic inheritance, *Weed Sci.*, 30, 281, 1982.
122. **Walker, R. H. and Buchanan, G. A.**, Crop manipulation in integrated weed management systems, *Weed Sci.*, 30(Suppl.), 17, 1982.
123. **Abel, A. L.**, The rotation of weedkillers, *Proc. British Weed Control Conf.*, Vol. I, 3, 249, 1954.
124. **Faulkner, J. S.**, Breeding herbicide-tolerant crop cultivars by conventional methods, in *Herbicide Resistance in Plants*, LeBaron, H. M. and Gressel, J., Eds., John Wiley & Sons, New York, 1982, 235.
125. **Warwick, D. D.**, Plant variability in triazine resistance of plants. pp. 56—64. E. Ebert and S. W. Dunford. *Residue Rev.*, 65, 1, 1976.
126. **LeBaron, H. and Gressel, J.**, Summary of accomplishments, conclusions, and future needs, in *Herbicide Resistance in Plants*, LeBaron, H. M. and Gressel, J., Eds., John Wiley & Sons, New York, 1982, 349.
127. **Chaleff, R. S.**, Isolation of agronomically useful mutants from plant cell cultures, *Science*, 219, 676, 1983.

Utilization of Crops

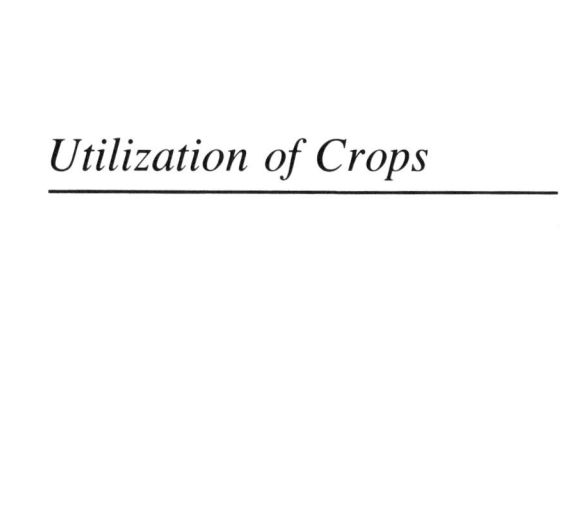

CEREALS

T. M. Starling

Cereals are crops grown primarily for their edible starchy seeds, and such crops constitute the world's major sources of food for humans and feed for livestock. It is estimated that cereal grains provide 56% of the food energy and 50% of the protein consumed on earth. In addition, cereals are widely used for industrial purposes — especially for the production of starch and industrial alcohols. Cereals also are used throughout the world for the production of a wide variety of distilled spirits, beers, ales, and some wines.

While parts of the plant other than the seeds find uses as feeds and for industrial purposes, the discussion herein shall be limited to uses of the seeds only. The crops which are used throughout the world for cereal purposes are listed alphabetically in Table 1. In addition to those listed, a few other species are used to a very limited extent. Table 2 gives the world, U.S., and Canadian production of major cereals during the 3-year period 1981 to 1983. On a worldwide basis, the major cereals in descending order are wheat, rice, maize, barley, and sorghum. Wheat, rice, and maize together make up $^3/_4$ of the world grain production. In the U.S., the major cereals are maize, wheat, sorghum, and barley. The quantity of production and relative cost of cereals affect their utilization, especially for industrial purposes.

The major uses of cereals vary widely throughout the world. Rice and wheat, although used to some extent for feed, are primarily considered as food grains. While sorghum and corn are primarily considered as feed grains in much of the western world, corn is considered a food grain in parts of Central and South America and sorghum as a food grain in Africa and parts of Asia. While oats, barley, buckwheat, and rye find some use as foods, they primarily are feed grains in most parts of the world. Despite being primarily a feed grain, rye is second only to wheat in importance as a grain for bread and other baked products. Triticale is a relatively new man-made crop and current production is very limited. There is very little commercial production of amaranth. Millet and sorghum are grown as food crops primarily along the edges of the deserts of Africa and Asia where lack of rainfall prohibits the production of other cereals. Wild rice is produced on a limited area in North America.

Characteristics of some cereals significantly affect their usage. Even within a species, there is significant opportunity to select types or cultivars which are more suitable for some uses than others. For instance, wheat is widely used throughout the world for bread and many other baked products because it is the only cereal having significant amounts of gluten, a component of wheat flour which permits the dough to adhere together and permits the baked product to increase in volume during the baking process. Hard wheats have more gluten than soft wheats, so flour from these two types of wheat is used according to the kind of product being baked. Even within these two types of wheat, cultivars vary widely in their milling and baking characteristics and wheat breeders must consider quality when developing new cultivars.

Rye has some gluten, but not enough that rye flour may be used alone in breads, so it is blended with wheat flour. Rye flour is included because of the unique flavor it imparts to "rye" bread. Where wheat is scarce or expensive, flour from other cereals often is mixed with the wheat flour as an extender. Oat flour has a property that retards development of rancidity in fat products. It is used as a coating for papers and containers used for packaging food products high in fat content and as an antioxidant and stabilizer in ice cream and other dairy products. While malt may be made from a variety of cereals, barley is the major cereal used for this purpose. In addition to tradition and cost of the grain, barley does have some characteristics that make it the choice of brewers as a source of malt. Hulls of barley are a

Table 1
CROPS USED FOR CEREAL PURPOSES

Common name	Latin name
Amaranth	*Amaranthus* species (several)
Barley	*Hordeum vulgare* L.
Buckwheat	*Fagopyrum sagittatum (esculentum)*
Tartary	*F. tartaricum*
Maize (corn)	*Zea mays* L.
Millet	
Pearl	*Pennisetum glaucum*
Proso	*Panicum miliaceum* L.
Finger	*Eleusine coracana* (L.) Gaertn.
Foxtail	*Setaria italica* Beavu.
Teff	*Eragrostis tef (abyssinica)*
Oats	*Avena sativa* L.
Rice	*Oryza sativa* L.
African	*O. glaberrima* Steud.
Rye	*Secale cereale* L.
Sorghum	*Sorghum bicolor* (L.) Moench.
Triticale	—
Wheat	
Common	*Triticum aestivum* L.
Club	*T. compactum* Host
Durum	*T. durum* Desf.
Polish	*T. polonicum*
Spelt	*T. spelta*
Emmer	*T. dicoccon (dicoccum)*
Wild rice	*Zizania palustris* L.

source of phenolic compounds which impart a particular flavor and "mouthfeel" to beer. However, too much of these phenolic compounds contributes to a nonbiological haze formation which is undesirable. The hulls also are desired in brewing beer because they serve as a filter during the brewing process.

There are several types of maize (corn) that vary widely in their endosperm characteristics, including dent, flint, flour, pop, sweet, and waxy. Some uses of maize are closely related to these endosperm characteristics. Dent maize consists of both corneous and soft starch and is suitable for many purposes. It is the most widely grown type in the U.S. Popcorn has a hard, corneous endosperm which is capable of expanding severalfold when heated, provided the kernels contain an appropriate amount of moisture. Sweet corn contains sugar as well as starch in the endosperm and is harvested at an immature stage and eaten fresh as a vegetable, or canned. Waxy maize has a soft endosperm which breaks with a wax-like fracture, having an amylopectin starch with a branched-chain molecular structure. This type starch makes a better adhesive than conventional starch. Flint and flour types are preferred in some parts of the world for food purposes.

Next to wheat, rice is the most nutritious cereal, provided that too much of the kernel is not removed in processing. Rice also has a nonsticky characteristic which permits the grains to remain free following cooking. Buckwheat flour is used in pancakes because of the flavor and texture it imparts to them.

Equally as important as characteristics of cereals in determining the relative importance of the various uses is the cost and availability of the different cereals. This obviously varies widely throughout the world and from season to season depending on production. Major cereals are those which are widely adapted and have a stable production, such as wheat, maize, and rice.

Table 2
WORLD, U.S., AND CANADIAN PRODUCTION OF MAJOR CEREAL GRAINS 1981—1983 (1000 MT)

Cereal crop	1981	1982	1983	Av	% of total
World					
All	1,653,608	1,707,250	1,638,847	1,666,568	100.0
Wheat	453,821	486,423	498,182	479,475	28.8
Rice	411,814	423,464	449,827	428,368	25.7
Maize	452,030	451,080	344,103	415,738	24.9
Barley	155,158	161,249	167,176	161,194	9.7
Rye	24,481	29,430	32,194	28,702	1.7
Oats	42,939	45,287	43,101	43,776	2.6
Millet	29,127	28,713	29,563	29,134	1.7
Sorghum	72,228	69,930	62,483	68,214	4.1
Others	12,010	11,674	12,218	11,967	0.7
U.S.					
All	333,805	338,515	208,903	293,741	100.0
Wheat	76,169	76,538	66,010	72,906	24.8
Rice	8,289	6,967	4,523	6,593	2.2
Maize	207,330	212,338	106,781	175,816	59.9
Barley	10,436	11,374	11,300	11,037	3.8
Rye	478	532	715	575	0.2
Oats	7,391	9,007	6,928	7,775	2.6
Sorghum	22,333	21,372	12,270	18,658	6.3
Others	379	385	376	380	0.1
Canada					
All	50,827	53,486	48,206	50,840	100.0
Wheat	24,802	26,790	26,914	26,169	51.5
Maize	6,673	6,513	5,875	6,354	12.5
Barley	13,724	14,074	10,616	12,805	25.2
Rye	927	913	831	890	1.7
Oats	3,188	3,684	2,773	3,215	6.3
Others	1,513	1,512	1,197	1,407	2.8

From FAO 1983 Production Yearbook, Vol. 37, Food and Agriculture Organization, Rome, 1984.

Statistics on the quantity of cereals used for various purposes on a worldwide basis are not readily available. However, statistics on the production and uses of major cereals grown in the U.S. and Canada in 1982 are given in Table 3. These two countries export considerably large quantities of grain to other countries and import relatively small quantities, largely from each other. The percentage of the domestically consumed cereal grains used for various purposes in these two countries in 1982 is given in Table 4.

Specific uses of cereals are extremely numerous and diverse and as previously indicated, vary widely throughout the world. Listed in Table 5 are some of the major uses of the grain from each of the cereal crops. The major use, whether as food, feed, or for industrial purposes, is italicized. Because of the many uses and names of the products, some uses and products are grouped in broad categories, such as baked products. Also flour from various cereals is blended in baking many products and in making breakfast cereals. For more specific details on uses, an extensive review of the literature would be required for each crop.

Table 3
PRODUCTION AND USES OF MAJOR CEREAL GRAINS IN THE U.S. AND CANADA IN 1982 (1000 MT)[a]

Crop and country

	Wheat	Rye	Barley	Oats	Maize	Sorghum	Rice	Cereals
U.S.								
Production	72,249	532	11,365	8,956	213,292	21,372	8,289	339,055
Export	42,564	5	1,437	73	49,126	5,433	3,929	102,577
Import	163	5	261	29	25	—	17	500
Total utilization	23,513	532	9,839	8,230	127,207	13,331	2,903	185,555
Feed	5,498	250	5,284	7,005	111,034	13,186	985	143,242
Seed	3,048	108	379	629	368	46	200	4,778
Processed	—		4,024	—	12,261	99	—	16,384
Food products	—		—	—	9,461	—	—	9,461
Industrial products[b]	—		4,024	—	2,800	99	—	6,923
Food	14,831	174	152	581	3,519	—	1,702	20,959
Waste	—	—	—	—	—	—	—	—
Other[c]	9,471	—	350	697	37,009	2,608	1,490	51,625
Canada								
Production	26,735	924	13,966	3,637	6,513	—	—	53,337
Export	21,368	314	5,648	105	511	—	—	27,950
Import	—	—	—	—	759	—	93	852
Total utilization	5,077	292	7,276	3,212	6,544	—	93	23,965
Feed	1,802	176	6,459	2,981	5,277	—	—	18,165
Seed	1,238	32	411	147	25	—	—	1,853
Processed	20	62	391	—	1,242	—	—	1,715
Food products	—	—	—	—	1,242	—	—	1,242
Industrial products[b]	20	62	391	—	—	—	—	473
Food	1,982	13	6	83	—	—	63	2,178
Waste	35	9	9	1	—	—	—	54
Other[c]	290	318	1,042	320	217	—	—	2,274

[a] Data from Food Consumption Statistics, 1973—1982, Organization for Economic Cooperation and Development, Paris, 1985.
[b] Includes grain for production of alcoholic beverages for human consumption.
[c] Includes stored grain carried over from the 1982 production.

Table 4
PERCENTAGE OF THE DOMESTICALLY CONSUMED CEREAL GRAINS IN THE U.S. AND CANADA USED FOR VARIOUS PURPOSES IN 1982[a]

	% Domestically consumed grain used for indicated purpose							
Use	Wheat	Rye	Barley	Oats	Maize	Sorghum	Rice	Total cereals
				U.S.				
Food	45	33	1	7	2	0	39	9
Feed	17	47	52	79	68	83	22	60
Seed	9	20	4	6	<1	<1	5	2
Other[b]	29	—	43	8	30	17	34	29
				Canada				
Food	37	2	<1	2	—	—	100	8
Feed	34	29	78	84	78	—	—	69
Seed	23	5	5	4	<1	—	—	7
Other[b]	6	64	17	9	22	—	—	15

[a] Calculated from data obtained from Food Consumption Statistics, 1973—1982, Organization for Economic Cooperation and Development, Paris, 1985.

[b] Includes processed grains, the end products of which find uses for food and industrial purposes.

Table 5
USES OF THE GRAIN (SEED) PORTION OF CEREAL CROPS

Amaranth
   Food[a]
      Whole grains — boiled, popped, flaked
      Flour — bread, chapaties
      Ground grain — beverages, porridges
   Feed — whole grains
Barley
   Feed[a] — whole grains, ground, distillery grains
   Industrial — malt for making beer, whiskey, alcohol, vinegar, syrups
   Food
      Hull-less or pearled grains — soups, stews, sauces, breakfast cereals, baby foods
      Malt — beverages, breakfast cereals, syrups, baked products
      Flour or grits — baby foods, breakfast cereals, porridges, mixed with wheat flour for baking
Buckwheat
   Feed[a] — whole grains, ground
   Food — pancake flour, breakfast cereal, porridges
Maize (Corn)
   Feed[a] — ground, whole, milling by-products, silage
   Industrial
      Starch — syrup, glucose, dextose, manufacture of paper and paper products, coffee whiteners, sizing in textile manufacture, adhesives, and many other industrial uses
      Starch (waxy) — adhesives for stamps, envelopes, gummed tape, paper boxes, plywood; as a substitute for tapioca starch in foods
      Oil — margarine, salad and cooking oils
      Distilled liquors — grain neutral spirits, corn whiskey, bourbon, beer
      Alcohol — industrial uses, use in gasohol
   Food
      Whole grains — popped, parched, vegetable purposes
      Bran removed — hominy, corn flakes
      Meal and flour — bread, tortillas, corn chips, snacks, other baked products, mixed with wheat flour in baked products
      Grits — boiled as a breakfast food
Millet
   Food[a]
      Whole grains — roasted, popped, boiled, puffed
      Ground grains — thick and thin porridges, nonalcoholic beverages, beer
      Flour — unleavened and leavened breads, cakes, puddings, chapaties, pancakes, mixed with wheat flour for baking
      Malt — syrups, beer
   Feed — whole grains (used for caged and wild birds)
Oats
   Feed[a] — ground, whole rolled, mill by-products, hulls
   Food
      Whole or flaked groats — rolled or flaked breakfast cereals
      Flour — prepared breakfast foods, baby foods, mixed with flour for baking, as an antioxidant or stabilizer for other fat-containing foods
   Industrial
      Flour — coating paper for wrapping fatty foods and meats
      Hulls — furfural, other fiber products
Rice
   Food[a]
      Whole grains (dehulled) — boiled, fried, breakfast cereals, puddings, soups, baby foods
      Flour — mixed with wheat flour in bread, bakery products, pancake mixes; snack foods, cookies, baby foods, breakfast foods; thickening agent in gravies, sauces, puddings
      Broken rice — breakfast cereals, baby food, brewing
      Bran — baked products, oil extraction

Table 5 (continued)
USES OF THE GRAIN (SEED) PORTION OF CEREAL CROPS

Rice (cont.)
  Industrial
    Starch — laundering purposes, face powders, textile manufacture, other industrial purposes
    Oil — cooking oil, margarine, wax manufacture, leather conditioning, other industrial purposes
    Alcohol — beer, wines
    Hulls — fuel, building boards, insulation, paper, packing material, wood alcohol, furfural
    Feed — bran, hulls, other milling by-products
Rye
  Feed[a] — whole ground
  Food — flour, generally mixed with wheat flour for rye breads, other bakery products
  Industrial — alcohol
Sorghum
  Feed[a] — whole, ground, milling by-products
  Food
    Whole grains — popped, puffed, malted, boiled
    Ground — thick and thin porridges, fermented beverages
    Flour — unleavened bread, tortillas, mixed with wheat flour for baked products
  Industrial — starch, dextrose, syrups, alcohol, edible oils, brewing, distilling
Triticale
  Feed[a] — whole, ground, milling by-products
  Food — similar to rye uses
  Industrial — similar to rye uses
Wheat
  Food[a]
    Flour — main ingredient throughout much of world for bread and other bakery products, and pasta products
      Hard wheat flour — bread, rolls, chapaties, baby foods, sweet goods, whole-wheat baked products, and used in combination with soft wheat flour in many baked products
      Soft wheat flour — crackers, cookies, biscuits, cakes, doughnuts, pancakes, waffles, muffins, pie crusts, ice cream cones, pretzels, soup and gravy thickeners, general-purpose flour
      Durum wheat flour — macaroni, spaghetti, noodles, other semolina products
    Whole, cracked, or flaked grains — puffed or flaked breakfast cereals, hot breakfast cereals, farina, bulgur
    Germs — used in baked products
    Bran — breakfast cereals, used in baked products
    Gluten — baked products, breakfast foods
  Feed — whole ground, and milling by-products; wheat germ used extensively in pet foods and food for fur-bearing animals
  Industrial
    Starch — food, fabric, and textile industries; adhesives
    Distilled alcohols — whiskey, gin, vodka, neutral spirits, beer
Wild rice
  Food[a] — boiled grains used similar to rice, alone or in casserole dishes; high prices and considered a delicacy

[a] Major use of the grain on a worldwide basis.

REFERENCES

1. **Briggs, D. E.**, *Barley*, Chapman and Hall, London, 1978.
2. **Bushuk, W.**, Ed., *Rye: Production, Chemistry, and Technology*, American Association of Cereal Chemists, Inc., St. Paul, 1976.
3. *Technology*, American Association of Cereal Chemists, Inc., St. Paul, 1976.
4. **Cole, J. N.**, *Amaranth: from the Past for the Future*. Rodale Press, Emmaus, Penna., 1979.
5. **Considine, D. M.**, Ed., *Foods and Food Production Encyclopedia*, Van Nostrand Reinhold, Co., New York, 1982.
6. **Dendy, D. A. V.**, Ed., *Proc. Symp. Sorghum and Millets for Human Food*, 9th Congr. Int. Assoc. Cereal Chemistry, Vienna, May 11 to 12, 1976, Tropical Products Institute, London.

7. **Grist, D. H.,** *Rice,* 5th ed., Longman, London, 1975.
8. **Houston, D. F., Ed.,** *Rice Chemistry and Technology,* American Association of Cereal Chemists, St. Paul, 1972.
9. **Inglett, G. E.,** *Corn: Culture, Processing, Products,* AVI, Westport, Conn., 1970.
10. **Inglett, G. E., Ed.,** *Wheat: Production and Utilization,* AVI, Westport, Conn., 1974.
11. **Inglett, G. E. and Munck, L.,** *Cereals for Food and Beverages, Recent Progress in Cereal Chemistry,* Proc. Int. Conf. Cereals for Food and Beverage, Copenhagen, August 13 to 17, 1979, Academic Press, New York, 1980.
12. **Kent, N. L.,** *Technology of Cereals,* 3rd ed., Pergamon Press, Oxford, 1983.
13. **Leonard, W. H., and Martin, J. H.,** *Cereal Crops,* Macmillan, New York, 1963.
14. **Luh, B. S.,** *Rice: Production and Utilization,* AVI, Westport, Conn., 1980.
15. **Martin, J. H., Leonard, W. H., and Stamp, D. L.,** *Principles of Field Crop Production,* Macmillan, New York, 1976.
16. **Pomeranz, Y., Ed.,** *Wheat Chemistry and Technology,* American Association of Cereal Chemists, St. Paul, 1971.
17. **Pomeranz, Y., Chairman,** *Industrial Uses of Cereals,* Proc. Symp. Held in Conjunction with 58th Annu. Meet. American Association of Cereal Chemists, American Association of Cereal Chemists, St. Paul, 1974.
18. **Pomeranz, Y. and Munck, L., Eds.,** *Cereals: A Renewable Resource,* American Association of Cereal Chemists, St. Paul, 1981.
19. **Rachie, K. O., and Majmudar, J. V.,** *Pearl Millet,* Pennsylvania State University Press, University Park, 1980.
20. **Rodale Press, Inc., Ed.,** *Proc. 2nd Amaranth Conf.,* Rodale Press, Emmaus, Penna., 1980.
21. **Stoskopf, N. C.,** *Cereal Grain Crops,* Reston Publishing, Reston, Va., 1985.
22. **Tsen, C. C., Ed.,** *Triticale: First Man-Made Cereal,* American Association of Cereal Chemists, St. Paul, 1974.
23. **Vogel, S. and Graham, M.,** Sorghum and Millet: Food Production and Uses, Report of a workshop held in Nairobi, Kenya July 4 to 7, 1978.
24. **Wall, J. S. and Ross, W. M., Eds.,** *Sorghum Production and Utilization,* AVI, Westport, Conn.
25. **Watson, S. A.,** Industrial utilization of corn, in *Corn and Corn Improvement,* Sprague, G. F., Ed., Monograph 18, American Society of Agronomy, Madison, 1977.
26. **Western, D. E. and Graham, W. R., Jr.,** Marketing, processing, uses, and composition of oats, in *Oats and Oat Improvement,* Coffman, F. A., Ed., Monograph 8, American Society of Agronomy, Madison, 1961.
27. **Yamazaki, W. T. and Greenwood, C. T., Eds.,** *Soft Wheat: Production, Breeding, Milling, and Uses,* American Association of Cereal Chemists, St. Paul, 1981.

SUGAR CROPS

G. A. Smith

INTRODUCTION

The term "sugar", from the Greek *Saccharis*, is vague. There are hundreds of different sugars and these are only a small section of the vast family of carbohydrates, which includes cellulose at one end of the scale and simple alcohols at the other. All are composed of carbon, hydrogen, and oxygen, although the alcohols contain these elements in different proportions than sugars.

Generally, when people speak of sugar, they refer to a disaccharide, sucrose, or X-D-glucopyranosyl B-D-fructofuranoside. This is a compound composed of two monosaccharides, glucose and fructose, into which it is split in the course of digestion. These two monosaccharides occur in fruits and in honey (together with sucrose). Most of the information presented herein will concentrate on sucrose, fructose, dextrose, and glucose, since these are the natural sugars most frequently used.

Sugar, being a product of photosynthesis, occurs in many plants. The principle crop sources of sugar as a commodity are sugarcane, sugar beet, corn, maple, and sorghum.

During its long history, sugar has been the cause and prize of wars, as well as the object of political activity. There are logical reasons for this. Sugar is an attractive commodity which has even provided a simple means of collecting taxes. Thousands of people throughout the world gain their livelihood from sugar. With a rapidly expanding world population, this is important because sugarcane and sugar beet are, respectively, the most efficient plant fixers of solar energy among tropical and temperate-zone vegetation. Sugarcane is four times as effective as any other tropical plant in terms of dry-matter production per hectare per year, and sugar beet is twice as productive as any temperate-zone plant.[1,16] It requires an average of only 0.07 ha to fix solar energy to the equivalent of 1 million kilocalories of energy in the form of sugar. All other forms of edible energy require more; beef is at the top end of the scale, needing 7.7 ha — over 100 times as much land as sugar.

Demand for sugar has increased with growing world population, and even faster perhaps because improving technology made its production cheap. In the 1830s, when the world population was 1 billion, recorded sugar production was 800,000 tons/year (from sugarcane and sugar beet). By 1975, with a world population of 4 billion, production of crystalline sugar was about 80 million tons, approximately equally divided between cane and beet.[1] For 1985/1986, world production is estimated at 99 million tons. Added to this production of crystalline sugar are the crop-produced syrup sweeteners. The most important of these are corn sweeteners, 7 million tons (dry basis), maple syrup (4 million gallons), and sweet sorghum syrup (2 million gallons).[2,21] Annual U.S. honey production of about 200 million pounds plus at least 200 million pounds from other countries add to the total crop-produced caloric sweetener supply.[18-20]

SUGAR QUALITIES AND CALORIC VALUE

To the biochemist, "sugar" is a broad term covering a large group of related organic compounds. Sugars, which are made by plant cells in the presence of sunlight, are the main reason why foods of plant origin are nutritionally useful. Starches, also made by plants, are dense complexes of sugar molecules. Starches and sugars make up the group of foodstuffs known as carbohydrates.

Carbohydrates are an essential component of the human diet. In 1974, the Food and

Nutrition Board of the National Academy of Sciences/National Research Council established revised Recommended Dietary Allowances (RDAs) for nutrients in the U.S. diet. The RDAs suggest that the average dietary energy intake (in calories) should consist of 10 to 15% protein, 35 to 40% fat, and 45 to 50% carbohydrates. Carbohydrates, therefore, contribute the major part of the available energy in the human diet. In less developed areas, it is not unusual to find 80 to 90% of available energy in the diet coming from carbohydrate sources.[4]

The digestion of carbohydrates beings in the mouth, with the action of salivary amylase on starch. Further digestion of polysaccharides and mono- and disaccharides is accompanied by the actions of numerous other enzymes, many of which are located at the brush border of the epithelium of the small intestine (e.g., sucrase, lactase, maltase, α-dextrinase). The monosaccharides (such as fructose, glucose, and galactose) which result from the actions of such enzymes, subsequently are absorbed into the bloodstream and transported to the liver and other tissues for utilization by individual cells. After absorption into an individual cell, the monosaccharide is either used directly for energy production, stored as glycogen or fat to meet future energy needs, or is converted to other metabolic intermediates needed for growth and/or maintenance of body tissue.[4]

In the cells of the human body, all usable carbohydrates are converted to the same basic fuel, pyruvic acid, which is then burned to release energy. Although proteins and fats can also be used as sources of energy, only sugars can yield pyruvic acid. That is why sugar is the principal and preferred fuel for the body's energy cycle.[6]

If fruits, vegetables, and cereals couldn't be broken down in the human digestive system to their component sugars, they would be nutritionally useless.

As well as its flavor, which was the original reason for its popularity, sugar supplies an important nutritional factor — 4 cal/g. One teaspoon of white table sugar (sucrose) weighs about 3.5 g.[14] The exceptional rapidity with which sugar is absorbed makes it particularly beneficial when energy is suddenly required.

Basic calorie output for maintaining life (respiration, circulation, muscle tone) varies between 750 cal for a 3-year old, 1630 for a man of 25, and 1300 for an adolescent of 13 or a woman of 55. But this man of 25, who needs 1630 calories in a state of complete rest will expend 7000 calories during the day if he is an athlete in training.[14] If it is absorbed in time, sugar not only permits the body to face intense muscular effort, but also to fight efficiently against sudden attacks of fatigue, accidents, or disruption of the circulation. For the nervous system, in a very short space of time, it revives sensory acuity, "presence of mind", and attention.

Naturally occurring sugars can generally be grouped into families, the members of which have the same empirical formula. The suffix "ose" identifies a sugar. In the hexose series, for example, in which dextrose is the most common member, all of the sugars have the formula $C_6H_{12}O_6$. All members of the group are therefore isomers of one another. The property of sweetness is directly related to the structural differences among the hexoses. D-Fructose, which is a ketohexose, has a relative sweetness in the range of 110 to 160 as compared to a value of 100 for sucrose. Dextrose, or D-glucose, has a relative sweetness of 70 to 80 as compared to sucrose and is an aldohexose. The enzymatic transformation of glucose to fructose was introduced to corn sweetener production in 1967.[4]

SWEETENER CONSUMPTION TRENDS

Table 1 presents the per capita caloric sweetener consumption trend in the U.S. Total caloric sweetener intake reached about 129 lb per capita in 1985. Overall use of caloric sweeteners rose to 14.0 million metric tons in 1985, an 8% increase from 1982. Refined sugars' (from sugarcane and sugar beet) share of total caloric sweetener use has declined from 67% (84 lb) in 1980 to 49% in 1985 (63 lb). The principal reason for this decline is

Table 1
POUNDS PER CAPITA OF U.S. CALORIC SWEETENERS, 1975—1985

Calendar year	Refined sugars	Corn sweeteners[a] HFCS	Glucose	Dextrose	Total	Total of all caloric sweeteners[b]
1975	89.1	5.0	17.5	5.0	27.5	118.0
1980	83.7	19.2	17.6	3.5	40.3	125.3
1981	79.5	23.3	17.8	3.5	44.6	125.5
1982	74.1	26.7	18.0	3.5	48.2	123.7
1983	71.0	30.8	18.0	3.5	52.3	124.7
1984	67.5	36.3	18.0	3.5	57.9	126.8
1985	63.4	43.5	18.0	3.5	65.0	129.8

[a] Dry wt basis.
[b] Includes honey and edible syrups.

From Sugar and Sweetener Outlook, Economic Research Service, U.S. Department of Agriculture, Washington, D.C., December 1982 and March 1986.

Table 2
U.S. CONSUMPTION OF CALORIC SWEETENERS AS PERCENT OF TOTAL CALORIC SWEETENERS, 1975—1985

Calendar year	Refined sugar	Corn sweeteners HFCS	Glucose	Dextrose	Total
1975	75.7	4.2	14.8	4.2	23.3
1980	66.8	15.3	14.1	2.8	32.2
1981	63.4	18.5	14.2	2.8	35.5
1982	59.9	21.6	14.6	2.8	39.0
1983	57.1	24.6	14.4	2.8	41.9
1984	53.2	28.6	14.2	2.8	45.7
1985	48.8	33.5	13.9	2.7	50.1

From Sugar and Sweetener Outlook and Situation, Economic Research Service, U.S. Department of Agriculture, Washington, D.C., December 1982 and March 1986.

the increased per capita consumption of corn sweeteners, especially high fructose corn syrup (HFCS).[4,10,20] The entry of aspartame as a low calorie industrial sweetener in 1981 and its approval for table-top use in 1982 is another reason for this decline. Table 2 presents the consumption trends for caloric sweeteners as percent of total caloric intake. Table 3 presents a comparison of noncaloric, low-calorie, and caloric sweetener use in the U.S.

MAJOR CROP SPECIES UTILIZED IN SWEETENER PRODUCTION

Although many fruit-bearing plants like the date palm and the carob produce sugar as a product of photosynthesis, the world's major supply of sugar is obtained from the cultivated or managed crops of sugarcane, sugar beet, corn, sugar maple, and sweet sorghum.

Sugarcane, corn, and sweet sorghum are cultivated grass plants. Sugar beet is an herbaceous broadleaf, whereas sugar maple is a hardwood tree. Table 4 presents the common

Table 3
POUNDS PER CAPITA CONSUMPTION OF NONCALORIC, LOW CALORIE, AND CALORIC SWEETENERS, 1978—1985

Calendar year	Saccharin	Aspartame	Total noncaloric and low calorie[a]	Caloric sweeteners	Total all sweeteners
1978	7.1	0	7.1	126.6	133.7
1979	7.4	0	7.4	127.1	134.5
1980	7.7	0	7.7	125.0	132.8
1981	8.0	0.2	8.2	125.1	131.4
1982	8.4	1.0	9.4	123.2	134.0
1983	9.5	2.5	12.0	124.6	138.8
1984	10.0	5.8	15.8	126.8	142.7
1985	6.0	11.0	17.0	129.8	146.8

[a] Sugar sweetness equivalent. Assumes saccharin is 300 times as sweet as sugar and aspartame is 200 times as sweet as sugar.

Table 4
MAJOR CROP SPECIES UTILIZED FOR SUGAR AND SWEETENER PRODUCTION

Crop	Genus and species	Plant part processed	Primary product
Sugarcane	*Saccharum* spp.	Stalk	Crystalized sugar (sucrose), molasses
Sugar beet	*Beta vulgaris* L.	Tap root	Crystalized sugar (sucrose), molasses
Corn	*Zea mays* L.	Grain	Maltodextrins, dextrose, high-fructose corn syrup, high-maltose corn syrup
Sugar maple	*Acer saccharum* Marsh. *Acer nigrum* Michx. F.	Sap	Syrup — sucrose, glucose
Sweet sorghum	*Sorghum bicolor* L. Moench	Stalk	Syrup (containing sucrose, dextrose, and levulose)

and scientific name, the plant parts utilized in sugar recovery, and the form and principal type of sugars produced. A partial listing of the principal commercial sugar products and by-products is presented under the description of each crop. A listing of products from sucrose in general is presented in Table 10.

DESCRIPTION OF THE MAJOR SWEETENER CROPS

Sugar Beet

Cultivated sugar beet *(Beta vulgaris)* is an herbaceous dicotyledon which normally completes its life cycle in 2 years. The biennial habit is variable and under certain conditions, the plant may function as an annual or even as a perennial. When sugar beet grows as a biennial, a fleshy tap root is produced the first year. It is this tap root which is harvested and processed for sugar recovery. *Beta vulgaris* and other closely related *Beta* species originated in the Mediterranean area. The beet root was a common part of the diet in Egypt during the building of the pyramids, but the potential of the sugar beet as a source of sugar was not discovered until the middle of the 18th century. Beginning in the second half of

Table 5
WORLD SUGAR PRODUCTION FROM SUGAR BEETS
1980—1985[a]

	Production (1000 metric tons)				
Region	1980/81	1981/82	1982/83	1983/84	1984/85
North America	2,964	3,129	2,651	2,560	2,710
South America	302	189	274	260	350
Western Europe	14,914	18,048	16,588	13,880	15,130
Eastern Europe	4,757	5,791	5,771	5,730	5,880
U.S.S.R.	7,196	6,400	7,300	8,700	7,800
Asia	1,259	1,231	1,270	1,540	1,460
North Africa	321	335	369	430	440
Middle East	1,001	1,479	1,591	2,300	2,280
World total	32,714	36,602	35,814	35,390	36,050

[a] Figures are raw value centrifugal sugar.

From Sugar and Sweetener Outlook and Situation, Economic Research Service, U.S. Department of Agriculture, Washington, D.C., December 1982 and December 1984.

the 19th century, plant breeders transformed the fodder beet into an efficient producer of sugar (sucrose).[7,11] The beet-sugar industry is concentrated in the temperate climates of Europe, North America, and the U.S.S.R. Sugar beets are also grown in parts of South America, North Africa, Japan, China, and the Near East. Table 5 presents the raw sugar production from sugar beet grown worldwide.

Sugar-Beet Products and By-Products

Numerous products are made from sugar in addition to food products (see Table 10). Specialty sugar products include granulated sugars produced with a specific grain size. Crushed grain sugars, which include the powdered sugars, have a sizable market in baking and candy. Brown or soft sugars are used by bakers and confectioners as a source of color and flavor in cakes, icing, and candies. Soft sugar is also used in the manufacture of syrup. Sugar is also molded into cubes and chipped slabs. A small volume of sugar is produced in vivid colors for both food and decorative purposes. Regrained sugar is a specialty sugar made by dissolving the original grain, subjecting the resulting syrup to some transformation, and repackaging for sale. Agglomerated sugar, developed by a Pillsbury Company patent, contains no starch. It is free flowing and is distributed in consumer-size packages. Liquid sugars, either liquid sucrose or liquid inverts which have equal parts of glucose and fructose, are marketed.[13]

The two principal by-products from sugar beet are molasses and dried pulp. About 60% of beet molasses produced in the U.S. is shipped to Steffen factories for the recovery of sugar by precipitation with lime. Other molasses is used for cattle feeding, either directly or mixed with feed products. Molasses, because of its high content of sugar and nitrogenous constituents, is a valuable raw material for the fermentation industries. Alcohol, yeast, glutamic acid, and citric acid are a few of the products manufactured from molasses.[13]

Dried pulp remaining after extraction of the sugar is a valuable cattle food, containing proteins, minerals, and carbohydrates.

Sugarcane

Cultivated sugarcane (*Saccharum* spp). is a vegetatively propagated perennial grass that

Table 6
WORLD SUGAR PRODUCTION FROM SUGARCANE, 1980—1985[a]

Region	\multicolumn{5}{c}{Production (1000 metric tons)}
	1980/81	1981/82	1982/83	1983/84	1984/85
North America	5,016	5,314	5,325	5,670	5,660
Caribbean	9,173	9,993	9,682	9,940	9,900
Central America	1,521	1,623	1,666	1,720	1,800
South America	12,916	12,981	14,034	14,230	13,840
Western Europe (Spain)	15	16	14	10	10
Africa	3,780	4,598	4,701	6,120	7,080
Asia	16,301	21,844	20,025	18,940	18,980
Oceania (Australia)	3,389	3,586	3,600	3,440	4,030
Other Countries	3,440	3,424	3,597	3,500	3,500
World total	55,551	63,379	62,644	60,170	61,490

[a] Figures are raw value centrifugal sugar.

From Sugar and Sweetener Outlook and Situation, Economic Research Service, U.S. Department of Agriculture, Washington, D.C., December 1982 and December 1984.

originated in Northern India and New Guinea. There are records of the utilization of the cane by civilized races in India and in China as far back as chronicled history goes and further evidence indicates that primitive man made use of it long before then.[16] He undoubtedly came to value the plant for the sweet taste of its stalks. Sugar is stored in the stem tissues of the plant and the stems devoid of leaves constitute the harvestable portion of the plant.

There are well-authenticated records of sugarcane and sugar production in both India and China several centuries before the Christian era.[5] The history of sugarcane improvement in early times is mainly one of variety substitution making use of naturally occurring types which gave improved yields due to better adaptation or greater resistance to disease. Improved varieties of sugarcane, known to be one of the oldest cultivated plants in the world, are grown commercially in the tropics and the subtropics. Sugarcane supplies about 55% of the annual world supply of granular sugar (sucrose). Table 6 presents the raw sugar production from sugarcane grown worldwide. Sugarcane accounts for about 63% of world raw sugar production.

Sugarcane Products and By-Products

Sucrose products and their uses are outlined in Table 10.

By-products from sugarcane include the following : (1) from bagasse, the fibrous residue of the cane stalk after crushing, the following products are recovered: poultry litter, fuel, methane, electricity, furfural, furfuryl alcohol, bleached pulp, carboard, newsprint, xylitol, fiberboard, particle board, and writing paper; (2) from molasses, the following products are obtained: animal feed, acetaldehyde, ethyl ether, butanol/acetone, feed yeast, ethyl acetate, ethanol, acetic acid, ethylene dichloride, rum, baker's yeast, monosodium glutamate, citric acid, and L-Lysine; and (3) from filter mud fertilizer, animal feed, and crude wax are obtained.[15]

Corn

Corn (*Zea mays* L.) is an annual grass plant indigenous to the western hemisphere. Columbus encountered corn on his first voyage in 1492, and subsequent explorers found

Table 7
**U.S. CORN SWEETENER CONSUMPTION,
1980—1985 (MILLION METRIC TONS)**

Calendar year	HFCS	Glucose	Dextrose	Total
1980	1.98	1.82	0.36	4.16
1981	2.40	1.85	0.36	4.64
1982	2.81	1.89	0.37	5.07
1983	3.26	1.91	0.37	5.56
1984	3.90	1.93	0.37	6.20
1985	4.72	1.94	0.38	7.04

From Sugar and Sweetener Outlook and Situation, Economic Research Service, U.S. Department of Agriculture, Washington, D.C., December 1982 and March 1986.

corn being grown by the Indians throughout the western hemisphere. It is now grown in every state of the U.S. and in every important agricultural area of the world. It is grown on more than 120 million hectares in the world, and is an important commodity in world trade. Approximately 6 billion bushels of corn are now grown annually in the U.S.[7] In drawing on the native grain as a source of industrial products for both feed and new food applications, the corn-refining industry has capitalized on a 165-year-old discovery which demonstrated that starch could be transformed into sweet substances by heating with dilute acid. This discovery by the Russian chemist Kirchoff and subsequent work by other scientists demonstrated that starch is a polymeric form of D-glucose, and that it could be broken down to D-glucose by hydrolysis or conversion and so the foundation was laid for a starch-derived sweetener industry.[4]

Liquid corn syrup technology advanced significantly with the post World War II commercial introduction of enzyme-catalyzed conversion. The enzymatic transformation of glucose to fructose was introduced to corn sweetener production in 1967.[4] By the early 1970s, a commercial process for the enzyme-catalyzed isomerization of glucose (dextrose) to the sweeter sugar fructose (levulose) was perfected. This process enables the production of both of the simple sugars that make up sucrose and thus match total invert syrup. Invert syrup is a partial or complete hydrolysis product of sucrose. The new product, high fructose corn syrup, is a significant development for the sweetener and food industry. In the 35 years from 1950 through 1985, while the per capita annual use of nutritive sweeteners in the U.S. varied from 116 to 129 lb, the corn sweetener share of the market rose from 13 to 50% (Table 2).[4,20] Since the introduction of high fructose corn syrup in 1967, the annual per capita use has risen from one-tenth of a pound to 43 lb at the end of 1985. Total per capita consumption of corn sweeteners totaled nearly 65 lb, dry basis, in 1985 (Table 1). Total U.S. consumption of corn sweeteners is presented in Table 7.

Corn Sweetener Products and By-Products

Fructose was previously an expensive sugar derived from sucrose. The dextrose industry had long suffered from low-level sweetness in their dextrose syrups; it is now possible to make sweeter dextrose syrups because of the presence of fructose introduced by glucose isomerization. Isomerized dextrose syrups are called high-fructose corn syrups (HFCS).

First-generation HFCS, introduced in the early 1970s, are composed of 42% fructose, 52% dextrose, and 6% higher saccharides. The 42% fructose syrup made from starch is used in soft drinks, yeast-raised baked goods, frozen desserts, canned fruits, jams, jellies, salad dressings, and confections. Second-generation HFCS contain fructose from 55 to 90%

Table 8
WORLD MAPLE SYRUP PRODUCTION AND EQUIVALENT SUGAR PRODUCTION

Country	Gallons × 1000				Sugar × 1000 lbs[a]			
	1977	1978	1979	1981	1977	1978	1979	1981
Canada	2,330	1,933	2,842	3,570	18,638	15,460	22,732	28,560
U.S.	1,221	1,154	1,219	1,409	9,768	9,231	9,751	11,272
Total	3,551	3,087	4,061	4,980	28,406	24,691	32,483	39,840

[a] Assuming that 1 gallon of syrup is equivalent to 8 pounds of sugar.

From Direction du Developpement Commercial — Ministere de L'Agriculture et de L'Alimentation du Quebec, and U.S.D.A. Agricultural Statistics, 1981, U.S. Government Printing Office, Washington, D.C., 1981.

of the total solids. A 55% HFCS contains 55% fructose, 40% dextrose, and 5% higher saccharides. The 90% HFCS contains 90% fructose, 7% dextrose, and 3% higher saccharides. The major uses of 55% HFCS are soft drinks, salad dressings, frozen desserts, capped fruits, jams, jellies, and breakfast cereals. The 90% HFCS is finding uses in "light" foods and beverages since a reduced quantity of 90% HFCS may be used to achieve the same level of sweetness in the final food. The 90% HFCS is used in soft drinks (full calorie and reduced calorie), salad dressings, jams, jellies, table syrups, wines, low-calorie frozen yogurts. and desserts.[10]

In addition to food products, corn sugars are a major source of alcohol produced by fermentation.

Sugar Maple

It is not known who first discovered how to make syrup and sugar from the sap of the maple tree. Both were well established items of barter among the Indians living in the area of the Great Lakes and the St. Lawrence River, even before the arrival of the white man.[22]

Maple syrup is a noncultivated, nonfertilized woodland crop. Since the trees grow best at altitudes of 600 ft and higher, maple syrup is usually produced in hilly country. Maple syrup and sugar are commercial crops from Maine to Minnesota and in adjacent areas of Canada. The trees on 1 acre will provide 160 tapholes and an average yield of 1 qt of syrup per taphole, or 40 gal of syrup per acre.[22,23]

Only 2 of the 13 species of maple *(Acer)* native to the U.S. are important in syrup production. *Acer saccharum* Marsh., commonly called the sugar maple, hard maple, rock maple, or sugar tree, furnishes three fourths of all sap used in the production of maple syrup. Although this tree grows throughout the maple-producing areas, the largest numbers are in the Lake States and the Northeast. Trees grow singly and in groups in mixed stands of hardwoods. The trunk of a mature tree may be 30 to 40 in. in diameter. The tree is a prolific seeder and endures shade well but unfortunately does not grow rapidly. *Acer nigrum* Michx. f., commonly known as black sugar maple, hard maple, or sugar maple, grows over a smaller range than does *A. saccharum*. It does not grow as far north or south but is more abundant in the western part of its range. This tree is similar to *A. saccharum* in both sap production and appearance.

Maple sap on a fresh basis typically contains about 2% sugar. It requires approximately 43 gal of 2% sap for 1 gal of standard-density syrup. Sucrose comprises 96% of the dry matter of the sap and 99.95% of the total sugar. The other 0.05% is composed of raffinose and several oligosaccharides.[22]

Total maple-syrup production for the U.S. and Canada is presented in Table 8. Vermont

Table 9
U.S. AND CANADA MAPLE SYRUP PRODUCTION BY STATE OR PROVINCE, 1979—1980

State or province	Gallons × 1000 1979	Gallons × 1000 1980	Sugar × 1000 lbs[a] 1979	Sugar × 1000 lbs[a] 1980
Maine	9	5	72	40
Massachusetts	30	18	240	144
Michigan	85	83	680	664
New Hampshire	76	55	608	440
New York	315	243	2,520	1,944
Ohio	90	88	720	704
Pennsylvania	57	56	456	448
Vermont	465	315	3,720	2,520
Wisconsin	92	110	736	880
Nova Scotia	16	N/A	128	N/A
New Brunswick	15	N/A	120	N/A
Quebec	2,594	2,883	20,752	23,064
Ontario	217	N/A	1,736	N/A
Total U.S.[b]	1,219	969	9,752	7,752
Total Canada	2,842	3,029	22,736	24,232

[a] Assuming that 1 gallon of syrup is equivalent to 8 pounds of sugar.
[b] Total may not sum to exact values due to differing data sources or rounding error. Total U.S. maple syrup production for 1984 was 1.36 million gallons.

is the leading maple-syrup state, producing about 500,000 gal in 1982. New York produced 320,000 gal in 1982.[21] The leading maple syrup province in Canada is Quebec which alone produced 2,593,975 U.S. gal in 1979 (Table 9). Production of maple syrup by state and province is presented in Table 9.

Maple Products and By-Products

Gross returns from the maple-syrup crop can be increased from 20 to 160% by converting syrup to sugar candies and maple spreads. The 8 lb of sugar in a gallon of syrup is worth $2.50/lb, based on syrup selling at $20/gal. This same weight of sugar, if converted to sugar products, can be sold for several times more than $2.50/lb. Numerous maple candies and other confections are produced utilizing maple syrup.[22,23]

Sweet Sorghum

The name "sweet sorghum" is used to identify varieties of sorghum, *Sorghum bicolor* L. Moench, that are sweet and juicy. These sweet-stalk varieties are also called "sugar sorghums". Sweet sorghum is grown for syrup or forage and is adapted to diverse climatic and soil conditions. Consequently, some sweet sorghum has been grown for syrup in 19 of the 50 states.[9]

Historically, sorghum syrup has been largely produced by open-kettle evaporative condensation of the juice crushed from the sorghum cane of varieties classified as "sweet" sorghum. About 90% of the sorghum syrup produced in the U.S. comes from the eight states of Alabama, Arkansas, Georgia, Iowa, Kentucky, Mississippi, North Carolina, and Tennessee.[3,9]

In many places, sorghum syrup continues to be produced by the old-style, open-kettle, batch method on farms where sorghum is grown. With the exception of one firm, most of

Table 10
THE PRODUCTS OF SUCROSE OF ACTUAL OR POTENTIAL IMPORTANCE PRESENTED BY THE TYPE OF REACTION LEADING TO THEIR FORMATION AND THEIR PRINCIPAL USE

Fermentation		Synthesis		Degradation	
Products	Use[a]	Products	Use	Products	Use
Dextran	B	Fatty acid esters of sucrose	A, C	Sorbitol, mannitol	B
Microbial alginate	A	Sucrose acetate isobutyrate	C	Glycerol	E
Xantham gum	C	Sucrose octaacetate	D	Propylene glycol	D
L-Lysine	A	Sucrose octabenzoate	D	Methylpiperazine	D
2,3 Butylene glycol	A	Sucrose polycarbonate	D	Fructose	B
Penicillin	B	Heptacyanoethylsucrose	E	Gluconic acid	C
Aureomycin	B	Polyhydroxyalkylsucrose	C	Oxalic acid	B
Terramycin	B	Sucrose polyurethane	D	Arabonic acid	B
Erythromycin	B	Sucrose xanthate	C	Hydroxymethylfurfural	D
Riboflavin (B$_2$)	B	Tetrachlorogalactosucrose	A	Levulinic acid	A
Cobamide (B$_{12}$)	B	Organotin sucrose pesticide	E	Lactic acid	A

[a] Use: A = food and feed improvers; B = pharmaceuticals, cosmetics, etc.; C = surfactants, viscosity improvers, etc.; D = resin intermediates, plasticizers, etc.; E = surface coatings, dielectrics, pesticides.

Adapted from Paturau, J. M., *By-products of the Cane Sugar Industry*, 2nd ed., Sugar Series, No. 3, Elsevier, Amsterdam, 1982.

the sorghum syrup for the commercial market is produced by relatively small producers, who sell in the local regions. The product is sold either as pure sorghum syrup or as a sorghum-flavored blend with a base of corn or cane syrup. Characteristically, sweet sorghum syrup is produced by small-acreage producers to satisfy local or regional consumer demand.

In 1960, the last year for which production statistics on sorghum syrup were reported by the U.S. Department of Agriculture, the output was 1,943,000 gal. This was a sharp drop from the 11.6 million gallons produced in 1945 and was only about one half of the 3.5 million gallons produced 10 years earlier in 1950.[17] The decline of sweet-sorghum syrup production in the U.S. has been attributed to the significant changes in the American way of life. Most notable among these changes is the migration from farm to city, improved food preservation and shipping practices, and reduced availability of farm labor. Sorghum syrup, formerly a basic food for many farm families is now a speciality product. The latest estimates indicate that approximately 2 million gallons of syrup are produced per year (about 250 gal/acre) by 14,000 growers in 14 states.[2,17]

Sweet-Sorghum Products and By-Products

The primary product from sweet sorghum is sorghum syrup which is usually sold as a pure product. Sweet sorghum can also be used to produce crystalline sugar using procedures similar to those used in sugarcane. However, commercial production of sugar from sweet sorghum has never become a reality.[3,9] The most recent interest in sweet sorghum has centered around its use for ethanol production. Owing to its wide area of adaptivity and low labor input for cultivation, further interest in sorghum as a source of ethanol may be seen.[2]

DERIVATIVES OF SUCROSE

There are some 10,000 chemicals and chemical intermediates which have been developed from sugar, and a large number of patents (probably well over 500) have been filed on the utilization of sugar.[15] Kollonitsch, in her book on sucrose chemicals, has covered a large number of these derivatives.[12] The book *Sucrochemistry* edited by Hickson gives up-to-date and fairly detailed data on the possible utilization and future development of sucrose-based chemicals.[8]

Table 10 presents only a very few sugar compounds which either are presently produced on an industrial scale or seem to show special promise for the future. Table 10 gives a simplified presentation of these products. They are grouped according to the type of reaction leading to their formation (fermentation, degradation, and synthesis) rather than according to their end usage as suggested in the excellent book *By-products of the Cane Sugar Industry* by Paturau.[15]

REFERENCES

1. **Anon.**, And this is the history of sugar!, *Sugarbeet Grower*, February, 1979.
2. **Broadhead, D. M.**, personal communication, 1983.
3. **Coleman, O. H.**, Syrup and Sugar from Sweet Sorghum, *Sorghum Production and Utilization*, 11, 416, 1968.
4. Nutritive Sweeteners from Corn, 2nd ed., Corn Refiners Assocation, Inc., Washington, D.C., 1979.
5. **Deerr, N.**, *The History of Sugar*, Vol. I., Chapman and Hall, London, 1949.
6. **Dirckx, J. H.**, The facts about sugar, *Sugro Info.*, 4, 3, 1982.
7. **Fehr, W. R. and Hadley, H. H.**, Eds., *Hybridization of Crop Plants*, American Society of Agronomy and Crop Science Society of America, Madison, 1980.
8. **Hickson, J. L.**, *Sucrochemistry*, American Chemical Society, Washington, D.C., 1977.
9. **Freeman, K. C., Broadhead, D. M., and Zummo, N.**, Culture of sweet sorghum for syrup production, in Agriculture Handbook No. 441, U.S. Department of Agriculture, Washington, D.C., May, 1973.
10. **Inglett, G. E.**, Sweeteners — a review, *Food Technol.*, March, 1981, 37—38, 40—41.
11. **Johnson, R. T., Alexander, J. T., Rush, G. E., and Hawkes, J. R.**, Eds., *Advances in Sugar Beet Production*, The Iowa State University Press, Ames, 1971.
12. **Kallonitsch, V.**, *Sucrose Chemicals*, International Sugar Research Foundation, Inc., Bethesda, Md., 1970.
13. **McGinnis, R. A.**, Ed., *Beet-Sugar Technology*, 3rd ed., Beet Sugar Development Foundation, Fort Collins, Colo., 1982.
14. **Morel, P.**, Sugar as a food, *Sugar J.*, February 1977, 13.
15. **Paturau, J. M.**, *By-products of the Cane Sugar Industry*, 2nd ed., Sugar Series, No. 3, Elsevier, Amsterdam, 1982.
16. **Stevenson, Y. C.**, *Genetics and Breeding of Sugar Cane*, Longman Green and Co., London, 1965.
17. Prospects for Revitalizing the Market for Sorghum Sirup, U.S.D.A.-ARS-Ne-66, October, 1975.
18. U.S.D.A., *Agricultural Statistics 1981*, U.S. Government Printing Office, Washington, D.C., 1981.
19. Sugar and Sweetener Outlook and Situation Report, U.S.D.A. Economic Research Service, Washington, D.C., December, 1982.
20. Sugar and Sweetener Outlook and Situation Report, U.S.D.A. Economic Research Service, Washington, D.C., December, 1984.
21. Sugar and Sweetener Outlook and Situation Report, U.S.D.A. Economic Research Service, Washington, D.C., March, 1986.
22. **Willits, C. O. and Hills, C. H.**, *Maple Sirup Producers Manual*, Agricultural Handbook No. 134, U.S. Government Printing Office, Washington, D.C., 1976.
23. **Winch, F. E., Jr. and Morrow, R. R.**, Production of Maple Sirup and Other Maple Products, Info. Bull. 95, State College of Agriculture and Life Sciences and U.S.D.A., New York, N.Y., 1978.

STARCH CROPS

D. L. Jennings

INTRODUCTION

Many plants store carbohydrate reserves in the form of starch in organs modified in various ways to accommodate it. Many of these organs have become important foods. In some societies, especially the developing countries of the tropics, they have become the staple food, while in more prosperous societies they have a less important role as vegetables.

In most situations a starchy root crop has become the chosen staple because it competes favorably on a cost basis with other sources of carbohydrates; and under traditional systems of farming, such crops give by far the highest yields of calories per hectare. However, though the relative economics of crops vary with locality, there is a general tendency to shift from a root crop to a cereal as the population becomes urbanized or becomes more prosperous and relies upon purchased foods. The root crop then becomes primarily a vegetable food that adds variety to the diet rather than the staple food. This is shown by the contrast between the average Nigerian, who is substituting rice and wheat for root crops, and consumers in poorer neighboring countries who are increasing their consumption of starchy root crops.[8] The situation is similar to that of the potato in Ireland: it was once the main staple there, but its contribution to the daily diet has declined in the past 40 years. This decline is true of all western countries except possibly for the poorer parts of Spain and Portugal. It is worth remembering that when potatoes, yams, cassava, and taro serve as the main providers of food calories, it is because they are the cheapest to produce and not because they are preferred: cost, not taste, is what matters. In tropical countries the root crops are preeminent because of their yield of food calories per unit of land and because the main cost is the physical effort of clearing and maintaining land in cultivation.

It is not surprising that the number of plant species whose storage organs are used as starch food is high (Table 1), but six kinds of crop have attained major importance and discussion is limited to these. The potato is by far the most important of them; it is primarily a temperate crop but is becoming important in parts of the tropics. Cassava is next in importance, closely followed by sweet potatoes and yams. Cocoyams and the other aroid species are of less but increasing importance.

For each of these crops the primary use is as a human food which is usually eaten boiled, roasted, or fried, but occasionally as a processed product. In addition, they are used for stock feed, commercial starch, flour, or industrial alcohol, and some of the processes involved yield valuable secondary products. Most commercial starch is produced from maize, but the similarity between starches and the ability of the chemist to tailor starches means that the market for a given starch is constantly changing in response to market and even political forces.[9] Hence starch from the developing countries is competing successfully with that produced in temperate countries; cassava starch in particular is preferred for the manufacture of newsprint, cardboard, glues for stamps and envelopes, and certain foods.

POTATOES

The potato tuber is still the world's leading starchy vegetable. As well as carbohydrates, potatoes provide useful amounts of other nutrients and are probably the source of some 8% of the riboflavin and 34% of the ascorbic acid of the British diet, besides significant amounts of protein and iron. Processed products are increasing in importance: small, uniform, peeled potatoes provide a useful convenience product when canned in 1.5 to 3.0% salt solution

Table 1
SOME SPECIES USED AS SOURCES OF STARCH

Common name	Species	Family	Primary product
African yam bean	*Sphenostylis stenocarpa* Hochst ex A. Ricl.	Leguminosae	Small, spindle-shaped tubers (5—7.5 cm)
Añu	*Tropaeolum tuberosum* Ruiz and Pav.	Tropaeolaceae	Small, conical or elipsoidal tubers (5—15 cm long, 3—6 cm wide)
Arracacha	*Arracacia xanthorrhiza* Bancroft	Umbelliferae	Slender lateral tubers, 6—10 per plant
Arrowhead	*Sagittaria sagittifolia* L.	Aponogetonaceae	Corms, 5 cm diam.
Arrowroot	*Maranta arundinacea* L.	Marantaceae	Cylindrical rhizomes 2.5 cm thick; 20—45 cm long
Cassava	*Manihot esculenta* Crantz	Euphorbiaceae	Tubers, 5—10 per plant; 30—45 cm long; 5—15 cm diam.
Chavar	*Hitchenia canlina* (Grah.) Baker	Zingiberaceae	Round tubers, 6 cm diam.
Chinese water chestnut	*Eleocharis dulcis* Trin.	Cyperaceae	Rounded corms, 1—4 cm diam.
Chufa	*Cyperus esculentus* L.	Cyperaceae	Tubers, 1.5—2.0 cm long, 1—2 cm diam.
Chuno	*Solanum juzepezukii* and *S. curtilobum*	Solanaceae	As potato
East Indian arrowroot	*Tacca contopetaloides* (L.) Kunze	Taccaceae	Many tubers, like potatoes, 10—15 cm diam., rarely 30 cm
Elephant yam	*Amorphophallus* spp.	Araceae	Depressed globose corms, 30 cm diam.
False yam	*Icacina senegalensis* A. Juss.	Icacinaceae	Tubers, 30—45 cm length 30 cm diam.
Giant taro	*Alocasia macrorrhiza* (L.) Schott	Araceae	Epigeous stem, 1 m high, 20 cm diam.
Hansa potato	*Solenostemon rotundifolius* (Poir.) J. K. Morton	Labiatae	Tubers, like small potatoes
Jerusalem artichoke	*Helianthus tuberosus* L.	Compositae	Tubers, 10—20 cm long, 2—7 cm diam.
Lotus root	*Nelumbo nucifera* Gaertn.	Nymphaeaceae	Rhizomes, 60—120 cm length, 5—10 cm diam.
Maco	*Lepidium meyenni* Walp.	Cruciferae	Root-hypocotyl
Oca	*Oxalis tuberosa* Molina	Oxalidaceae	Tubers, like potatoes. 5—7 cm length, 1—4 cm diam.
Potato	*Solanum tuberosum* L.	Solanaceae	Tubers, thickened underground stems
Queensland arrowroot	*Canna edulis* Ker-Gawl	Cannaceae	Tubers, variable shape usually 5—9 cm diam., 10—15 cm length
Sago palm	*Metroxylon sagus* + *M. rumphii*	Palmae	Tree trunk
Shoti	*Curcuma zedoaria* (Berg.) Roscue	Zingiberaceae	Rhizomes 15 cm length, 2—3 cm diam.
Swamp taro	*Cyrtosperma chamissomis* (Schott.) Merr.	Araceae	Spherical corms 2 m length, 0.6 m diam.
Sweet potato	*Ipomoea batatas* L.	Convolvulaceae	Tubers from adventitious roots variable size and shape
Tannia	*Xanthosoma* spp. (L.) Schott.	Araceae	Central plus lateral corms. 12—25 cm length, 12—15 cm diam.
Taro	*Colocasia esculenta* (L.) Schott.	Araceae	Central plus lateral corms; variable size and shape.
Topee tambo	*Calathea allouia* (Aubl.) Lindl.	Marantaceae	Tubers, like potatoes; 4—6 cm length, 2—4 cm diam.

Table 1 (continued)
SOME SPECIES USED AS SOURCES OF STARCH

Common name	Species	Family	Primary product
Ulluco	*Ullucus tuberosus* Caldas	Basellaceae	Tubers, like small potatoes
Yacón	*Polymnia sonchifolia* Poepp and Endl.	Compositae	Tuberous fusiform roots; 20 cm length, 3—10 cm diam.
Bitter yam	*Dioscorea dumetorum* (Kunth) Pax	Dioscoreaceae	Tubers, variable
Chinese yam	*Dioscorea opposita* Thunb.	Dioscoreaceae	Tubers, variable
Cush-cush yam	*Dioscorea trifida* L.	Dioscoreaceae	Many small tubers 15—20 cm length
Greater yam	*Dioscorea alata* L.	Dioscoreaceae	Tubers, normally single and large
Intoxicating yam	*Dioscorea hispida* Denst	Dioscoreaceae	Tubers
Lesser yam	*Dioscorea esculenta* (Lour.) Burk.	Dioscoreaceae	Ovoid tubers, 4 per plant, length 10—20 cm
Potato yam	*Dioscorea bulbifera* L.	Dioscoreaceae	Bulbils or aerial tubers in leaf axils; 7—10 cm length, 2—5 cm diam.
White yam	*Dioscorea rotundata* Poir.	Dioscoreaceae	Usually one large tuber
Yellow yam	*Dioscorea cayenensis* Lam.	Dioscoreaceae	Tubers
Yam bean	*Pachyrrhisus erosus* (L.) Urban	Leguminosae	Tubers or fleshy roots 10—30 cm diam.

Based on data from Kay, D. E., Root crops, *TPI Crop and Product Digests*, Tropical Products Institute, London, 1973.

with calcium salts added to improve texture. For this purpose, tubers with a sugar content below 2% are required and it is important to choose a cultivar and storage conditions which ensure this. Frozen chips (french fries) and crisps are other well-known products, while dehydrated flakes, granules, or dice are less well-known ones made by dehydrating cooked mashed potatoes. Potato flour has been processed for longer than these products, particularly in America, and is used by the baking industry to make certain types of bread and confectionery.

As a stock feed, potatoes, especially culls, are fed fresh to cattle and sheep, or they may be dried and used as meal, or stored as silage. With a total digestible nutrient content of about 17%, potatoes are equal in value to about a quarter of that of an average grain mixture.

Relatively small amounts of potatoes are used as starch because it cannot normally be produced at a price competitive with maize starch. Small amounts are used in the food, paper, and textile industries. In some European countries potato alcohol is produced extensively from pulped and fermented potatoes.

Among the secondary products of potato processing are potato pulp and potato peels, which are products of starch manufacture and are used for livestock feed, and potato processing-water effluent, which is the effluent water of potato processing and is used as a source of high-grade starch or high-grade protein.[5]

Chuno is a processed product made from *Solanum juzepezukii* or *S. curtilobum,* the bitter potatoes, which are not fit for consumption fresh because of the presence of a bitter glycoalkaloid. These potatoes are important because they grow at altitudes of up to 4500 m in the Andes where frost damage to other crops leaves chuno to make up 80% of the diet. The processing involves utilizing natural freezing and thawing for 3 days prior to extracting the glycoalkaloid; the product can be stored for several years.[2]

CASSAVA

Cassava is the most important starchy root crop of the tropics. About 65% of total production is used for direct human consumption, about 20% for animal feed, and lesser amounts for industrial starches. The method of use is very variable; fresh roots deteriorate rapidly and so they are often stored as sun-dried chips and then eaten after being ground into a flour. In West Africa, cassava is made into a fermented meal known as *gari*. Fresh roots are also eaten as a vegetable after roasting or boiling or they may be boiled and pounded into a paste and added to soups. Because of the low protein content of cassava (usually less than 1%), the exclusive use of cassava food has become associated with the nutritional disorder *kwashiorkor*, but this is largely a problem of urban areas where the diet includes less of the leafy vegetables and legumes available in rural parts.

A modern development is to use a cassava substrate in a fermentation process for the production of a single-cell protein. The microorganisms used include *Aspergillus fumigatus* and *Rhizopus chinensis*, and a product containing up to 37% crude protein has been made experimentally.[6,10]

A major problem for human consumption is the presence of the glycoside linamarin which is converted to hydrocyanic acid (HCN) by the enzyme linamarase. The HCN is concentrated in the rind and fibrous core in the center of the root and its content varies considerably with cultivar and growing conditions. In most cultivars the level is not high in freshly harvested roots and rises as the glycoside comes into contact with the enzyme after lifting. The HCN then has to be removed by a diversity of methods, notably heating, sun-drying, fermenting, or leaching. These methods are effective, and cases of acute poisoning arising from cassava consumption are rare, but recent evidence suggests that chronic toxicity due to the regular intake of low levels of HCN is more common and can be responsible for diseases associated with goiter, cretinism, tropical ataxic neuropathy, and tropical diabetes, especially in parts of Africa where the diet is low in protein and iodine.[7]

The processed food for which cassava is best known is tapioca. Most of this comes from the Malayan peninsula and consists of partly gelatinized cassava starch; it can take the form of flakes, seeds, or pearls, depending on the processing method used.[5] However, cassava flour is also used to a small extent in the manufacture of bread and confectionery and even in macaroni and spaghetti. Another use is as an adulterant of cereal flour. Indeed the very extensive experimentation of recent years on the mixing of wheat and noncereal flours has probably involved cassava flour more than any other. The problem has been to discover the proportion of "extender" flour that can be added to wheat flour without losing too much quality. This has been found to vary from 3 to 30% according to the bakery product; for example, a higher proportion can be used in biscuits than in bread because there are fewer problems of volume and texture. However, the use of new emulsifiers and other additives to substitute for the role of wheat gluten in facilitating gas retention in the dough and gelatinization during baking has brought good success, though mostly from the best grades of cassava starch. Unfortunately, very few of the findings from this research have been utilized.[1,10]

Large quantities of peeled dried cassava roots are used for livestock feeding.[5,6,9] They are usually available as small chips, (pieces of dried root up to 5 cm long), broken roots (like chips but larger), or pellets made by compressing the powder of dried roots. These goods contain some 70% of starch, and their popularity fluctuates in Europe with the relative cost of barley grains; in South America their price compares favorably with that of imported alfalfa meal. A cassava meal for livestock is also made from the residue left after the extraction of starch or tapioca. Roots can also be stored as silage.

One of the advantages of the cassava plant is that its leaves are a rich source of protein and can be used after detoxification to augment both human and animal foods to give a more balanced diet. Foliage can be cut from 4-month-old plants and thereafter every 2 months to produce 4 tons of crude protein per hectare per annum. Thus, a balanced animal ration can be based upon about 70% of cassava products produced from a small area.[6]

The main industrial outlet for cassava is in the manufacture of starch for use as a foodstuff, textiles, adhesives, or paper. For the human market, starches with a high content of amylopectin and low content of amylose are preferred, but this is not important for livestock. Most cultivars of cassava produce starch low in amylose, which is a disadvantage for many commercial usages.

Industrial ethanol is also made from cassava roots, especially in Brazil where economic circumstances are unusual and there is a government policy to expand the use of ethanol as a motor fuel. The use of sugarcane alcohol is more economical and constitutes the major proportion of current production, but the proportion of cassava alochol is expected to increase following the impetus received from agronomic improvements and better processing technology. Cassava-alcohol distilleries have been set up and production is projected to reach 1 million m³ by 1985 with yields in the order of 180 ℓ of alcohol per ton of root.[10] There are considerable waste products from this process and, as with starch manufacture, there are problems in using them. Without further processing their HCN content makes them difficult to use as a growth medium for microbial protein production.

YAMS

Many species of *Dioscorea*, the yam genus, are used as food plants in tropical countries, but the most important are *D. alata,* the greater yam, which is the most important species in Asia, and *D. rotundata,* the white yam, which is the most important species in Africa along with *D. cayenensis,* the yellow yam. *D. esculenta,* the lesser yam, and *D. trifida,* the cush-cush yam, are also important, but the other species are less so and used mostly as famine food or medicinally.[3,5] The bitter principle present in species like *D. dumetorum* and *D. bulbifer* can be removed by the same methods as used for cassava.

Both *D. alata* and *D. rotundata* belong to the Enantiophyllum section of the genus, whose members are characterized by a single large tuber and only limited formation of subsidiary tubers. The tuber is organographically regarded as a stem tuber, but it has been shown to arise from the hypocotyl in some of the African species.[3]

The greater yam is usually a large individual tuber of from 5 to 10 kg. It is widely used as a staple in Asia but very little is processed. In some countries a small amount is used to produce dehydrated yam flakes or yam powder suitable for blending in food products. Some of the cultivars with colored tubers are used to produce natural coloring or flavoring agents.

The greater yam, like other yams, can be used to produce starch or alcohol but it is usually cheaper to use other species. Similarly, apart from cullings and waste products, it is not common to use yams for stock feed. All yams are costly to grow and these costs are leading to the partial replacement of yams by cassava for human food, even where the preference is for the yam.

The tubers of the white yam are similar to those of the greater yam, but they are preferred in West Africa because of their suitability for the preparation of *fufu*. This is because the starch has large granules (35 to 50 μ) and forms gels of considerable strength and viscosity. Fufu is produced by boiling sliced and peeled tubers until they are soft, and then pounding the resultant dough until it becomes stiff and glutinous and can be cut into slices.

The yellow yam is also widely grown in Africa, possibly because it is high yielding, but it is not suitable for making fufu and is therefore not so popular: this is unfortunate because it is superior in respect of proteins and vitamins. The lesser yam is also high yielding but unsuited for making fufu and used only as a vegetable.

The cush-cush yam has excellent flavor and is one of the more popular yams grown in the Caribbean and northern parts of South America. The Chinese yam is grown mostly in Japan, where the custom is to use yams largely to produce raw material for the preparation of a wide range of food products.[5]

SWEET POTATOES

Sweet potatoes are essentially fleshy enlargements of adventitious roots; a single plant may produce 40 to 50 of them. Cultivars can be conveniently classified into three groups: one where the flesh is dry and mealy when cooked, one where the flesh is soft and watery when cooked because of a tendency to convert much of the starch to sugar, and one with a coarse flesh suitable only for stock feed or processing. They are essentially a second-world crop and their primary use is as a vegetable for human use. Where they are eaten fresh, the white-fleshed types are preferred. In India and parts of Africa they may be peeled, sliced, and sun-dried to produce a flour, and in the U.S. a high proportion are canned, frozen, or used to produce dehydrated flakes for use in a variety of food products. The yellow-fleshed types have a high carotene content and are preferred for canning, for which whole or sliced tubers or puree may be used.

Large quantities of sweet potatoes, mostly culls, are used as livestock feed, particularly in the U.S. Outbreaks of poisoning sometimes occur if moldy potatoes are included in the feed, due apparently to the presence of the metabolites ipomeanorone and isomeamaronol, which have been isolated from tubers showing only minor blemishes.

Commercial starch is an important secondary product, though it is regarded as of low grade, partly because of its color. A high proportion of the Japanese production is used in this way in a cottage-based industry. However, production costs are high and maize starch is a better commercial prospect.

Sweet-potato flour made from peeled, sliced tubers is used as an extender of wheat flour in much the same way as described for cassava flour.

Among the waste products, the vines are used as a green vegetable or for livestock feed, often as silage, and proteins are obtained from the skins and other residues.[5]

COCOYAMS

Cocoyam is the name given to several aroid species including *Colocassia esculenta*, also known as taros, and species of *Xanthosoma*, also known as tannias. Their overall importance is minor compared to the other tropical starch crops, but they are of considerable local importance, especially in West Africa, because they are adapted to varied conditions, from tropical wet swamps to dry uplands. There are two main types of taro: the eddoes, which have a relatively small corm surrounded by large well-developed cormels, and the dasheens, which have large central corms and fewer, smaller, and more compactly clustered cormels. These two types are often given specific rank, but it is preferable to regard *C. esculenta* as a polymorphic species and to denote the eddoe and dasheen variants as *C. esculenta* var. *antiquorum* and *C. esculenta* var. *esculenta*, respectively.

In most areas the corms and cormels are prepared in ways similar to those used for potatoes to give mealy products with a characteristic mealy flavor. The small size of the starch grain makes the food easy to digest, but also makes taro an unsuitable source of industrial starch. An interesting variation in preparation occurs in Hawaii, where grated corms are fermented and combined with coconut to produce *poi*. Taro flour is made in some parts and the rich mucilage of the corms is used in making paper and also medicinally.

Four or more species of *Xanthosoma* (tannias) are grown for food, but it is common to refer to them all as *X. sagittifolium*. They all produce a central corm surrounded by ten or more cylindrical lateral corms. Only the lateral corms are eaten because the main one is usually acrid; yellow-fleshed cultivars are preferred and are eaten in ways similar to taro. They are more nutritious than taro but not so easily digested because their starch grains are larger.

These species are not closely related to either the giant taro or the swamp taro. The former is *Alocasia macrorrhiza* and, unlike most of the edible aroids, is grown for its stem, though its underground cormels are sometimes used as well. The swamp taro is *Crytosperma chamiosonis* and is important because it is one of the few crops that flourish on swampy saline soils. It requires 2 to 3 years to produce a reasonably sized tuber and so other crops are normally preferred in other situations. An advantage is that it can be left for 10 years or more; the resultant corm may then weigh 90 kg but would probably be dry and stringy.[5]

OTHER STARCHY FOODS

Many of the other species listed in Table 1 have local importance because they have characteristics which enable them to be grown in special situations. Although the primary product in most instances is a root tuber there are some interesting exceptions. One is the sago palm, which is important on the freshwater swamps and peaty soils of Malaysia. This plant accumulates starch, known as sago, in its trunk and it is estimated that a hectare containing some 130 palms can yield about 24 tons of dry starch a year. The dry matter composition of these trunks closely resembles that of cassava roots and the sago is used for the same purposes, mainly human and animal foods and industrial starches. Palms are not normally harvested until their first fruits are formed 8 years after planting. The trunks are then cut and debarked, and the pith (often 1000 kg per trunk) is then rasped to obtain the starch. The long waiting period before harvest is probably the main reason why this very useful crop is not more widely used.[4]

REFERENCES

1. **Chatelanat, R. P.**, The use of cassava starch in the development of 'Composite Flour' bakery products, *Proc. 2nd Symp. Int. Soc. Tropical Root Crops, 1970*, 2, 14, 1970.
2. **Christiansen, J. A. and Thompson, N. R.**, The utilisation of "bitter" potatoes in the cold tropics of Latin America, *Proc. 4th Symp. Int. Soc. Trop. Root Crops*, 4, 212, 1976.
3. **Coursey, D. G. and Martin, F. W.**, The past and future of the yams as crop plants, *Plant Foods Hum. Nutr.*, 2, 133, 1972.
4. **Flack, M.**, The sago palm: a potential competitor to root crops, *Proc. 3rd Symp. Int. Soc. Trop. Root Crops*, 1973.
5. **Kay, D. E.**, Root crops, *TPI Crop and Product Digests*, Tropical Products Institute, London, 1973.
6. **Nestel, B. and Graham, M., Eds.**, Cassava as Animal Feed, Monograph IDRC-095e, International Development Research Centre, Ottawa, 1977.
7. **Nestel, B. and MacIntrye, R., Eds.**, Chronic Cassava Toxicity, Monograph IDR-010e, International Development Research Centre, Ottawa, 1973.
8. **Nweke, F. I.**, Consumption patterns and their implications for research and production in tropical Africa, Proc. 1st Symp. Int. Soc. Tropical Root Crops — Africa Branch, 1980.
9. **Phillips, T. P.**, Cassava Utilisation and Potential Markets, Monograph IDRC-020e, International Development Research Centre, Ottawa, 1974.
10. **Weber, E. J., Cock, J. H., and Choninard, A., Eds.**, Cassava Harvesting and Processing, Monograph IDRC-114e, International Development Research Centre, Ottawa, 1978.

OILSEED CROPS

J. R. Wilcox and R. C. Leffel*

Oilseed crops are important sources of edible and industrial fats and oils. Production of all oilseeds in 1981 and 1982 exceeded 170 million metric tons (Table 1).[31] Over 40 million metric tons of fats and oils were extracted from these oilseeds (Table 2).[30] Corn is not usually considered an oilseed crop. However, corn oil, produced in the embryo of the seed, is one product of the wet-milling industry and has become an important edible vegetable oil. Oilseed crops produced nearly 75% of the edible and industrial fats and oils consumed and processed in the world. Of the vegetable oils produced, about 90% were consumed as edible oils and 10% used in industrial processes.

Production of various oilseeds varies by country (Table 3).[32] The U.S. produces over half the soybeans in the world and is a major producer of cottonseed, peanuts, sunflowers, and flax. India and China are the leading producers of peanuts and, with the addition of Canada, of rapeseed. Major producers of sunflower seed, in addition to the U.S., are the U.S.S.R. and Argentina. Most of these oilseed crops are grown primarily as a source of oil. Cottonseed could be considered a byproduct of the cotton industry, and the supply of this oilseed is dependent upon the acreage of cotton grown for fiber in the producing countries.

Oilseed constituents are used in a wide variety of products (Table 4). The fleshy fruit may be an important food, such as the olives that are consumed as green or ripe fruits. The dehulled seed may be consumed as a food product with little or no processing as the sunflower seed, peanuts, and coconuts. More commonly the oil is extracted from the embryo, including cotyledons, and is used in salad oils, as cooking oils, in margarines or shortenings, or in industrial products.

Fats and oils make up one of the three major classes of foods, with the others being carbohydrates and proteins.[35] They are a concentrated food material having more than twice the net heat value as the same weight of carbohydrates and proteins. Fats and oils serve as carriers of fat-soluble vitamins and furnish the fatty acids essential for normal growth and development of animals. Besides their direct nutritive value, fats and oils make other foods more appetizing. They are indispensible in practical cooking and baking since many foods cannot be made palatable without fats or oils.[8]

Industrial uses of oils include lubricants and uses as constitutents of paints, varnishes, waxes, plastics, and pharmaceuticals. Linseed and tung oils are commonly used in paints and varnishes because of their rapid drying properties. With current technology, properties of many of the vegetable oils can be altered to provide characteristics needed in industrial products. This has altered the potential industrial uses of specific vegetable oils.

Specific physical characteristics of the major vegetable oils are given in Table 5. The iodine number refers to the number of grams of iodine absorbed by 100 g of fat. Oils having iodine numbers below 100 are considered as nondrying. Iodine numbers from 100 to 130 indicate semidrying oils and those above 130 as drying oils.[13] The melting point is determined by chilling (4 to 10°C) a sample of oil in a sealed capillary tube, then heating the tube in a water bath. The melting point is the temperature at which the sample of oil becomes uniformly clear and transparent.[35] The saponification value is defined as the weight in milligrams of potassium hydroxide needed to completely saponify 1 g of fat. It is inversely related to the average molecular weight of the fat.[35]

Typical oil contents and the fatty-acid composition of the major vegetable oils are given in Table 6. Fatty-acid composition of vegetable oils is the major factor determining the use

---

* Tables follow the text.

of the oil. High-erucic-acid oils, such as rapeseed and crambe, can be used as lubricants in the continuous casting of steel, in formulated lubricants, and in the manufacture of rubber additives.[21] Salad and cooking oils commonly have palmitic, oleic, and linoleic acids as their major constituents. The linolenic acid in soybean, rapeseed, flax, and tung oils is an important constituent of drying oils, such as those used in paints and varnishes. However, linolenic acid in edible soybean oil has been associated with objectionable flavors and poor stability of the oil. Industrial processes have been developed to hydrogenate and deodorize soybean oil to improve flavor and stability.[4]

The fatty-acid contents shown in Table 6 for these vegetable oils should be considered only as typical values, since plant breeders have developed cultivars of oilseeds with very different fatty acid compositions for most of the crops. One notable example is the zero erucic acid strain of rapeseed, *Brassica napus*.[10] As well as having no erucic acid, this strain has higher oleic and linoleic acid contents than those of commonly produced rapeseed. Another example is that of soybean (*Glycine max*), where lines have been developed with altered fatty acid composition of the oil. These include lines with high stearic acid (45.4%),[11] high oleic acid (51%),[37] low linoleic acid (31.9%), and low linolenic acid (3.5%).[12,36] These examples for rapeseed and soybean demonstrate that fatty acid composition of the oil of specific oilseed crops can be genetically altered to increase their usefulness for specific edible and industrial products.

An important byproduct of oilseed crops is the residual meal after the oil is extracted from the fruit or seed. Potential production of 44% equivalent oilseed meal was in excess of 91 million metric tons in 1981—1982 (Table 7).[31] The meal of most oil-seed crops contains an excellent balance of amino acids essential for animal growth and development (Table 8). Therefore, the meal is commonly used as a high-protein feed supplement for livestock and poultry.

Oil and protein constituents are synthesized at different rates and times during oilseed development. Tables 9 through 19 give the fatty-acid composition of the oil, and the nitrogen and amino-acid composition of the seed for various oilseeds at specific stages of development and maturation. Changing environmental conditions during seed development and maturation affect the synthesis of fatty acids and, therefore, their final proportions in the oils of mature seeds. Since not only oil composition but oil content as well are affected by environment, this can affect oil utilization and the value of specific oilseed crops.

Potentially new oilseed crops are being evaluated as alternate sources of energy or as sources of oils with unique characteristics. Jojoba is one example of such a crop (Tables 20 and 21). The oil, or liquid wax, that can be extracted from the seeds is of a type that is difficult to synthesize commercially and can substitute for the complete range of uses of sperm oil.[20]

Oilseed crops will continue to be a major source of both edible and industrial oils. As the supply of fossil fuels decreases or becomes more expensive, oilseed crops will become an increasingly important source of industrial oils. Since most oilseeds supply a major proportion of high-protein feed supplements, they will also play a significant role in maintaining a productive livestock industry.

Table 1
WORLD PRODUCTION OF OILSEEDS: 1979—1980 TO 1981—1982

Oil seed[a]	1979—1980[b]	1980—1981[b]	1981—1982[b]
Soybeans	93,770	80,714	86,221
Cottonseed	25,200	25,570	27,890
Peanuts	16,999	15,896	18,838
Sunflower	15,316	13,067	14,240
Rapeseed	10,084	11,386	12,466
Copra	4,826	4,901	4,848
Flaxseed	2,687	2,082	2,107
Sesame seed	1,820	1,630	1,945
Palm kernel	1,469	1,523	1,871
Safflower seed	1,150	944	984
Castor beans	868	804	836
Total	174,189	158,517	172,246

[a] Data unavailable on world production of corn used specifically for oil.
[b] Values given indicate thousands of metric tonnes.

From Foreign Agriculture Circular: Oil Seeds and Products, Foreign Agricultural Service, U.S. Department of Agriculture, Washington, D.C., Dec., 1982.

Table 2
WORLD PRODUCTION OF FATS AND OILS FROM 1979—1980 TO 1981—1982 (IN THOUSAND METRIC TONS)

Commodity	1979—1980[a]	1980—1981[a]	1981—1982[a]
Edible vegetable oils	32,617	30,137	31,691
Soybean	14,376	12,172	13,154
Sunflower	5,577	4,763	5,075
Rapeseed	3,420	3,863	4,115
Cottonseed	3,209	3,213	3,506
Peanut	3,083	2,802	3,154
Olive	1,473	1,925	1,269
Sesame	610	589	651
Corn	512	518	525
Safflower	357	292	242
Palm oils	8,535	9,150	9,517
Palm	4,702	5,041	5,414
Coconut	3,028	3,283	3,238
Palm kernel	677	696	735
Babussa kernel	128	130	130
Industrial oils	1,431	1,275	1,245
Flaxseed	808	643	622
Castor	375	349	393
Olive residue	134	179	136
Tung	100	90	90
Oiticica	14	14	14
Animal fats	15,014	14,801	14,804
Marine oils	1,317	1,252	1,234
Total — all oils	58,914	56,615	58,491

[a] Values given indicate thousands of metric tonnes.

From United States Department of Agriculture, Agricultural Statistics, U.S. Government Printing Office, Washington, D.C., 1982.

Table 3
AVERAGE AREA, YIELD, AND PRODUCTION OF MAJOR OILSEEDS BY MAIN PRODUCERS: 1977—1978 TO 1981—1982

Country	Area (1000 ha)	Yield (kg/ha)	Production (1000 metric tons)	Country	Area (1000 ha)	Yield (kg/ha)	Production (1000 metric tons)
Soybeans				Sunflower seed			
U.S.	26,409	1,998	52,777	U.S.S.R.	4,411	1,177	5,190
Brazil	8,300	1,517	12,594	Argentina	1,653	938	1,550
China	7,298	1,084	7,909	U.S.	1,507	1,382	2,082
Argentina	1,721	2,045	3,520	Romania	512	1,618	828
Paraguay	370	1,435	531	Bulgaria	242	1,689	408
Others	4,246	1,121	4,758	Others	3,391	1,086	3,682
Total	48,344	1,698	82,090	Total	11,715	1,172	13,740
Cottonseed				Rapeseed			
U.S.S.R.	3,087	1,579	4,874	India	3,810	488	1,861
China	4,866	995	4,839	China	2,845	837	2,380
U.S.	5,306	904	4,797	Canada	2,233	1,182	2,640
India	7,891	337	2,658	Poland	303	1,783	540
Pakistan	2,028	635	1,287	France	323	2,204	712
Others	9,585	736	7,052	Others	1,623	1,415	2,297
Total	32,763	778	25,506	Total	11,137	936	10,429
Peanut				Flaxseed			
India	7,175	845	6,061	India	1,835	243	446
China	2,068	1,410	2,915	Argentina	829	811	673
U.S.	602	2,704	1,627	U.S.S.R.	1,247	187	233
Senegal	1,084	700	759	Canada	615	960	590
Sudan	965	935	902	U.S.	332	775	257
Brazil	265	1,473	390	Others	414	632	261
South Africa	231	1,085	250	Total	5,271	466	2,460
Others	5,217	880	4,592	Total (above crops)	126,837		151,722
Total	17,607	993	17,592				

From Foreign Agricultural Circular: Oil Seeds and Products, Foreign Agricultural Service, U.S. Department of Agriculture, Washington, D.C., Dec., 1982.

Table 4
UTILIZATION OF OILSEEDS AND OILSEED PRODUCTS

Oilseed	Plant part	Component	Products	Ref.
Soybean *Glycine max* L. Merr.	Seed	Whole seed	Full fat soy flours, soy concentrates, vegetable beans, soybean sprouts, snack foods, speciality foods from processed or fermented beans including soy milk, tufu, miso, sufu, tempeh, hamanatto, natto, soy sauce	2
		Oil	Salad oil, salad dressing, margarines, cooking oil, shortening stocks, dairy-product substitutes Lecithin as a food emulsifier, wetting agent, antioxidant Paints, varnishes, resins, and plastics	
		Meal	High-protein livestock and poultry feed, pet foods, soy-protein isolates used in simulated meat products	
Sunflower *Helianthus annuus* L.	Seed	Whole seed	Confectionary, bird feed	8
		Oil	Salad oil, cooking oil, margarines, mayonnaise, lubricants	
		Meal	Livestock and poultry feed	
Rape *Brassica napus*	Seed	Oil	Salad oil, cooking oil, margarine, lubricants, fuel, soaps	8, 21
		Meal	Livestock and poultry feed	
Cotton *Gossypium hirsutum* L.	Seed	Oil	Salad oil, mayonnaise, cooking oil, shortening stock, soap stocks	8
		Meal	Livestock and poultry feeds (raw cottonseed contains gossypol, a pigment toxic to some animal species)	
Peanut *Arachis hypogaea* L.	Seed	Whole seed	Edible whole nuts or whole nut products such as peanut butter	18
		Oil	Salad and cooking oil, shortenings, soaps	
		Meal	Livestock and poultry feed, flakes, grits, flours, high-protein concentrates and isolates	
Olive *Olea europaea*	Fruit	Whole fruit	Edible as green or ripe fruit	8
		Pulp	Virgin olive oil	
		Oil	Salad and cooking oils, canning oil, pharmaceuticals, lubricants, soaps	
		Meal	Livestock feed	
Sesame *Sesamum indicum* L.	Seed	Whole seed	Baked goods, confectionary products	17
		Oil	Salad oil, cooking oil, shortenings, margarine, soaps, pharmaceuticals	
		Meal	High-protein livestock and poultry feed	

Table 4 (continued)
UTILIZATION OF OILSEEDS AND OILSEED PRODUCTS

Oilseed	Plant part	Component	Products	Ref.
Corn *Zea mays* L.	Seed	Oil Meal	Salad or cooking oil, margarines Ingredient in swine, ruminant, and poultry feeds; germ flour	33
Safflower *Carthamus tinctorius*	Seed	Oil Meal	Salad oil, cooking oil, mayonnaise, margarines Paints, varnishes, linoleum, plastics Livestock and poultry feed	8
Palm *Elaeis guineensis*	Fruit	Pulp (oil) Seed or kernel (oil)	Source of oil used as cooking oil in margarine and shortenings Source of palm kernel oil used in margarines, salad oils, cooking oils, confectionary coatings, soaps	8
Coconut *Cocus nucifera* L.	Seed	Oil Copra cake Dessicated coconut Coir (husk) Shell	Margarines, salad oils, cooking oils, dairy-product substitutes, soap, fuel oil, synthetic detergents, lubricating oil additive, corrosion inhibitor, cosmetics Livestock and poultry feed Confectionary, coconut flour Fiber used in mats, rugs, carpets, rope, yarn, insulation; bristle fiber used in brushes Fuel, charcoal, plastics, carved articles	1
Babussa *Arbignya speciosa*	Seed	Oil	Confectionary and baking trade similar to coconut oil, margarine, cooking oil, soap	8
Flax *Linum usitatissimum* L.	Seed	Whole seed Oil Meal	Specialty edible products: cereal and bread flour ingredients Coatings: paint, varnish, lacquers, oil cloth, tarpulins; industrial chemicals: fatty acids, soaps, glycerine; antispalling and curing treatments for concrete Printing inks, core oils, asbestos, binders, shoe polish, pesticide carriers, edible oil (not in U.S.) High-protein animal feed (toxic to poultry in high proportions unless soaked or pyridoxin supplemented)	5
Castor *Ricinus communis*	Seed	Oil	Medicinal and technical purposes; lubricant, dyes, soaps, imitation leather, inks, plasticizer; dehydrated castor oil used in enamels, paints, and varnishes	29
Tung *Aleuritus fordii*	Seed	Oil	Quick-drying industrial oil for paints, varnishes, enamels, lacquers; used in pressed fiber boards, linoleum, and inks	29
Oiticica *Licania rigida*	Seed	Oil	Drying oil used in paints and varnishes	8

Table 4 (continued)
UTILIZATION OF OILSEEDS AND OILSEED PRODUCTS

Oilseed	Plant part	Component	Products	Ref.
Crambe *Crambe abyssinica* Hochst.	Seed	Oil	A high-erucic-acid oil used in formulating lubricants in continuous steel casting; plasticizers, adhesives; wax esters used in commercial waxes	15
		Meal	High-protein livestock and poultry feed (contains glucosinilates associated with unpalatability and goitrogenicity)	
Jojoba *Simmondsia chinensis*	Seed	Oil	An unsaturated liquid wax with potential use in lubricants, paper coatings, polishes, electrical insulation, carbon paper, textiles, leather, precision casting, and pharmaceuticals	20
		Meal	Potential use as livestock feed (meal contains simmondsin, a toxic factor)	

Table 5
CHARACTERISTICS OF SELECTED VEGETABLE OILS[8,13,29,35]

Vegetable oil	Iodine value (no.)	Melting point (°C)	Saponification value (no.)
Soybean	120—141	−23—−20	189—195
Sunflower	125—136	−18—−16	188—194
Rapeseed	81.4	−9	170—180
Cottonseed	96.8—111.6	−2—2	189—198
Peanut	84—100	−2	188—195
Olive	80—88	0—6	188—196
Sesame	106—130	−5—0	188—193
Corn	124.4	−12—−10	187—193
Safflower	143.3	−18—−16	190
Palm	48—56	27—50	196—202
Coconut	7.5—10.5	23—26	250—264
Palm kernel	14—23	24—26	245—255
Babussa kernel	14—16	22—26	247—251
Flaxseed	170—204	—	189—196
Castor	81—91	—	176—187
Tung	157—172	—	189—195
Oiticica	140—152	—	186—195
Crambe	93	61—63	—
Jojoba	82	11—12	92

Table 6
TYPICAL OIL CONTENTS AND FATTY ACID COMPOSITION OF SELECTED VEGETABLE OILS (%)

Composition	Soybean[14]	Sunflower[25]	Rapeseed[3]	Cottonseed[35]	Peanut[38]	Olive[8]	Sesame[9]	Corn[23]	Safflower[15]
Oil content	19.3	46.3	38.0	19.5	52.6	27.0	46.4	5.0	30.0
Fatty acid									
Caprylic $C_8H_{16}O_2$	—	—	—	—	—	—	—	—	—
Capric $C_{10}H_{20}O_2$	—	—	—	—	—	—	—	—	—
Lauric $C_{12}H_{24}O_2$	—	—	—	—	—	—	—	—	—
Myristic $C_{14}H_{28}O_2$	—	—	—	1.0	—	—	—	—	—
Palmitic $C_{16}H_{32}O_2$	12.0	5.3	3.2	25.0	11.0	10.2	9.7	11.1	6.7
Eleostearic	—	—	—	—	—	—	—	—	—
Stearic $C_{18}H_{36}O_2$	3.2	3.9	1.5	2.8	3.1	2.2	4.8	2.0	2.7
Oleic $C_{18}H_{34}O_2$	20.9	30.9	23.9	17.1	46.9	79.1	41.2	24.1	12.9
Ricinoleic	—	—	—	—	—	—	—	—	—
Linoleic $C_{18}H_{34}O_3$	55.2	59.3	15.6	52.7	31.5	7.6	44.4	61.9	77.5
Linolenic $C_{18}H_{32}O_2$	8.5	—	8.8	—	—	—	—	0.7	—
Licanic $C_{18}H_{30}O_2$	—	—	—	—	—	—	—	—	—
Arachidic $C_{18}H_{30}O_3$	—	—	—	—	1.4	—	—	—	—
Eicosenoic $C_{20}H_{40}O_2$	—	—	11.8	—	1.2	—	—	—	—
Behenic $C_{20}H_{38}O_2$	—	—	—	—	2.6	—	—	—	—
Erucic $C_{22}H_{44}O_2$	—	—	34.6	—	—	—	—	—	—
Lignoceric $C_{22}H_{42}O_2$	—	—	—	—	1.0	—	—	—	—
Tetracosenoic $C_{24}H_{48}O_2$	—	—	—	—	—	—	—	—	—
$C_{24}H_{46}O_2$	—	—	—	—	—	—	—	—	—

Table 6 (continued)
TYPICAL OIL CONTENTS AND FATTY ACID COMPOSITION OF SELECTED VEGETABLE OILS (%)

Composition	Palm[35]	Coconut[1]	Babussa[13]	Palm kernel[35]	Flax[5]	Castor[29]	Tung[29]	Oiticica[13]	Crambe[15]	Jojoba[19]
Oil content	50.0	64.5	—	49.5	36.0	50.0	17.5	60.0	32.0	47.5
Fatty acid										
Caprylic $C_8H_{16}O_2$	—	7.5	6.5	1.4	—	—	—	—	—	—
Capric $C_{10}H_{20}O_2$	—	7.4	2.7	2.9	—	—	—	—	—	—
Lauric $C_{12}H_{24}O_2$	—	47.6	45.8	50.9	—	—	—	—	—	—
Myristic $C_{14}H_{28}O_2$	1.2	15.8	19.9	18.4	—	—	—	—	—	—
Palmitic $C_{16}H_{32}O_2$	46.8	9.0	6.9	8.7	6.5	2.0	4.0	7.0	2.0	1.0
Eleostearic $C_{18}H_{30}O_2$	—	—	—	—	—	—	77.0	—	—	—
Stearic $C_{18}H_{36}O_2$	3.8	2.1	—	1.9	3.5	1.0	1.0	5.0	—	—
Oleic $C_{18}H_{34}O_2$	37.6	6.6	18.1	14.6	17.0	8.0	6.0	6.0	15.0	10.0
Ricinoleic $C_{18}H_{34}O_3$	—	—	—	—	—	85.0	—	—	—	—
Linoleic $C_{18}H_{32}O_2$	10.0	1.8	—	1.2	21.0	3.0	9.0	—	10.0	—
Linolenic $C_{18}H_{30}O_2$	—	—	—	—	51.5	—	3.0	—	—	—
Licanic $C_{18}H_{30}O_3$	—	—	—	—	—	—	—	78.0	—	—
Arachidic $C_{20}H_{40}O_2$	—	—	—	—	—	—	—	—	—	—
Eicosenoic $C_{20}H_{38}O_2$	—	—	—	—	—	—	—	—	7.0	70.0
Behenic $C_{22}H_{44}O_2$	—	—	—	—	—	—	—	—	2.0	—
Erucic $C_{22}H_{42}O_2$	—	—	—	—	—	—	—	—	58.0	13.0
Lignoceric $C_{24}H_{48}O_2$	—	—	—	—	—	—	—	—	—	—
Tetracosenoic $C_{24}H_{46}O_2$	—	—	—	—	—	—	—	—	3.0	2.0

Table 7
WORLD POTENTIAL PRODUCTION OF
44% PROTEIN MEAL EQUIVALENT
CALCULATED FROM ASSUMED EXTRACTION
RATES

Meal	1979—1980	1980—1981	1981—1982
Soybeans	65,687	55,815	59,376
Cottonseed	7,448	7,457	8,239
Sunflower seed	4,937	4,239	4,626
Peanut	4,141	3,728	4,447
Rapeseed	3,753	4,245	4,672
Linseed	1,138	882	894
Copra	763	774	766
Sesame seed	645	569	692
Safflower seed	433	356	370
Palm kernel	267	277	340
Fish	6,957	6,911	6,898
Total	96,168	85,252	91,320

[a] Values are given in thousands of metric tonnes.
From Foreign Agricultural Circular: Oil Seeds and Products, Foreign Agricultural Service, United States Department of Agriculture, Washington, D.C., Dec., 1982.

Table 8
CRUDE PROTEIN AND AMINO-ACID CONTENT OF FARM FEEDS DERIVED FROM OIL-SEED MEALS (%)

Product	Crude protein	Arginine[a]	Histidine[a]	Isoleucine[a]	Leucine[a]	Lysine[a]	Methionine[a]	Phenylalamine[a]	Threonine[a]	Tryptophan[a]	Valine[a]
Soybean meal	45.29	7.46	2.49	5.50	7.69	6.17	1.39	4.86	4.03	1.69	5.40
Sunflower seed meal	21.02	7.76	2.19	4.52	5.95	3.81	2.19	5.12	3.43	1.38	4.90
Cottonseed meal	39.61	11.02	2.70	4.01	6.20	4.20	1.49	5.25	3.47	1.59	4.98
Peanut meal	38.95	10.33	2.16	4.33	6.68	3.53	1.04	4.97	2.98	1.22	4.82
Sesame meal	46.08	11.91	2.21	4.27	6.92	2.76	2.65	4.73	3.64	1.91	5.06
Corn oil cake meal	22.69	6.65	3.44	4.23	8.68	5.24	1.90	4.19	4.14	1.28	6.61
Safflower meal	22.10	7.78	1.99	3.85	5.52	2.71	1.54	5.25	2.94	1.18	4.93
Copra meal	22.22	11.43	1.71	4.00	6.30	3.06	1.59	4.23	3.78	0.94	5.49
Palm kernel meal	19.16	13.26	1.62	3.97	6.42	3.44	2.14	4.28	3.13	1.04	5.38
Babussa meal	22.72	14.04	1.81	3.87	6.17	4.31	2.33	5.94	3.13	1.06	5.24
Linseed meal	36.73	9.28	1.80	4.57	5.96	3.62	1.66	4.49	3.78	1.74	5.55
Castor pomace	39.79	10.03	1.68	4.65	5.65	3.02	1.46	4.68	3.24	1.11	5.40
Tung meal	20.88	10.87	2.30	4.93	7.57	4.17	2.24	7.28	4.60	1.68	8.33

From Lessman, K. J. and Anderson, W. P., in *New Sources of Fats and Oils*, Pryde, E. H., Princin, L. H., and Mukherjee, K. D., Eds., American Oil Chemists Society, Champaign, Ill., 1981. With permission.

Table 9
OIL AND FATTY ACID COMPOSITION OF 'HAROSOY 63' SOYBEAN SEED AT VARIOUS STAGES OF DEVELOPMENT

	Seed					Oil composition				
Days after flowering	Fresh wt (mg/seed)	Dry wt (mg/seed)	Moisture (%)	Oil (mg/seed)	Dry (%)	Palmitic (%)	Stearic (%)	Oleic (%)	Linoleic (%)	Linolenic (%)
24	5.5	0.9	83.0	0.03	3.5	19.0	8.2	7.5	35.0	30.0
25	14.8	2.3	84.5	0.1	5.3	19.6	6.5	10.1	34.6	29.1
26	35.9	5.5	84.5	0.4	7.9	19.0	5.2	11.3	39.1	25.4
30	84.0	15.5	81.5	1.9	12.3	16.1	4.0	21.3	44.8	13.8
32	146.0	30.8	78.9	4.8	15.5	14.5	3.2	24.0	46.2	12.1
37	224.5	57.5	74.4	9.9	17.2	12.0	2.9	25.6	49.4	10.1
39	214.1	64.6	69.8	12.6	19.5	10.1	2.8	26.3	51.0	9.2
40	225.3	70.3	68.8	14.1	20.1	10.8	3.2	25.8	51.4	9.1
43	272.4	82.5	69.5	16.6	20.2	10.6	3.1	26.2	52.0	8.6
45	295.5	95.8	67.6	17.7	19.0	10.7	3.2	25.9	51.8	8.0
46	324.0	100.5	69.0	20.7	20.1	10.6	3.2	25.8	52.7	7.7
50	328.8	119.3	63.7	24.3	20.4	10.0	2.8	28.3	52.2	6.8
53	366.6	134.6	63.3	27.4	20.4	10.4	2.8	25.4	53.8	6.8
57	385.1	154.1	60.0	31.7	20.6	10.1	3.8	26.3	54.8	5.1
58	442.7	162.1	63.4	33.6	20.7	10.2	3.4	27.5	52.2	5.9
63	452.0	179.0	60.4	39.9	22.3	10.4	3.1	26.1	54.1	6.3
64	450.8	187.0	58.5	44.9	24.0	10.4	2.6	26.3	54.8	5.9
67	397.4	186.2	53.1	44.3	23.8	10.4	2.9	26.1	54.7	5.9
72	311.8	186.6	40.2	44.0	23.6	10.4	2.9	26.4	54.1	5.8

Reproduced from *Crop Science*, Volume 12, 1972, pages 739—741 by permission of the Crop Science Society of America.

Table 10
NITROGEN COMPOSITION AND PROTEIN CONTENT OF 'ACME' SOYBEAN SEED FROM NODE 1 DURING VARIOUS STAGES OF DEVELOPMENT

	Seed			Nitrogen fraction			
Days after flowering	Fresh wt (mg/seed)	Dry wt (mg/seed)	Moisture (%)	Nonprotein nitrogen (%)	Protein nitrogen (%)	Protein (%)	Dry (mg/seed)
25	22.2	3.8	83.0	2.48	5.15	32.2	1.2
27	52.1	9.2	82.4	2.30	4.86	30.4	2.8
28	52.0	10.7	79.4	1.80	4.82	30.1	3.2
29	73.5	13.2	82.0	1.91	5.07	31.6	4.2
30	88.4	16.2	81.6	1.55	5.12	32.0	5.2
30	88.9	16.7	81.2	1.41	5.01	31.1	5.2
31	109.0	21.1	80.7	1.30	4.92	31.0	6.5
32	124.2	27.3	78.0	1.16	5.08	31.8	8.7
32	134.9	29.6	78.0	1.11	4.88	30.6	9.1
33	122.2	31.2	74.5	1.10	5.16	37.2	11.6
35	175.3	37.8	78.4	1.09	5.23	32.6	12.3
36	159.8	39.6	75.2	1.17	5.31	33.2	13.1
37	198.5	54.5	72.5	0.88	5.26	32.9	17.9
38	227.0	61.3	73.0	0.84	5.63	35.2	21.6
41	266.3	75.7	71.6	0.76	5.47	34.1	25.8
42	271.4	77.4	71.5	0.54	5.36	33.5	25.9
45	289.7	102.9	64.5	0.52	5.19	32.4	33.3
47	310.8	112.3	63.9	0.41	5.47	34.2	38.4
50	370.8	122.2	67.0	0.50	5.57	34.8	42.5
52	352.3	131.6	62.6	0.38	5.54	34.6	45.5
60	359.4	165.6	53.9	0.38	5.77	36.0	59.6
74	238.5	200.6	15.9	0.29	5.60	35.0	70.2

Reproduced from *Crop Science*, Volume 12, 1972, pages 739—741 by permission of the Crop Science Society of America.

Table 11
MEAN VALUES[a] AND STANDARD ERROR OF THE MEAN FOR AMINO ACIDS FROM 'HAROSOY 63' SOYBEAN SEED

	\multicolumn{7}{c	}{Days after flowering}	
	10	20	30	40	50	60	70	$S_{\bar{x}}$
Lysine	6.7	5.4	6.0	6.1	6.6	6.3	6.7	0.3
Histidine	5.7	3.1	2.5	2.2	2.6	2.7	2.7	0.1
Arginine	6.0	8.5	8.5	8.2	8.4	8.3	8.3	0.4
Aspartic	10.8	14.2	12.3	11.9	12.0	12.1	13.6	0.4
Threonine	4.8	4.3	4.9	4.4	4.4	3.9	5.1	0.3
Serine	5.4	6.0	5.5	5.8	6.0	6.1	6.0	0.1
Glutamic	17.8	14.9	18.0	19.4	20.0	19.9	20.2	0.9
Proline	4.2	3.9	4.7	4.5	5.6	3.8	3.8	0.4
Glycine	4.0	4.1	4.3	4.1	4.3	4.5	4.3	0.1
Alanine	6.0	6.3	5.7	5.6	4.6	4.5	4.6	0.3
Cystine	0.7	1.2	1.0	1.1	1.4	1.2	0.9	0.2
Valine	4.7	4.9	4.9	4.9	5.0	5.6	5.3	0.1
Methionine	1.6	1.5	1.2	1.3	1.2	1.2	1.3	0.1
Isoleucine	3.8	4.0	4.9	4.8	4.6	4.5	4.6	0.2
Leucine	6.3	6.7	7.0	7.2	7.3	7.8	7.3	0.3
Tryrosine	3.9	3.6	4.1	3.6	3.8	4.0	4.0	0.4
Phenylaine	4.2	5.3	4.6	4.6	4.8	5.3	5.1	0.3
Unknown	3.4	3.8	3.3	2.4	1.4	0.3	0.5	0.4

[a] Mean values given are in mg/16 mg N.

Reproduced from *Agronomy Journal,* Volume 69, 1977, pages 481—486, by permission of the American Society of Agronomy.

Table 12
EFFECT OF STAGE OF MATURITY ON OIL AND FATTY-ACID COMPOSITION OF 'SUN GRO 380' SUNFLOWER SEED

Days after flowering	Moisture (%)	Total oil (%; dry basis)	Dry wt (g/100 seed)	\multicolumn{9}{c	}{Fatty-acid composition (%)}
				16:0	16:1	18:0	18:1	18:2	18:3	20:0	20:1	22:0
7	83.0	1.6	1.3	20.8	0.5	3.7	11.7	47.8	10.7	1.5	—	0.8
14	76.7	16.6	2.4	7.5	0.1	7.7	59.6	22.6	0.8	0.8	Tr[a]	0.9
21	64.3	38.9	3.1	5.7	Tr	5.0	51.7	36.1	0.1	0.4	0.1	0.9
28	47.9	50.6	4.8	5.4	Tr	3.7	43.3	46.4	Tr	0.3	0.2	0.7
35	36.2	53.5	5.6	5.0	Tr	3.3	38.5	52.1	Tr	0.3	Tr	0.6
42	21.7	53.3	5.5	5.0	Tr	3.2	37.3	53.7	Tr	0.2	Tr	0.6
49	14.6	53.2	5.5	5.3	Tr	3.1	32.7	57.5	Tr	0.5	0.2	0.7
56	9.6	53.6	5.1	5.8	Tr	2.9	31.4	59.2	Tr	0.2	Tr	0.4
63	8.2	53.9	5.0	5.7	Tr	3.1	32.3	57.9	Tr	0.4	Tr	0.5
70	8.9	53.5	5.4	5.2	Tr	2.7	37.2	53.8	Tr	0.3	0.1	0.7

[a] Tr, trace; less than 0.1%.

From Robertson, J. A., Chapman, G. W., Jr., and Wilson, R. L., *J. Am. Oil Chem. Soc.*, 55, 266, 1978. With permission.

Table 13
FATTY-ACID CONTENT OF *BRASSICA NAPUS* FIELD-PRODUCED SEED IN PERCENT OF TOTAL FATTY ACIDS AT SIX STAGES OF DEVELOPMENT

Fatty acid	Zero-erucic-acid strain A (days after pollination)						Nugget rapeseed (days after pollination)					
	7	14	21	28	35	42[a]	7	14	21	28	35	42[a]
Palmitic	13.6	14.6	7.1	4.9	4.4	4.1	15.6	15.4	8.0	4.8	3.7	3.8
Palmitaleic	1.6	1.6	0.9	0.6	0.4	0.7	0.7	1.2	1.1	0.5	0.6	0.5
Stearic	9.0	7.2	4.2	2.0	2.0	1.5	9.3	7.4	4.6	2.3	2.1	1.8
Oleic	2.2	8.9	53.5	61.1	63.8	63.5	2.9	5.0	38.3	28.1	25.4	25.2
Linoleic	48.6	48.5	23.8	21.1	19.1	20.0	48.5	49.5	23.4	15.9	14.2	13.5
Linolenic	22.3	17.4	7.6	7.8	7.7	7.8	20.2	19.1	8.2	7.3	6.4	6.0
Arachidic	2.7	1.8	1.8	1.2	1.3	1.1	2.8	2.4	1.7	1.6	1.0	1.3
Eicosenoic	0.0	0.0	1.1	1.3	1.3	1.3	0.0	tr	8.4	14.5	13.6	14.4
Erucic	0.0	0.0	0.0	0.0	0.0	0.0	0.0	0.0	6.3	25.0	33.0	33.5

[a] Mature

Reproduced from Fowler, D. B. and Downy, R. K., *Can. J. Plant Sci.*, 50, 233, 1970. With permission.

Table 14
PEANUT CHARACTERISTICS USED TO ESTIMATE KERNEL AGE

Stage	Pericarp	Kernel
4	Very watery, soft, spongy	Very small, flattened, completely white, mostly seed coat
5	Still soft, not as watery, inner pericarp fleshy — no cracks	Larger than 4 weeks, flat; white or just turning pink at one end
6	Inner pericarp tissue beginning to show cracks	Torpedo-shaped; generally pink at embryonic-axis end of kernels
7	Inner pericarp beginning "cottony" appearance	Torpedo- to round-shaped; embryonic-axis end of kernel pink; other end white to light pink
8	Inner pericarp beginning to dry out — cracks more numerous	Round, light pink all over
9	Inner pericarp white but beginning to show brown splotches	Dark pink at embryonic axis end, light to dark pink elsewhere
10	Many dark-brown splotches on inner pericarp	Large, generally dark pink all over; seed coat beginning to dry out
11	Inner pericarp almost completely brown	Dark pink, may show imprint of pericarp on seed coat in places; seed coat drying out
12	Black splotches appearing on inner pericarp	Dark pink, may show imprint of pericarp on seed coat in places; seed coat drying out

From Pattee, H. E., Jones, E. B., Singleton, J. A., and Saunders, T. A., *Peanut Sci.*, 1, 57, 1974. With permission.

Table 15
INFLUENCE OF MATURITY ON QUANTITY AND COMPOSITION OF PETROLEUM ETHER-EXTRACTED OIL FROM 'FLORUNNER' PEANUTS

Maturity stage	Dry weight per seed (g)	Oil (% dry weight)	Hydrocarbon sterol ester[a]	Triacyl-glycerol[a]	Free fatty acid[a]	Diacyl-glycerol[a]	Sterol[a]	Monoacyl-glycerol	Polar lipid[a]
5	0.03	25.3	0.6	85.3	4.5	4.7	2.4	0.5	2.0
6	0.08	30.8	0.4	89.3	3.1	3.5	1.8	0.6	1.4
7	0.15	34.4	0.5	88.3	2.5	3.6	1.3	0.8	1.9
8	0.32	42.8	0.6	90.8	1.8	3.0	1.4	0.8	1.6
9	0.45	45.6	0.6	92.6	1.3	2.2	1.0	0.9	1.3
10	0.57	46.7	0.3	94.3	0.9	2.0	0.8	0.6	1.0
11	0.59	48.4	0.4	94.8	0.7	1.9	0.9	0.5	0.7
12	0.59	48.2	0.4	95.8	0.7	1.7	0.5	0.3	0.6

[a] Relative weight percent. Each value is the mean of three replications.

From Sanders, T. H., *J. Am. Oil Chem. Soc.*, 57, 8, 1980. With permission.

Table 16
EFFECT OF MATURITY ON FATTY-ACID COMPOSITION OF TRIACYLGLYCEROL, COMPOSITION OF FREE FATTY ACIDS, AND COMPOSITION OF SN-1,3-DIACYLGLYCEROLS IN 'FLORUNNER' PEANUT OIL

Maturity stage	16:0	18:0	18:1	18:2	20:0	20:1	22:0	24:0
\multicolumn{9}{c}{Fatty-acid composition of triacylglycerol[a]}
6	16.6	2.3	41.7	28.6	1.3	1.8	5.9	1.6
7	14.4	2.0	44.8	28.6	1.4	1.9	5.2	1.5
8	13.0	1.9	45.9	31.1	1.2	1.7	3.5	1.6
9	12.4	2.0	47.7	31.1	1.2	1.4	2.6	1.6
10	12.4	2.0	49.4	29.3	1.1	1.3	2.4	1.7
11	12.6	2.0	50.3	29.2	1.1	1.2	2.3	1.3
12	13.2	2.3	50.7	28.3	1.3	1.1	2.1	1.1
\multicolumn{9}{c}{Composition of free fatty acids[a]}
6	17.5	1.9	46.4	29.2	0.7	2.9	1.0	0.4
7	14.3	1.9	55.0	24.1	0.7	2.8	0.7	0.4
8	13.0	2.0	51.8	28.4	0.8	2.5	0.9	0.6
9	12.3	2.0	52.0	29.1	0.9	2.1	0.9	0.6
10	12.7	2.2	52.1	28.2	1.0	1.8	1.2	0.8
11	12.5	2.1	52.6	28.1	1.0	1.8	1.1	0.7
12	13.3	2.5	52.6	27.2	1.1	1.5	1.1	0.6
\multicolumn{9}{c}{Fatty-acid composition of sn-1,3-diacylglycerols[a]}
6	8.9	3.1	46.1	25.6	2.9	5.6	6.8	0.9
7	10.2	2.1	54.4	24.9	1.4	3.7	2.2	0.9
8	9.7	2.4	52.5	26.4	1.7	4.3	2.2	0.9
9	10.6	1.9	53.4	27.5	1.5	2.9	1.4	0.8
10	11.7	2.1	55.4	25.3	1.0	2.0	1.7	0.9
11	12.4	2.1	54.9	24.7	1.1	2.0	1.8	1.0
12	12.5	2.2	55.4	25.1	1.1	1.6	1.7	1.1

[a] Values given are in mol %.

From Sanders, T. H., *J. Am. Oil Chem. Soc.*, 57, 12, 1980. With permission.

Table 17
ACCUMULATION OF OIL AND FATTY ACIDS IN DEVELOPING KERNELS OF HIGH (IHO), MEDIUM (H51), AND LOW (K6) OIL-PRODUCING CORN LINES

Strain	Days after pollination	100 kernels Wet wt (g)	100 kernels Dry wt (g)	Oil (%)	Triglycerides Total	16:0	18:0	18:1	18:2	18:3	Polar lipids Total	16:0	18:0	18:1	18:2	18:3
IHO	10	8.3	1.1	3.0	10.1	19.1	2.1	10.9	51.7	16.2	70.8	27.0	1.5	8.7	49.8	13.0
	15	15.6	2.5	5.6	41.1	14.0	1.7	35.7	46.2	2.4	45.1	24.5	1.3	8.5	58.0	7.7
	30	26.7	11.5	10.9	78.4	13.9	1.3	40.4	44.1	0.3	9.6	22.1	0.9	12.1	59.7	5.2
	45	31.6	18.0	13.7	84.0	12.8	1.7	35.2	49.7	0.6	6.0	20.6	1.2	12.9	60.9	4.4
	60	30.6	19.0	13.8	88.1	14.2	1.7	35.2	48.5	0.4	4.5	22.7	1.8	15.9	55.2	4.4
	75	33.4	23.4	13.4	92.0	11.5	1.3	32.0	54.2	1.0	3.9	21.4	1.4	23.3	50.5	3.4
	90	31.8	23.8	13.8	92.4	12.6	2.1	36.1	48.6	0.6	3.9	21.5	1.3	28.4	46.9	1.9
H51	10	6.6	0.8	3.3	13.0	20.4	2.0	19.6	50.0	8.0	69.7	25.4	1.3	5.4	57.2	10.7
	20	9.7	1.5	3.9	24.0	20.3	2.1	23.7	49.6	4.3	59.3	24.8	0.9	5.2	60.4	8.7
	30	30.5	11.9	5.6	68.8	19.1	1.7	31.7	46.5	1.0	18.7	23.8	1.2	11.4	58.9	4.7
	45	28.7	15.5	6.1	75.4	17.9	1.5	31.7	47.9	1.0	10.0	20.0	0.8	12.0	61.0	6.2
	60	19.3	16.0	5.3	79.0	17.7	1.7	30.8	48.6	1.2	8.7	22.4	1.2	26.6	46.7	3.1
	75	22.4	19.9	5.7	80.8	17.4	1.4	31.9	48.3	1.0	6.9	20.3	1.4	28.3	47.6	2.4
	85	23.3	21.9	5.2	80.7	17.6	1.5	31.1	48.2	1.6	8.1	21.2	1.5	27.6	46.9	2.8
K6	10	6.0	0.7	3.4	17.3	15.0	1.7	10.5	58.2	14.6	72.4	27.4	1.0	8.7	52.1	10.8
	15	10.0	1.7	3.6	27.8	14.9	1.2	19.8	57.6	6.5	63.1	25.5	1.5	9.5	55.2	8.3
	30	26.9	11.0	2.9	56.3	12.7	1.0	28.1	56.2	2.0	33.1	24.5	1.3	9.2	58.6	6.4
	45	31.1	17.7	3.9	74.4	11.7	1.2	24.5	60.7	1.9	13.9	20.2	1.2	10.8	59.4	8.2
	60	30.7	20.3	2.4	75.8	11.7	1.4	21.4	63.4	2.1	13.0	20.2	1.7	11.5	58.9	7.7
	75	21.2	16.5	2.6	74.9	12.5	1.1	21.2	63.1	2.1	12.0	24.8	1.5	18.8	51.6	3.3

From Weber, E. J., *J. Am. Oil Chem. Soc.*, 46, 485, 1969. With permission.

Table 18
AMINO ACID CONTENT OF OIL-FREE MEAL FROM CRAMBE SEED AND SEED PARTS (g/16 g N)

Amino acid	Pericarp	Seed	Seedcoat	Hypocotyl	Cotyledon
Lysine	1.5	5.1	7.4	5.6	5.5
Methionine	0.3	1.6	1.2	1.8	1.5
Arginine	1.3	5.7	4.4	6.5	6.1
Glycine	1.8	5.2	5.0	5.8	5.1
Histidine	0.4	2.4	2.0	2.5	2.9
Isoleucine	1.4	3.7	3.7	3.9	3.4
Leucine	2.4	5.9	5.1	6.5	5.1
Phenylalanine	1.5	3.4	3.2	4.5	3.4
Tyrosine	0.7	3.0	3.2	2.7	2.5
Threonine	1.7	4.2	5.0	4.0	4.9
Valine	1.8	4.5	5.0	4.7	3.8
Alanine	1.7	4.0	4.0	4.4	3.6
Aspartic acid	3.1	6.0	5.9	6.2	6.8
Glutamic acid	3.6	14.2	11.2	16.7	14.8
Hydroxyproline	1.1	0.9	5.9	0.0	0.0
Proline	1.7	5.5	6.7	5.7	5.7
Sorine	1.9	3.5	4.2	3.5	3.5
Nitrogen as amino acids (% of N)	23	74	72	76	70
Nitrogen as ammonia (% of N)	26	13	14	15	15

From Earle, F. R., Peters, J. E., Wolff, I. A., and White, G. A., *J. Am. Oil Chem. Soc.*, 43, 330, 1966. With permission.

Table 19
CHANGES IN FATTY ACID COMPOSITION OF MATURE SEEDS OF C.I. 1303 FLAX AS INFLUENCED BY TEMPERATURE

Temp. (°C)	Seed age (days)	Palmitic	Oleic	Linoleic	Linolenic
15	9	13.2a	23.2a	19.5a	41.7a
	16	11.8b	23.5a	21.3a	41.0a
	23	8.1c	27.0b	14.1b	48.5b
	37	6.5d	28.0b	14.6b	48.7b
20	9	13.6a	24.9a	20.6a	38.6a
	16	11.0b	29.7b	17.9b	39.0a
	23	7.6c	35.2c	10.5c	43.9b
	37	6.8d	35.7c	11.5c	43.7b
25	9	13.9a	27.6a	20.1a	36.3a
	16	10.0b	35.2b	12.9b	39.6b
	23	7.8c	37.7b	10.1c	42.2b
	37	7.6c	39.9b	9.9c	40.5b
30	9	13.1a	41.8a	17.0a	25.5a
	16	9.0b	46.6b	11.3b	30.4b
	23	8.3c	47.1b	11.0b	31.0b
	37	8.3c	47.4b	10.6b	30.9b

Fatty acid composition (% of total fatty acids)

Note: When values for individual fatty acids at the four seed ages within a temperature group are arranged in order of magnitude, adjacent means followed by a common letter do not differ significantly at the 5% level. Stearic acid content averaged 2.4% and was unaffected by temperature and age.

Table 20
WAX CONTENT AND PROTEIN CONTENT OF WAX-FREE MEAL AND MOISTURE CONTENT AND SEED WEIGHT (g) OF JOJOBA SEED[a,b]

Harvest dates	Wax (%)	Protein (%)	Moisture (%)	Seed dry wt
June 20	13.5	22.3	71.8	0.06
June 27	20.4	24.3	65.9	0.08
July 4	27.8	25.7	60.3	0.14
July 11	40.5	29.0	56.2	0.25
July 18	43.6	29.8	45.6	0.25
July 25	47.6	30.8	41.3	0.26
August 3	49.8	31.6	24.6	0.29
August 15	49.4	32.6	10.3	0.34

[a] Seeds harvested in Aguanga, Calif., on different dates. Data represent means from 15 single plant seed samples.
[b] Data on percent of wax and percent of protein are given on a moisture-free basis.

From Yermanos, D. M., J. Am. Oil Chem. Soc., 52, 115, 1975. With permission.

Table 21
GAS-CHROMATOGRAPHIC COMPOSITION OF EXPELLER-PRESSED JOJOBA OIL

Wax esters	%	Free alcohols	%	Sterols	%
C-33	0.02	C-16	0.01	Campesterol	0.05
C-34	0.08	C-18	0.04	Stigmasterol	0.08
C-35	0.04	C-20	0.49	Sitosterol	0.21
C-36	1.16	C-22	0.49	Others	0.52
C-37	0.02	C-24	0.07		
C-38	6.23	C-26	0.01		
C-39	0.04		1.11		
C-40	30.56				
C-41	0.10				
C-42	49.50	Free acids	%		
C-43	0.06	C-16	0.08		
C-44	8.12	C-18	0.23		
C-45	0.03	C-19	0.01		
C-46	0.86	C-20	0.60		
C-48	0.16	C-21	0.03		
C-50	0.06	C-22	0.03		
	97.05	C-24	0.02		
			1.00		

From *Products from Jojoba: A Promising New Crop for Arid Lands*, National Academy of Sciences, Washington, D.C., 1975, 14.

REFERENCES

1. **Child, R.,** *Coconuts,* 2nd ed., Longman Group Ltd., London, 1974.
2. **Cowan, J. C.,** Processing and products, in *Soybeans: Improvement, Production, and Uses,* Caldwell, B. E., Ed., American Society of Agronomy, Madison, 1973.
3. **Craig, B. M.,** Varietal and environmental effects on rapeseed. III. Fatty acid composition of 1958 varietal tests, *Can. J. Plant Sci.,* 41, 204, 1971.
4. **Dutton, H. J., Lancaster, C. R., Evans, C. D., and Cowan, J. C.,** The flavor problem of soybean oil. VIII. Linolenic acid, *J. Am. Oil Chem. Soc.,* 28, 115, 1951.
5. **Dybing, C. D. and Lay, C.,** Flax: *Linum usitatissimum,* in *CRC Handbook of Biosolar Resources,* Vol. 2, *Resource Materials,* McClure, T. A. and Lipinsky, E. S., Eds., CRC Press, Boca Raton, Fla., 1981.
6. **Earle, F. R., Peters, J. E., Wolff, I. A., and White, G. A.,** Compositional differences among crambe samples and between seed components, *J. Am. Oil Chem. Soc.,* 43, 330, 1966.
7. **Dybing, C. D. and Zimmerman, C. D.,** Fatty acid accumulation in maturing flaxseeds as inflenced by environment, *Plant Physiol.,* 41, 1465, 1966.
8. **Eckey, E. W.,** *Vegetable Fats and Oils,* Reinhold, New York, 1954.
9. **El Tinay, A. H., Khattab, A. H., and Khidir, M. O.,** Protein and oil compositions of sesame seed, *J. Am. Oil Chem. Soc.,* 53, 648, 1976.
10. **Fowler, D. B. and Downey, R. K.,** Lipid and morphological changes in developing rapeseed, *Brassica napus, Can. J. Plant Sci.,* 50, 233, 1970.
11. **Graef, G. L., Miller, L. A., Fehr, W. R., and Hammond, E. G.,** Fatty acid development in a high stearic acid mutant of soybean, *Agron. Abstr.,* p. 91, 1983.
12. **Hammond, E. G., Fehr, W. R., and Snyder, H. E.,** Improving soybean quality by plant breeding, *J. Am. Oil Chem. Soc.,* 49, 33, 1972.
13. **Jamieson, G. S.,** *Vegetable Fats and Oils,* Reinhold, New York, 1943.
14. **Kleiman, R. R.,** Personal communication on analysis of soybean germplasm accessories, 1983.
15. **Lessman, K. J. and Anderson, W. P.,** Crambe, in *New Sources of Fats and Oils,* Pryde, E. H., Princin, L. H., and Mukherjee, K. D., Eds., American Oil Chemists Society, Champaign, Ill., 1981.
16. **Lyman, C. M., Kuiken, K. A., and Hale, F.,** Essential amino acid content of farm feeds, *J. Agric. Food Chem.,* 4, 1008, 1956.
17. **Lyon, C. K.,** Sesame: current knowledge of composition and use, *J. Am. Oil Chem. Soc.,* 49, 245, 1972.
18. **McWatters, K. H. and Cherry, J. P.,** Potential food uses of peanut seed proteins, in *Peanut Science and Technology,* Pattee, H. E. and Young, C. T., Eds., American Peanut Research and Educational Society, Inc., Yoakum, Tex., 1982.
19. **Miwa, T. K.,** Jojoba oil wax esters and derived fatty acids and alcohols: gas chromatographic analysis, *J. Am. Oil Chem. Soc.,* 48, 259, 1971.
20. National Academy of Sciences, *Products from Jojoba: A Promising New Crop for Arid Lands,* National Academy of Sciences, Washington, D.C., 1975, 14.
21. **Nieschlag, H. J. and Wolff, I. A.,** Industrial uses of high erucic oils, *J. Am. Oil. Chem. Soc.,* 48, 723, 1971.
22. **Pattee, H. E., Jones, E. B., Singleton, J. A., and Sanders, T. A.,** Composition changes of peanut fruit parts during maturation, *Peanut Sci.,* 1, 57, 1974.
23. **Reiners, R. A. and Gooding, C. M.,** Corn oil, in *Corn: Culture, Processing Products,* Inglett, G. E., Eds., AVI, Westport, Conn., 1970.
24. **Robertson, J. A., Chapman, G. W., Jr., and Wilson, R. L., Jr.,** Relation of days after flowering to chemical composition and physiological maturity of sunflower seed, *J. Am. Oil Chem. Soc.,* 55, 266, 1978.
25. **Robertson, J. A., Morrison, W. H., III, and Wilson, R. L.,** Effects of planting and temperature on the oil content and fatty acid composition of sunflower seeds, *USDA-SEA-ARS Southern Series No. 3,* 1979.
26. **Rubel, A., Rinne, R. W., and Canvin, D. T.,** Protein, oil, and fatty acid in developing soybean seeds, *Crop Sci.,* 12, 739, 1972.
27. **Sanders, T. H.,** Effects of variety and maturity on lipid class composition of peanut oil, *J. Am. Oil Chem. Soc.,* 57, 8, 1980.
28. **Sanders. T. H.,** Fatty acid composition of lipid classes in oils from peanuts differing in variety and maturity, *J. Am. Oil Chem. Soc.,* 57, 12, 1980.
29. **Swern, D.,** Composition and characteristics of individual fats and oils, in *Bailey's Industrial Oil and Fat Production,* 3rd ed., Swern, D., Ed., Interscience, New York, 1964.
30. United States Department of Agriculture, Agricultural Statistics, U.S. Government Printing Office, Washington, D.C., 1982, 142.
31. Foreign Agricultural Circular: Oil Seeds and Products, Foreign Agricultural Service, United States Department of Agriculture, Washington, D.C., December, 1982, 4.
32. Foreign Agricultural Circular: Oil Seeds and Products, Foreign Agricultural Service, United States Department of Agriculture, Washington, D.C., August, 1983, 10.

33. **Watson, S. A.,** Industrial utilization of corn, in *Corn and Corn Improvement,* Sprague, G. F., Ed., American Society of Agronomy, Madison, 1977.
34. **Weber, E. J.,** Lipids of maturing grain of corn (*Zea mays* L.). I. Changes in lipid classes and fatty acid composition, *J. Am. Oil Chem. Soc.,* 46, 485, 1969.
35. **Weiss, T. J.,** *Food Oils and Their Uses,* AVI, Westport, Conn., 1979.
36. **Wilcox, J. R., Cavins, J. F., and Nielsen, N. C.,** Genetic alteration of soybean oil composition by a chemical mutagen, *J. Am. Oil Chem. Soc.,* 61, 97, 1984.
37. **Wilson, R. F., Burton, J. W., and Brim, C. A.,** Progress in the selection for altered fatty acid composition in soybeans, *Crop Sci.,* 21, 788, 1981.
38. **Worthington, R. E. and Hammons, R. O.,** Genotypic variation in fatty acid composition and stability of *Arachis hypogaea* L. oil, *Oleogineux,* 11, 695, 1971.
39. **Yazdi-Samadi, B., Rinne, R. W., and Seif, R. D.,** Components of developing soybean seeds: oil, protein, sugars, starch, organic acids and amino acids, *Agron. J.,* 69, 481, 1977.
40. **Yermanos, D. M.,** Composition of jojoba seed during development, *J. Am. Oil Chem. Soc.,* 52, 115, 1975.

PROTEIN CROPS

Alfred E. Slinkard

INTRODUCTION

Protein crops are high in protein and are members of the legume family (Leguminosae). Most legumes are high in protein by virtue of their ability to fix atmospheric nitrogen symbiotically when infected with an appropriate strain of *Rhizobium*. Those legumes utilized by humans, primarily as dry seed, are termed grain legumes or pulses, except that soybeans and groundnuts (peanuts) are given priority status as oil seeds.

Over 70% of the protein in the human diet worldwide is of plant origin — 50% from cereals; 13% from pulses, oilseeds, and nuts; 5% from starchy roots; and 3% from vegetables and fruits.[1] The nutritional significance of protein relates to the availability and proportion of essential amino acids. Legume proteins are deficient in the sulfur amino acids, methionine and cystine. The limiting amino acids in cereal protein are lysine and threonine. Consequently, legume and cereal proteins are nutritionally complementary. As a broad generalization, a diet containing about 65% cereal protein and 35% legume protein approaches nutritional adequacy for humans, provided adequate sources of energy, vitamins, and minerals are included.

Early man recognized the complementary effects of cereal and legume proteins in that many early civilizations used such combinations; soybean and rice in the Orient, pea and lentil with wheat and barley in the Fertile Crescent, *Vigna* bean and sorghum in Africa, and *Phaseolus* bean and maize in the New World.[2] Many areas of the modern world are predominantly vegetarian because of religion or the excessive cost of animal protein. Thus, the complementary relationship between legume and cereal proteins is as important today as it ever was.

The following discussion of pulse crop utlization in the world will cover both utilization in the conventional way and as processed components, hulls (seed coats), starch, and protein. A short section on toxic constituents of pulses will conclude the discussion.

CONVENTIONAL UTILIZATION OF PULSE CROPS

Pulse crops are utilized primarily for their dry seed. In addition many pulse crops are utilized for their immature seed and immature pods. The seven major pulse crops of the world are presented in Table 1, along with statistics on world population of the dry seed, the major producers, parts consumed directly by man, and a typical protein content for the dry seed. The two most important pulse crops are the dry bean (*Phaseolus vulgaris* and *Vigna* species other than *V. unguiculata*) and the dry pea, which together account for over 50% of the world production of pulse crops. Dry bean and dry pea are also the most important pulse crops in international trade, and the major pulse-exporting nation is the U.S. Pulse crop production is particularly important in developing nations, where they are utilized primarily for home consumption, with small surpluses sold at local markets. As such, pulse crops have little value as cash crops or as exports, except as noted above. However, they are essential as sources of protein and energy for the local populace.

Production and utilization of most pulse crops are localized in a few nations or a geographical area. This is particularly evident for the 27 minor pulse crops listed in Table 2. The low moisture content and the hard seed coat of the dry seeds facilitate storage under adverse conditions such as found in many developing nations.

Little genetic improvement has been made in most pulse crops other than *Phaseolus* bean

Table 1
MAJOR PULSE CROPS OF THE WORLD —
PRODUCTION, MAJOR PRODUCERS, PARTS CONSUMED,
AND PERCENT PROTEIN IN THE DRY SEED

Pulse crop	1980 world production of seed (millions of tonnes)[3]	Major producers[3]	Parts consumed directly by man	Protein in dry seed (%)[4]
Dry bean (*Phaseolus vulgaris*)	14.7[a]	Brazil, U.S., Mexico, China,[a] India[a]	Dry seed, immature seed, immature pods, young leaves	22.0
Dry pea (*Pisum sativum*)	11.1	Russia, China, India	Dry seed, immature seed, immature pods	22.5
Faba (broad) bean (*Vicia faba*)	6.7	China, Ethiopia, Egypt	Dry seed, immature seed	25.4
Chickpea (*Cicer arietinum*)	4.8	India, Pakistan, Mexico	Dry seed, immature seed, young shoots	17.1
Lentil (*Lens culinaris*)	1.2	India, Turkey, Syria	Dry seed, immature seed, sprouts	25.0
Pigeon pea (*Cajanus cajan*)	—	India	Dry seed, immature seed, immature pod, sprouts, young leaves and shoots	19.2
Cowpea (*Vigna unguiculata*)	—	Nigeria	Dry seed, immature seed, immature pods, sprouts, young leaves and shoots	23.4

[a] Includes *Vigna* species (except *V. unguiculata*) grown primarily in China, India, and Southeast Asia.

and dry pea. However, local farmers have selected within various pulse crops for adaptation and palatability over the past centuries. There is tremendous potential for genetic improvement in all pulse crops.

UTILIZATION OF PROCESSED COMPONENTS

In developing nations, processing of pulse seed refers to removal of the seed coat (decortication) by pounding or grinding or both and splitting. Tempering the seed by soaking in water (cold or hot) facilitates decortication. The split seeds are then utilized directly in various foods or they may be ground into a flour which is then used in other food preparations.

Boiling of the seed (whole or split) is the most common method of preparation. Other methods include roasting and parching, frying, puffing, steaming, germination, fermenting, and agglomeration.[10]

Processing of leguminous seeds into their components has been pioneered with the oilseed legume, the soybean. It has been processed into oil and meal, which is primarily used as a protein supplement in livestock rations. However, increasing amounts of soybean meal have been processed into food-grade components such as protein concentrates and protein isolates which are utilized in meat extenders and analogues.

Processing of dry peas and dry beans has also been initiated, but the protein products cannot compete price-wise with soybean protein products. Some progress has been made in processing dry peas into pea hulls, pea starch, and pea protein concentrate. A dry milling process has been developed and is being used commercially by one plant in Saskatchewan, Canada. The dehulled peas are finely ground and the pea flour is processed through an air classifier which separates the heavier starch granules from the lighter protein fraction.[11]

Table 2
MINOR PULSE CROPS OF THE WORLD, MAJOR PRODUCERS, PARTS CONSUMED, AND PERCENT PROTEIN IN THE DRY SEED

Pulse crop	Major producers	Parts consumed directly by man	Protein in dry seed (%)	Ref.
Jack bean (*Canavalia ensiformis*)	Tropics	Dry seed, immature seed, immature pods, young leaves, shoots, flowers	23.7	5
Sword bean (*C. gladiata*)	Tropics	Dry seed, immature seed, immature pods	32.0	5
Ye-eb (*Cordeauxia edulis*)	Somalia(shrub)	Dry seed (13.4% oil)	12.1	5
Cluster bean (guar) (*Cyamopsis tetragonoloba*)	India	Immature pods; seed is a major source of galactomannan	33.3	6
Horsegram (*Dolichos uniflorus*)	India	Dry seed, sauce from fermented seeds	24.9	7
Kersting's groundnut (*Kerstingella geocarpa*)	Tropics	Dry seeds	21.5	5
Hyacinth bean (*Lablab niger*)	India, Sudan	Dry seed, immature seed, immature pods, sprouts	25.1	5
Louvana (*Lathyrus ochrus*)	Cyprus	Dry seed, immature seed (8)	—	
Grass pea (flat pea); (*L. sativus*)	India	Dry seed, immature seed, young leaves and shoots	28.2	4
White lupine (Egyptian lupine); (*Lupinus albus*)	Egypt, Ethiopia	Dry seed	40	7
Tarwi (*L. mutabilis*)	Peru, Ecuador, Bolivia, Chile	Dry seed (17.8% oil)	47.8	5
Velvet bean (*Macuna pruriens*)		Dry seed, immature seed, immature pods, young leaves	24	7
Yam bean (*Pachyrrhizus erosus*)	Southeast Asia	Tubers	9.9 tuber dry weight	6
Potato bean (*P. tuberosus*)	Western South America	Tubers	9.9 tuber dry weight	9
Tepary bean (*Phaseolus acutifolius*)	Southwestern U.S., Mexico	Dry seed, immature seed	24.5	5
Runner bean (*P. coccineus*)	Temperate zone, subtropics	Dry seed, immature seed, immature pods	23.1	5
Lima bean (*P. lunatus*)	Tropical, subtropical America	Dry seed, immature seed, immature pods, young leaves	25.0	5
Winged bean (goa bean); (*Psophocarpus tetragonolubus*)	Southeast Asia	Dry seed (8.8% oil), immature seed, immature pods, young leaves, shoots and flowers, tubers	36.4 / 11.4 tuber dry weight	5 / 5
African yam bean (*Sphenostylis stenocarpa*)	Western Africa	Dry seed, tubers	21.1 / 10.8 tuber dry weight	5 / 5
Fenugreek (*Trigonella foenumgraecum*)	India, Ethiopia, Northern Africa	Dry seed, immature seeds, immature pods, young leaves (primarily a condiment)	27.4	7
Marama bean (*Tylosema esculentum*)	Southwestern Africa	Dry seed (42.8% oil), tubers	29.5 / 8.1 tuber dry weight	5 / 5
Mat bean (moth bean or pillipesara); (*Vigna aconitifolia*)	India	Dry seed, immature seed, immature pods	23.0	4
Adzuki bean (*V. angularis*)	China, Japan	Dry seed	21.0	
Mung bean (green gram) (*V. aureus*)	India, China	Dry seed, immature seed, sprouts	23.6	4
Black gram (urd) (*V. mungo*)	India	Dry seed, immature seed, immature pods	23.4	4
Rice bean (*V. umbellata*)	Southeast Asia	Dry seed, immature seed, immature pods, young leaves	21.7	4
Bambara groundnut (*Voandzeia subterraea*)	Zambia	Dry seed, immature seed	16—21	4

Likewise a wet milling process has been developed and is being used commercially in Manitoba, Canada. The pea hulls (seed coats) are finely ground and used in the production of high-fiber bread. The pea starch is used in the production of carbonless carbon paper and as a desliming agent in potash refining. The pea protein concentrate has potential in the production of high protein bread and in pea protein isolates which may have various food and feed applications. Unfortunately, to date pea protein concentrates and pea protein isolates are unable to compete with comparable soybean products, so neither processing plant has been operating at capacity. These processing plants produce about twice as much starch as protein concentrate and, thus, the development of a high-value specialty starch market could carry the cost of the operation. Then both processing plants would quickly start operating at capacity. Perhaps the changing freight-rate picture will make starch use in the local potash refining industry more attractive in a few years. The dry pea production potential, the dry processing plant, and the potash refineries are all located in Saskatchewan. Only the right economic incentives are lacking to make production and local utilization of processed dry pea products viable.

TOXIC CONSTITUENTS

Although pulses are high in protein and, thus, of great nutritional value, they also contain various antinutritional or toxic constituents. Fortunately, over time man has devised ways and means of reducing or eliminating the deleterious effects from most pulse crops before ingesting them. In many instances these toxic constituents are water soluble or heat labile or both, in which case discarding the cooking water twice reduces the hazard to a minimum. Heating largely eliminates these toxic effects. Air classification of pulse flour concentrates the toxic constituents in the protein fraction.[12,13]

Protease inhibitors inhibit the proteolytic activity of certain enzymes (thus reducing protein digestibility) and are widely distributed in the pulses. The most widely studied are trypsin inhibitors in soybean. Up to six different trypsin inhibitors have been reported in soybean, and not all are heat labile.[14] Protease inhibitors of one type or another probably occur in all pulses, but are usually in low concentrations and destroyed during cooking.

Phytohemagglutinins or lectins have the property of agglutinating red blood cells and are widely distributed in the pulses. They have varying levels of toxicity.[15] Lectins impede the absorption of the products of digestion in the intestine.[4] They have also been associated with host specificity of the nitrogen-fixing *Rhizobium* and provide some protection against certain plant diseases.

Cyanogenetic glycosides occur in most pulses, but in toxic amounts only in some tropical types of the lima bean.[4] They produce hydrocyanic acid, a strong poison, when hydrolyzed by β-glucosidase.

Saponins are glycosides characterized by a bitter taste and form foams in aqueous solution.[16] High concentrations have been reported in *Canavalia* (jack bean and sword bean).[4]

Lathrogen, a neurotoxin, has been isolated from the grass pea *(Lathyrus sativus)*. This neurotoxin causes paralysis of the leg muscles and other symptoms in man. However, it is rarely a problem except in years of severe drought, when seed of the grass pea may comprise a major portion of the protein in the diet in certain parts of India.[4]

The faba (broad) bean can induce hemolytic anemia (favism) in susceptible individuals, namely, those deficient in the enzyme glucose-6-phosphate dehydrogenase.[17] The agent is heat labile and the most severe cases occur from ingesting raw seeds or inhaling pollen.

Polyphenols (tannins) are concentrated in seed coats of most pulses, particularly those with dark-colored seed coats or those that darken with age such as faba (broad) bean and lentil. They are bitter and interfere with protein digestion. Decortication results in removal of most of these compounds.[10]

Alkaloids are heat-stable toxic constituents of lupines. While sweet (low-alkaloid) mutants have been found in the white lupine, only the bitter (high-alkaloid) types are known in tarwi *(Lupinus mutabilis)*. They have to be soaked for a long time in salt water to make them palatable.[18]

DOPA (dihydroxyphenylalanine) is found in faba (broad) beans. Maturing plants turn black as they dry, but DOPA is not toxic as such.[17]

There are also two groups of indigestible materials in pulses — unavailable carbohydrates and protein conjugates.[4] Unavailable carbohydrates (raffinose and stachyose) are not digested and pass into the large intestine where they fermented anaerobically to produce gas (flatus).[19]

Air classification of finely ground pulse flour concentrates these unavailable carbohydrates in the protein fraction, leaving little in the starch fraction.[19] The formation of protein conjugates with phytin and hemicellulose adversely affects protein digestibility.

REFERENCES

1. **Roberts, L. M.**, The Food Legumes: Recommendations for Expansion and Acceleration of Research, Mimeographed report to the Rockefeller Foundation, 1970.
2. **Isely, D.**, Leguminosae and *Homo sapiens, Econ. Bot.*, 36, 46, 1982.
3. FAO, *Production Yearbook, 1980,* Vol. 34, Food and Agriculture Organization, Rome, 1980.
4. **Smartt, J.**, *Tropical Legumes,* Longman Group, London, 1976.
5. National Academy of Sciences, *Tropical Legumes: Resources for the Future,* National Research Council, Washington, D.C., 1979.
6. **Rachie, K. O. and Roberts, L. M.**, Grain legumes of the lowland tropics, *Adv. Agron.*, 26, 1, 1974.
7. **Westphal, E.**, *Pulses in Ethiopia. Their Taxonomy and Agricultural Significance,* Centre for Agricultural Publications and Documentation, Wageningen, The Netherlands, 1974.
8. **Photiades, J. and Alexandrou, G.**, Food legume research and production in Cyprus, in *Food Legume Improvement and Development,* Hawtin, G. C. and Chancellor, G. J., Eds., International Development Research Centre, Ottawa, 1979, 75.
9. **Purseglove, J. W.**, *Tropical Crops: Dicotyledons,* Longman Group, London, 1968.
10. **Siegel, A. and Fawcett, B., Eds.**, *Food Legume Processing and Utilization,* International Development Research Centre, Ottawa, 1976.
11. **Reichert, R. D.**, Air classification of peas *(Pisum sativum)* cultivar Trapper varying widely in protein content, *J. Food Sci.*, 47, 1263, 1982.
12. **Elkowicz, K. and Sosulski, F. W.**, Antinutritive factors in eleven legumes and their air classified protein and starch fractions, *J. Food Sci.*, 47, 1301, 1982.
13. **Marquardt, R. R., McKirdy, J. A., Ward, T., and Campbell, L. D.**, Amino acid, hemagglutinin and trypsin inhibitor levels, and proximate analysis of faba beans *(Vicia faba)* and faba bean fractions, *Can. J. Anim. Sci.*, 55, 421, 1975.
14. **Liener, I. E. and Kakade, M. L.**, Protease inhibitors, in *Toxic Constituents of Plant Foodstuffs,* 2nd ed., Liener, I. E., Ed., Academic Press, New York, 1980, 7.
15. **Jaffe, W. G.**, Hemagglutinins (lectins), in *Toxic Constituents of Plant Foodstuffs,* 2nd ed., Liener, I. E., Ed., Academic Press, New York, 1980, 73.
16. **Birk, Y. and Peri, I.**, Saponins, in *Toxic Constituents of Plant Foodstuffs,* 2nd ed., Liener, I. E., Ed., Academic Press, New York, 1980, 161.
17. **Majer, J., Chevion, M., and Glaser, G.**, Favism, in *Toxic Constituents of Plant Foodstuffs,* 2nd ed., Liener, I. E., Ed., Academic Press, New York, 1969, 266.
18. **Nowacki, E.**, Heat stable antinutritional factors in leguminous plants, in *Advances in Legume Science,* Summerfield, R. J. and Bunting, A. H., Eds., Royal Botanic Gardens, Kew, 1980, 171.
19. **Sosulski, F. W., Elkowicz, L., and Reichert, R. D.**, Oligosaccharides in eleven legumes and their air-classified protein and starch fractions, *J. Food Sci.*, 47, 498, 1982.

PLANT FIBERS

Joshua A. Lee

INTRODUCTION

The terms food and fiber are often linked in references to agriculture, suggesting an importance of fibers second only to that of food. Without fibers — plant and animal — there is doubt that human societies could have advanced much beyond the paleolithic mode.[16,17] There would have been no fishing lines and nets, no snares, no bindings for weapons and housing, no textiles, and no paper for recording the annals of human progress and history.

In commerce, the term fiber, when applied to vegetable products, is very loose, ranging from multicellular roots and stems to unicellular "seed hairs", such as cotton. Three commercial groupings of fibers are commonly used.[10,16] These are based upon the kind of fiber and the kind of plant from which the material is taken. The first group, *surface* fibers, is a classification derived more for commercial convenience than from any sense of botanical cohesion. The grouping includes all useful fibrous materials growing from the surfaces of plants, and, as used, include roots, leaves, and stems. By far the most important kinds of fibers taken from plant surfaces are the floss fibers, mostly cotton and kapok. These fibers are mostly unicellular outgrowths from the seed coat or the inner wall of the ovary.[9,16] Other surface fibers include filling materials collected as stems, roots, and leaves, and various braiding materials. The second group, *bast* fibers, develop in the cortex, phloem, and pericycle of many dicotyledonous plants.[16] These structural cells are regarded as true botanical fibers.[14] Typically, bast fibers occur as clusters of cells, each cell cemented to neighboring cells by the pectic middle lamella. The individual fibers, or in some materials bundles of fibers, are most often freed from enclosing tissues by retting, a process of rotting away unwanted tissues by microbial action in a moist environment. Bast fibers are sometimes decorticated by mechanical means followed by chemical digestion of pectic materials adhering to the fibers. The third group, *hard,* or leaf, fibers are extracted from the leaves of various monocotyledonous plants, and a "fiber" is frequently an intact vascular bundle.[10—14] The individual fibers within the vascular matrix are very short cell — one to a few millimeters in length — and cemented end-wise into lengthy strands. The extracted fibers consist often of both xylem and phloem, along with various ensheathing cells, and are found scattered through the pithy matrix of the leaf. The cells are lignified to varying degrees and are thus described as hard.[14,16,21] Hard fibers are most often decorticated by mechanical means.

Plant fibers are often grouped into four classes according to end use as follows: (1) textile and fabric fibers, (2) cordage fibers, (3) brush and braiding fibers, and (4) filling fibers.[10,16] These categories are imprecise because many fibers are interchangeable in end use. Thus hemp — a fiber similar to flax, a premier textile fiber — was once the most important cordage fiber of Europe, but has come to be used more and more for textiles as the hard fibers have become prominant in cordage markets. Cotton, the most important textile in the world, is also used for filling purposes and for weak cordage. Nevertheless, end-use category is a convenient, and generally meaningful, way to discuss plant fibers, and the system will be used herein to provide information on some of the most important fibers.

There are currently no international standards for grading fibers, with most nations having legally sanctioned domestic standards for assessing fiber quality. However, the standards of all nations are based upon such properties as length, strength, fineness, appearance, and content of foreign matter. Therefore, the standards of a given nation are usually transposable into the relevant metrics of others.[5]

Through the ages fibers have been taken from the sources available and adapted to the needs at hand. Hundreds of plant genera have been exploited.[10,14,16] According to Kirby[9] more than 1000 species have been utilized, and Schery[16] states that no fewer than 300 species have been used in East Africa alone. Still, only about 24 species have a history of commercial importance.[9] A fairly exhaustive list of fiber plants can be gleaned from Kirby[9] and Mauersberger.[14]

Economic consideration more often than comparative quality of products have determined the success of a particular fiber. Thus some of the best fibers — ramie, for example — have not been great commercial successes because of the costs entailed in processing the raw fiber. The following criteria are important in determining the commercial potentialities of a particular fiber[9]: (1) there must be sufficient quantity in a particular part of the plant to make extraction worthwhile, (2) there must be an efficient and economical means of extraction, (3) the fiber must be suitable for a particualr end use, (4) the fiber must be available regularly in sufficient quantities to satisfy a particular demand, and (5) the fiber must be competitive in price.

TEXTILE AND FABRIC FIBERS

The arts of spinning and weaving antedate history in both the Old and New Worlds.[1,8,17,18,21] Woven fabrics ventilate better than animals skins in warm weather, and provided a way to meet clothing needs as populations increased in neolithic and early historical times.[15,16] The finer, softer, and more pliable fibers are required for textiles. Fineness is important because there should be no less than 150 individual fibers at any one point along a skein of textile yarn. Fewer would decrease adhesion and increase breakage during spinning and weaving. Flexibility decreases abrasion during carding and weaving, and high tensile strength reduces breakage.[9] Textile fibers of vegetable origin remain very important in world commerce in spite of recent inroads from petroleum-based synthetics.

Cotton

Cotton, *Gossypium* spp. (Malvaceae), is the most important textile fiber offered on world markets today, with modern production per year amounting to between 70 and 80 million bales of about 218 kg each. The U.S. and the U.S.S.R. usually lead in production with the annual crop of either nation fluctuating between 12 and 15 million bales. However, the People's Republic of China produced about 20 million bales in 1985 and has thus an increasing interest in growing the crop.

Cotton is an ancient crop; use of the commodity dates beyond 2700 BC for the Indus Valley,[8] to 2500 BC for coastal Peru,[1,9] and a somewhat earlier date for the Tehaucan Valley of Mexico.[18] There seems to be little doubt that cotton was domesticated for the seed floss (lint), although the seeds might have been useful as livestock feed.[3] A population of the linted diploid species *Gossypium herbaceum* L. grows wild in southern Africa, and wild races of the linted tetraploid species, *Gossypium hirsutum* L. and *Gossypium barbadense* L. grow as rare, widely dispersed populations in tropical America, and on some islands of the Pacific. These cottons bear spinnable lint, although none approach modern cultivars in fiber yield potential nor in quality of lint. In fact, modern domesticated cottons have diverged totally in all aspects of plant development and properties of the fiber. Moreover, changes in reproductive responses to climate have permitted the crop to be moved from tropical to subtropical and warm temperate zones where modern cotton production is centered. The crop is grown as far north as the 42nd parallel in Asia.[12]

A cotton fiber results from the outgrowth of a cell from the epidermis of the ovule. As the cell elongates, cellulose is laid down in transverse bands about the lumen. At maturity the protoplast dies and the cell wall collapses inward to form a convoluted strand.[2] Flattening and convolution provide the kind of surfaces and baffles that promote adhesion during the spinning process. Cotton is unique in that it is the only floss fiber that has the combination of length and morphology required for efficient spinning.[2]

Cotton is marketed on the basis of fiber quality and grade, the latter being a measure of fiber appearance and trash content. Fiber length is crucial in determining the best end use of a particular lot of cotton; longer fibers are needed for finer-count yarns.

In the U.S. cotton is measured as upper-half mean and as mean fiber length.[20] The former is the average length of the longer half of the fibers as determined on a digital fibergraph.[6] The latter is the average length of all fibers in a given sample. The ratio of the two measures gives an index of the uniformity of length stated as a percent.

Tensile strength is measured in centi-Newtons per Tex, a Tex being the weight of 1 km of processed yarn.[20] The jaws of the breaking device, the stelometer,[6] are set 3.2 mm apart, and a measure of the elasticity of the fiber bundle relates to the distance the jaws travel before breakage. Fibers breaking at 25 cN Tex are considered strong.

Fiber perimeter, a measure of fineness, is important in assessing maturity of cotton lint, a necessity when matching growths for blending, or for blending with other fibers. Fiber perimeter is estimated with a micronaire device,[6] with the principal being that air at a constant pressure flows at a more rapid rate through a compressed sample of coarse fibers than through an equal weight of fine fibers. Cottons in the range of 3.5 to 5 are considered premium grade for fineness.

About 90% of the current world production of cotton is from *G. hirsutum*. Fiber length of most modern uplands, the annual form of *G. hirsutum*, range from 25.4 to 28.6 mm for the upper-half-mean. The chief end uses of upland fiber are for cotton textile yarns, in blends with polyester for textile yarns, and for the production for weak cordage. Linters, the short fibers cut from seed before crushing, are an important source of industrial cellulose.

Gossypium barbadense accounts for about 9% of the current world supply of cotton.[12] These have upper-half-mean lengths mostly in the 28 to 38 mm range, and the fibers are very strong. The Sea Island cottons, now obsolete, ranged upward to 63 mm. The fibers of *G. barbadense* are used mostly for the production of sewing thread and luxury fabrics.

Gossypium arboreum L. and *G. herbaceum* account for the remainder of the world cotton production. These species are grown mostly in southern Asia in environments where the tetraploid species do not thrive. Neither species has been subjected to much improvement, so yields are low and fiber quality generally below that of upland. Fiber length, for example, rarely exceeds 25 mm, and the fibers are frequently coarse. The lint of the diploid cottons is used for textile purposes — often in a cottage industry, and somewhat for filling and quilting purposes.

Flax

Flax, *Linum usitatissimum* L., Linaceae, is one of about 50 species in the genus distributed about the holarctic region. Several wild species of the Old World are similar to cultivated flax and cross readily with the species. Still, the wild ancestor of domesticated flax has not been identified.[16]

Flax is believed to be the most ancient of domesticated fiber crops, with cultivation of the plant dating from neolithic times. The Swiss lake dwellers of 3000 BC grew flax, as did the Egyptians of 4000 BC. Hutchinson et al.[11] theorized that cotton was domesticated in Old World communities that already had a technology for the spinning and weaving of flax and wool.

Flax grows best in a cool, moist climate, and the culture of the crop is similar to that for small grains. The plants are pulled when mature and retted either by emersion or by atmospheric moisture in the field (dew retting). Retted flax is dried and the fibers separated from dross tissues on breaking and skutching devices, machines that crush the pithy core of the stem and remove the nonfibrous materials from the pericycle.

Table 1
RELATIVE FITNESS OF SOME BAST FIBERS FOR TEXTILE PURPOSES

	Rank 1	Rank 2	Rank 3	Rank 4
Durability	Ramie	Flax	Hemp	Jute
Tensile strength	Ramie	Hemp	Flax	Jute
Fiber length	Ramie	Flax	Hemp	Jute
Cohesiveness	Flax	Hemp	Jute	Ramie
Fineness	Ramie[a]	Flax	Hemp	Jute
Uniformity	Flax	Ramie	Hemp	Jute
Pliability	Flax	Ramie	Jute	Hemp
Color	Ramie	Flax	Hemp	Jute
Consensus	Flax	Ramie	Hemp	Jute

[a] Ramie varies from the finest to the coarsest fiber in the chart. See Table 2.

Adapted from Weindling, L., *Long Vegetable Fibers*, Columbia University Press, New York, 1947.

Table 2
PHYSICAL CHARACTERISTICS OF SOME INDIVIDUAL FIBERS OF TEXTILE MATERIALS

Fiber	Length (mm)	Width (μ)	Individual fiber tensile strength (g)
Ramie	60—250	50	25—50
Flax	25	20	18.0
Hemp	20	22	15.6
Cotton	13—63	18—27	6.2

Adapted from Dempsey, J. M., *Fiber Crops*, University of Florida Press, Gainesville, 1975.

The ultimate (individual) fibers of good-quality flax average 27 mm, whereas fiber strands can be more than 1 m in length. Mean diameter of flax fiber is about 23 μm, and the lumen is very narrow. Flax is stronger than cotton (Table 1) but much less elastic.

Retted flax contains 64% cellulose, 17% hemicellulose, and 2% lignin.[9] Flax is regarded as having the highest utility of any fiber for producing fine textiles. Table 2 ranks bast fibers as textile materials. The chief problem with flax, from the standpoint of modern commerce, is the high cost entailed in processing the fiber.

The U.S.S.R. accounts for more than one half of the recent world production of flax which has run as high as 650,000 metric tonnes (t).[16] Other leading producers are Poland, Belgium, the Netherlands, and France. Most of the fiber is spun within Europe and is used mostly for linen fabrics and sewing thread. Short and waste fibers from the hackling (combing) process are used for cordage, stuffing, and high-grade paper.

Hemp

Hemp, *Cannabis sativa* L. (Moraceae), is apparently native to eastern Asia, the plant having been grown as a crop in China for at least 5000 years.[16] Hemp was cultivated in preclassical times in Greece, and was an important seed and fiber crop in Europe during the Middle Ages.

The hemp plant can be either monoecious or dioecious, with the male plants producing the best fiber. The fiber strands of hemp are up to 130 cm in length. However, the ultimate fiber cells vary from 5 to 55 mm in length, and diameter from 16 to 50 μm. The cell lumen is broader than that of flax, and the fibers are less flexible. Hemp contains 65% cellulose, 16% hemicellulose, and 8% lignin.[9]

Hemp was the first important cordage fiber of the Old World. During the last century the cordage market was largely lost to hard fibers, and more recently synthetics. Still, a large amount of hemp is used locally for cordage in Europe, with cordage grades deriving mostly from dew-retted stems.

Other than dew retting, i.e., allowing the stems to rot on the ground, hemp is retted in tanks or ponds. Tank-retted hemp is finer and of better quality than cordage-grade hemp and is used for textile purposes, especially in the U.S.S.R., Italy, and the Far East.

Current world production of hemp has been estimated at 260,000 t, mostly in eastern and southern Europe, with the U.S.S.R. the leading producer.[5] India, China, and Korea produce substantial amounts of hemp for local use.

Ramie

Ramie, *Boehmeria nivea* (L.) Guad. (Urticaceae), is a premium bast fiber that has not lived up to its potential. The chief difficulty has been in extracting the fiber from the gummy ensheathing cells. Once extracted, good-quality ramie has an outstanding luster and is one third to twice as strong as good-quality flax (Table 2). Although Dempsey[4] quotes ramie as being finer than either flax or hemp, the fiber is very variable in fineness, ranging from finer than flax to much coarser. Ramie, thought to be native to China, is a perennial shrub with stems up to 3 m in length.

The ensheathing gums of ramie cannot be removed by retting and have been traditionally removed by pounding the ribbon — pericycle stripped from the stem — followed by hand scraping. Recently chemical degumming has come into prominance, i.e., boiling in caustic soda, followed by bleaching and treating with an acidic solution.[14]

Ultimate fibers of ramie range up to 250 mm, and the fiber strands can be 2 m in length. Ramie is resistant to water and fungi, and is well adapted to most of the uses of cotton and flax. The fiber is particularly suited for production of high-pressure water hoses. In 1975 world production of ramie was estimated at about 35,000 t with the People's Republic of China the leading producer.[4] Export trade in ramie fiber is increasing.

Jute

Two species of jute are cultivated, *Corchorus capsularis* L., lowland jute, and *Corchorus olitorius* L., upland jute (Tiliaceae). Jute has been grown in India for at least 2800 years.[16,21] The plants grow best in deep soils containing an ample supply of moisture. Jute of either species grows to 5 m tall in from 3 to 5 months.

Jute stems are retted in ponds, and the fibers are floated free of the bark by agitating the stems on the surface of the pond. The cleaned fibers vary in color from silvery white to dark brown, with the lighter fibers being the best grade.[4] Ultimate fiber length averages 2.03 mm with strands exceeding 1 m. Fiber diameter averages 23 μm, and tensile strength 24.1 kg for a 38.1-cm bundle weighing 0.324 g.[14] An equal quantity of flax would have a breaking strength of about 29.6 kg. Chemically, jute fibers consist of 75% cellulose, 11% lignin, and 12% xylan. Annual production amounts to about 2.7 million t, two thirds of which is grown in India and Bangladesh. The remainder is grown mostly in Africa and Brazil.[16]

Virtually 80% of the world jute production is used for packaging fabrics, particularly in the manufacture of gunny and burlap bagging.[4] Other important uses for jute fabrics are furniture webbing, backing for carpets, linoleum frames, and weak cordage.

MALVACEOUS FIBERS

Kenaf and roselle are extracted from *Hibiscus cannabinum* L. and *Hibiscus sabdariffa* L. (Malvaceae), respectively. The fibers are similar to jute, though somewhat coarser and less pliable. The uses of kenaf and roselle are the same as for jute. India and Thailand are prominant in the production of kenaf and roselle, virtually all of which is processed locally.

Urena lobata L. and *Abutilon theophrasti* Medic. (Malvaceae), are of considerable importance locally as producers of packaging fibers. The former is a tropical perennial grown mostly in west Africa and southern Asia, and the latter, known as China jute, is a temperate species grown mostly in the U.S.S.R., China, Korea, and Japan. The lignified fibers of these crops are frequently mixed with hemp or jute for cordage, or for making coarse fabrics.

Wilson[23] lists 13 species in the genus *Hibiscus* that yield fiber, and Mauersberger[14] no less than 17. There are fiber plants in several other malvaceous genera. World production of malvaceous fibers is under 1 million t annually.[9]

Sunn Hemp

Sunn hemp, *Crotalaria juncea* L. (Fabaceae), is grown mostly in India with an annual production of about 97,000 t.[16] The fiber is resistant to deterioration when wet and so is used for fish nets and fabrics that come in contact with water. Small amounts of sunn hemp are exported to the U.S. and to Britain for the manufacture of cigarette papers, and the plant as been investigated as a source of paper pulp.[22]

CORDAGE FIBERS

In modern times most cordage has been made from hard fibers extracted from the leaves of monocotyledonous plants. Typically these fibers are lengthy, coarse, and rigid, but have, for most classes, greater breaking strength in relation to cross-section than the bast fibers. Although hard fibers have been used since prehistoric times,[17] production of hard fibers on a commercial scale commenced with the development of the East African sisal industry in the first half of the nineteenth century.[13] Hard fibers remain important in world commerce, although there is now severe competition from synthetics for some end uses. Virtually all classes of hard fibers are extracted through mechanical scraping.

Agave Fibers

Sisal and henequen, leaf fibers of *Agave sisalana* Perrine and *Agave fourcroydes* Lem. (Agavaceae), respectively, are the most important in terms of total volume of production of the cordage fibers. Both species are native to Mesoamerica, growing there in seasonally xeric environments. In recent times interspecific hybrids of *Agave* have been explored for fiber production with some success.[13]

About one half of the world total production of hard fibers is accounted for by sisal. This amounts to about 370,000 t. Henequen accounts for about 100,000 t, and cantala, from *Agave cantala* Roxb., about 3000 t. Sisal is the strongest and most lustrous of the agave fibers. The breaking strength of a 0.324-g bundle 38.1 cm in length was given by Mauersberger[14] at 28.1 kg. Henequen was somewhat weaker at 26.2, and cantala very weak at 10.2 kg.

Sisal is produced mostly in east Africa with some production in west Africa, Indonesia, the West Indies, and Brazil; henequen in Mexico and the West Indies; and cantala in the Philippines.

Table 3
RELATIVE BREAKING STRENGTHS OF FIVE
HARD CORDAGE FIBERS

Abaca[a]	Sisal	Sansevieria	Henequen	Phormium	Cantala
100.0	79.0	73.5	70.5	54.7	28.0

[a] Breaking strength of abaca estimated at 35.6 kg/0.324-g bundle 38.1 cm in length.

Adapted from Mauersberger, H. R., *Matthew's Textile Fibers*, John Wiley & Sons, New York, 1954.

Table 4
RELATIVE FITNESS OF SOME HARD FIBERS FOR
CORDAGE PURPOSES

	Rank 1	Rank 2	Rank 3	Rank 4
Durability	Abaca	Sisal	Istle	Mauritius
Tensile strength	Abaca	Sisal	Istle	Mauritius
Strand length	Abaca	Mauritius	Sisal	Istle
Fineness	Mauritius	Abaca	Sisal	Istle
Uniformity	Mauritius	Abaca	Sisal	Istle
Pliability	Mauritius	Abaca	Sisal	Istle
Consensus	Abaca	Sisal	Mauritius	Istle

Adapted from Weindling, L., *Long Vegetable Fibers*, Columbia University Press, New York, 1947.

The chief use of sisal is for cordage, especially for twine used in baling hay. Henequen is used mostly in Mexico for cordage and for rough bagging. Plantation production of sisal is usual in east Africa.[13]

Several species of *Agave* yield fiber known as istle, especially *Agave heteracantha* Zucc. and *Agave lecheguilla* Torr. Istle is softer and weaker than sisal and is thus suited for rough fabrics and close-weave bagging. At one time istle was used in the U.S. for making cheap brushes. Istle is produced only in Mexico.

Abaca

Abaca, the premier rope fiber of the world (Tables 3 and 4) is extracted from the leaf stalks of *Musa textilis* Nee (Musaceae), a close relative of the common banana. The species is endemic to the Philippines, with the islands producing about 95% of the current world crop of about 96,000 t.[16] The fiber is sometimes used for coarse fabrics in the Philippines, but most is exported for the production of high-quality cordage, particularly marine hawsers.[21]

Abaca fibers are 2 to 4 m long, light in weight, very strong (Table 3), and very resistant to deterioration in sea water. Abaca is now largely a plantation crop where the decortication of fiber, as with the agave fibers, is largely mechanized.

Coir

Coir is extracted from the unripe husks of coconut, *Cocos nucifera* L. (Palmaceae), and is one of the few fibers taken from the fruit of a plant. The fibers are retted free of enclosing tissues in sea water.

The chief use of coir is for making sennit braid, the standard cordage of coastal south Asia and the South Seas. Total annual production of coir has been estimated at 95,000 t, about half from India.[9] Fifty percent of the India crop is exported to Europe for making doormats and for filling purposes.[16]

Miscellaneous Hard Fibers

Several species of *Samuela* and *Yucca* (Liliaceae), furnish istle of local importance in northern Mexico, and *Phormium tenax* J. R. & G. Forst. (Liliaceae), New Zealand flax, yields a lustrous fiber of moderate strength (Table 3) that is used mostly in New Zealand for the production of bags and cordage. Several species of *Sansevieria* (Liliaceae), have been explored for use as cordage fibers. However, none have achieved commercial importance.

Furcraea foetida (L.) Haw. (Amaryllidaceae), although native to Brazil, is best known as Mauritius hemp. The fiber is weaker and more pliable than sisal and is thus less suited for cordage (Table 4). The fiber is produced only on the island of Mauritius for the manufacture of sugar bags.

BRUSH AND BRAIDING FIBERS

Virtually any fiber with the required stiffness can be used for making brushes. Thus, various hard fibers — particularly sisal and istle — have been used for making cheap brushes of the kind usually made from animal bristles.[16] A few kinds of fibers are collected especially for the brush trade. Very few plants are cultivated especially for making brushes.

Piassava

Piassava fibers have set the standards for stiff-brush manufacture for over a century.[9] These fibers are taken from the stalks of various species of palms of the genera *Attalea*, *Leopoldinia*, *Raphia*, *Caryota*, *Borassus*, and *Dictyosperma*. The fibers are collected from weathered materials, or else are removed by retting followed by pounding and scraping.[9] Individual fiber strands of piassava can be more than 1 m in length.

As much as 18,000 t of piassava have been collected annually, mostly from Brazil and West Africa. In recent decades synthetic materials have displaced much of the market for piassava.

Broomcorn

Broomcorn, a variety of *Sorghum bicolor* (L.) Moench., Poaceae, is the favored material for soft brooms, particularly whisk brooms.[8] The fibers, up to 40 cm in length, are taken from unripe infloresences.

Braiding Materials

Braiding materials are classified as fibers in commerce. Many species of plants are used locally for mats and basketry; important are stems of sedges and rushes, leaves of tough grasses and palms, and wood splints. Leghorn straw hats are woven from the stems of wheat and rye.[16] One of the most famous of braiding fibers is strips of leaf of *Carludovicia palmata* Ruiz & Pav. (Cyclanthaceae), the panama hat palm. Only young leaves are used. Curiously, panama hats are not products of Panama but are woven in Ecuador as a cottage industry. Several thousand hats are exported annually.[16]

FILLING FIBERS

A great variety of fibers have been used for stuffing and filling purposes, the lower grades and wastes of textile fiber serving as well as any for many purposes. Only a few fibrous materials have been gathered especially for use as fillers. The use of vegetable and animal fibers for filling purposes have declined in recent years in favor of synthetics.

Kapok

Kapok, the pod floss of the silk cotton tree, *Ceiba pentandra* (L.) Gaertn. (Bombacaceae), is the most important of the natural filling materials. Although the silk cotton tree is thought to be a native of the New World tropics, kapok production is centered in southern Asia, particularly Indonesia.

Kapok is about 64% cellulose, is resilient, and very resistant to wetting. Compressed kapok has about one sixth the weight of an equal volume of cotton, and loses only about 10% of its buoyancy after a month of immersion in water.[9]

Kapok is used widely for stuffing and insulating purposes, and is especially useful for making life jackets and belts. Production of kapok has run as high as 30,000 t with about 19,000 t exported in 1959.[9] Although kapok remains very useful in the tropics, particularly for stuffing mattresses, use of the fiber has declined in nations where synthetic stuffing materials are available.

The floss of the red silk cotton tree, *Bombax malabaricum* D.C. (Bombacaceae), of India has properties similar to those of kapok. Akund, a floss inferior to kapok, is collected from *Calotropis* spp. (Asclepiadaceae), small trees of southern Asia, and the floss of various species of milkweeds — especially *Asclepias syriaca* L. — was collected in North America as a substitute for kapok during World War II.

Crin Vegetal

The leaf fibers of the dwarf palm, *Chamaerops humilis* L. (Palmaceae), of the Mediterranean basin have been collected and used as a substitute for horsehair for centuries. The center of the industry in recent years has been Morocco.[9]

Spanish Moss

At one time spanish moss, *Tillandsia usneoides* L. (Bromeliaceae), was a favorite filling material for furniture. The plant is not a true moss but an epiphyte related to the pineapple. The "moss" was collected mostly in the southeastern U.S. during warm weather, retted in piles to remove the epidermis, dried, and baled for shipment. The collection of spanish moss has declined with the use of synthetic stuffing materials.[16]

THE FUTURE OF VEGETABLE FIBERS

There is little reason to suppose that vegetable fibers will diminish in importance in the less developed regions of the world where synthetic fibers would be too costly for most of the population, if available at all. Indeed, there is reason to believe that the advantage lies with fibers of plant origin in the more developed nations in the long run, as the mineral stocks used in the production of synthetic fibers grow scarcer and more dear. Even so, future markets and end uses will likely continue to sort out those fibers that can be mass produced most efficiently, perhaps as fibrous materials in bulk rather than the classes of fibers known today. Thus there is the likelihood that the markets of tomorrow will look to fibers synthesized from plant hydrocarbons, particularly cellulose. For that reason wood fibers — not covered herein — could loom larger in the textile and cordage markets of the future.

An old cliche in fiber trade states that cost determines quality in the long run; that is to say, any fiber reasonably adequate for a particular end use that can compete in price usually captures the markets. The foregoing probably explains why much of the world cotton crop is below desired quality standards, why jute, a mediocre fiber when compared with flax or hemp, enjoys a larger share of the world fiber market than either, and why 75% of the textile fibers spun in the U.S. are polyester. Costs will always be the foremost consideration in choosing a fiber for a particular end use, whether synthetic or natural.

Of the fibers currently offered, cotton seems the most assured of a place in the fiber needs of the future. Production and processing of the crop is largely mechanized, and the fiber can be used as such or as a source of high-quality cellulose. Bast fibers might wane in importance as extracted fiber, but certain classes, particularly the malvaceous fibers, offer potential as a source of cellulositic pulp. The future of the hard and filling fibers seems the most uncertain of all.

REFERENCES

1. **Bird, J. and Mahler, J.**, America's oldest cotton fabrics, *Am. Fabrics*, 20, 73, 1951—52.
2. **Brown, H. B. and Ware, J. O.**, *Cotton*, McGraw-Hill, New York, 1958.
3. **Chowdhury, K. A. and Burth, G. M.**, Cotton seeds from the neolithic of Egyptian Nubia and the origin of Old World cotton, *J. Linn. Soc. London Biol.*, 3, 303, 1971.
4. **Demspey, J. M.**, *Fiber Crops*, University of Florida Press, Gainesville, 1975.
5. **Doberczak, A., St. Dowgielewicz, and Zurek, W.**, *Cotton, Bast, and Wool Fibers*, Centralny Instytut Informacji Naukowo-Technicznej I Ekonomicznej, Warsaw, 1964.
6. **Duckett, K. E.**, Cotton Fiber Instrumentation Research in Tennessee, University of Tennessee Agricultural Experimental Station Bulletin, 536, 1974.
7. **Fryxell, P. A.**, *The Natural History of the Cotton Tribe*, Texas A & M University Press, College Station, 1979.
8. **Gulatti, A. M. and Turner, A. J.**, A note on the early history of cotton, Indian Central Cotton Committee Technical Laboratory Bulletin No. 17, 1928.
9. **Kirby, R. H.**, *Vegetable Fibres*, Leonard Hill, London, 1963.
10. **Hill, A. F.**, *Economic Botany*, McGraw-Hill, New York, 1937.
11. **Hutchinson, J. B., Silow, R. A., and Stephens, S. G.**, *The Evolution of Gossypium and the Differentiation of the Cultivated Cottons*, Oxford University Press, London, 1947.
12. **Lee, J. A.**, Cotton as a world crop, in *A Cotton Monograph*, Crop Science Society of America Press, Madison, 1983.
13. **Lock, G. W.**, *Sisal*, John Wiley & Sons, New York, 1962.
14. **Mauersberger, H. R.**, *Matthew's Textile Fibers*, Wiley, New York, 1954.
15. **Scherer, J. A. B.**, *Cotton as a World Power*, F. A. Stokes, New York, 1916.
16. **Schery, R. W.**, *Plants for Man*, Prentice-Hall, Englewood Cliffs, N.J., 1972.
17. **Smith, C. E., Jr.**, Plant fibers and civilization — cotton, a case in point, *Econ. Bot.*, 19, 71, 1965.
18. **Smith, C. E., Jr. and Stephens, S. G.**, Critical identification of Mexican archaeological cotton remains, *Econ. Bot.*, 25, 160, 1971.
19. **Stephens, S. G. and Moseley, M. E.**, Early domesticated cottons from aracheological sites in central coastal Peru, *Am. Antiq.*, 39, 109, 1974.
20. **ASTM**, *The 1965 Book of ASTM Standards*, Part 25, The American Society for Testing and Materials, Philadelphia, 1965.
21. **Weindling, L.**, *Long Vegetable Fibers*, Columbia University Press, New York, 1947.
22. **White, G. A. and Haun, J. R.**, Growing *Crotalaria juncea*, a multipurpose legume for paper pulp, *Econ. Bot.*, 19, 175, 1965.
23. **Wilson, F. D.**, An evaluation of kenaf, roselle and related *Hibiscus* for fiber production, *Econ. Bot.*, 21, 132, 1967.

VEGETABLE CROPS

R. J. Hilton

INTRODUCTION

In much of the continents of North America and Europe, the commercial production of the major crops listed as vegetables is carried out as a specialized type of farming, and often on a mass-production basis. However, in many countries a very significant proportion of the fresh vegetable supply is produced in small operations and either is used directly by the producer or marketed in village, town, or city produce markets. For this reason it is difficult to obtain reliable estimates of vegetable production for any one country.

Although the global areas of land and sea differ widely in their capacity to provide food for the human population of the earth, the urge to obtain sufficient food is constant. And of the sources of food, that product of plants which we call vegetables represents the greatest diversity (Tables 1 and 2). The land area producing vegetables in developing (mainly tropical) countries (4830×10^3 ha) is greater than that in the temperate countries (2798×10^3 ha), but the volume of vegetable produce in the former is about 12% lower.

Vegetable crops are especially important in developing countries where protein from livestock and fish products is often unavailable. Thus, in many tropical and subtropical regions, the dietary needs for starches and proteins are supplied by sweet potatoes, plantains, yams, taro, cassava, and an important group of warm-season species of peas, beans, and other pulse (legume) crops.

It is not possible to compute a reliable value relationship of vegetable produce compared with wheat, corn, and rice, but a reasonable estimate would be that vegetables are worth about 70% of wheat, 32% of corn, and 80% of rice. These proportions will vary widely among different regions by reason of widespread differences in population densities, climatic factors, and economic status.

ECONOMICS OF VEGETABLE PRODUCTION

During the past three decades, several important factors have combined to result in a reduction in the area utilized for vegetable growing and this is associated with an increase in tonnage of produce harvested. These changes are more marked in vegetable production than in most other agricultural pursuits (fruit and livestock enterprises, for example), but they obviously parallel the increased mechanization coupled with improved cultivars (varieties) and fertilization practices that have brought about a similar phenomenon in cereal grain production. Global statistics for vegetable production are crude estimates, and not provided here. However, a good picture of the effect noted above can be seen in recent U.S. Department of Agriculture figures (Table 3).

Developed countries have emphasized mass production of vegetables on large and highly mechanized land units, whereas most developing countries still have greatest production from small (under 1 ha) farms. These require low capital input but they are very labor intensive.[1]

Generally speaking, processing of vegetable crops is of minor importance in the developing countries, but in most of North America and Europe, for example, the area devoted to production of processed vegetable crops is slightly greater than that used for growing fresh market vegetables. The processing crops are grown usually in an extensive fashion whereas fresh market produce is a much more intensive operation. As a result, the value of fresh market produce is about 2.5 times that of processed produce. Again using USDA figures,

Table 1
BOTANICAL CLASSIFICATION OF SELECTED MAJOR VEGETABLE KINDS[2,3]

Family	Species	Common name
Gramineae — grass family	*Zea mays* var. *rugosa*	Sweet corn
	Z. mays var. *exerta*	Popcorn
Liliaceae — lily family	*Asparagus officinalis*	Asparagus
Araceae — arum family	*Calocasia esculenta*	Taro (eddo, dasheen)
Amaryllidaceae — amaryllis family	*Allium cepa*	Onion
	A. porrum	Leek
	A. sativum	Garlic
	A. escalonicum	Shallot
	A. schoenoprasum	Chive
Dioscoraceae — yam family	*Dioscorea alata*	Greater (Asiatic) yam
	D. trifida	Cush-Cush yam
Polygonaceae — buckwheat family	*Rheum rhaponticum*	Rhubarb
	Rumex acetosa	Sorrel
Chenopodiaceae — goosefoot family	*Beta vulgaris*	Beet
	B. vulgaris var. *cicla*	Chard
	Spinacea oleracea	Spinach
Aizoaceae — carpetweed family	*Tetragonia expansa*	New Zealand spinach
Cruciferae — mustard family	*Brassica oleracea* var. *acephala*	Kale
	B. oleracea var. *gemmifera*	Brussels sprouts
	B. oleracea var. *capitata*	Cabbage
	B. oleracea var. *botrytis*	Cauliflower
	B. oleracea var. *italica*	Sprouting broccoli
	B. cauloraps	Kohlrabi
	B. napus var. *napobrassica*	Rutabaga (Swede turnip)
	B. campestris var. *rapa*	Turnip
	B. pekinensis	Pe-tsai (Chinese cabbage)
	B. chinensis	Pak-choi (Chinese cabbage)
	Nasturtium officinale	Watercress
	Amoracia rusticana	Horseradish
	Raphanus sativus	Radish
Leguminosae — pea (pulse) family	*Pisum sativum*	Pea
	P. sativum var. *macrocarpon*	Edible-podded pea
	Vicia faba	Broad (fava) bean
	Phaseolus vulgaris	Kidney (snap, French) bean
	P. lunatus	Lima (butter) bean
	P. aureus	Mung (green gram) bean
	P. coccineus	Scarlet runner bean
	Vigna sesquipedalis	Asparagus bean
	V. sinensis	Cowpea
	Glycine max	Soybean
	Cajanus cajan	Pigeon pea
	Cicer arietinum	Chick pea
Euphorbiaceae — spurge family	*Manihot esculenta*	Cassava (manioc, tapioca)
Malvaceae — mallow family	*Hibiscus esculentus*	Okra
Umbelliferae — parsley family	*Daucus carota*	Carrot
	Petroselinum crispum	Parsley
	Apium graveolens	Celery
	A. graveolens var. *rapaceum*	Celeriac
	Pastinaca sativa	Parsnip
Convolvulaceae — morning glory family	*Ipomea batatas*	Sweet potato
Sapindaceae — soapberry family	*Blighia sapida*	Akee (ackee)

Table 1 (continued)
BOTANICAL CLASSIFICATION OF SELECTED MAJOR VEGETABLE KINDS[2,3]

Family	Species	Common name
Solanaceae — nightshade family	*Solanum tuberosum*	Potato
	S. melongena	Eggplant (aubergine)
	Lycopersicon esculentum	Tomato
	L. esculentum var. *cerasiforme*	Cherry tomato
	Capsicum annuum	Sweet pepper
	C. frutescens	Hot pepper
Cucurbitaceae — gourd family	*Cucurbita pepo*	Pumpkin
	C. pepo var. *condensa*	Summer squash
	Citrullus vulgaris	Watermelon
	Cucumis sativus	Cucumber
	C. melo	Muskmelon
	C. melo var. *reticulatus*	Netted melon
	C. melo var. *cantaloupensis*	Cantaloupe
	C. melo var. *inodorus*	Casaba melon
Compositae — daisy or sunflower family	*Cichorum indivia*	Endive
	Tragopogon porrifolius	Salsify
	Taraxacum officinale	Dandelion
	Lactuca sativa var. *crispa*	Curled lettuce
	L. sativa var. *capitata*	Head lettuce
	Lactuca sativa var. *longifolia*	Cos lettuce
	Helianthus tuberosum	Jerusalem artichoke
	Cynara scolymus	Artichoke

it is noted that when vegetable crop production is compared with that of all other food crops in the U.S., the area for vegetables is only 1.5% of the total, whereas the vegetable produce accounts for 8.5% of the value of all nonmeat food crops.[4] FAO projections as noted by Burger[5] were for annual expansion in vegetable production of 3.6% between 1970 and 1980. The actual growth levelled at 2.4%/year for the period. To some degree the increases in mass production in recent years have led to food volumes in developed countries that are close to the saturation point. The many developing countries that are unable to produce foodstuffs adequate for their own needs find it uneconomic to import because of transport and/or storage costs.

VEGETABLES AND CLIMATE

The diverse nature of the many vegetable kinds, as well as their genetic variability, have led to a very general but useful classification based on plant reaction to growing-season temperatures (Table 4).

In evaluating the cool-season/warm-season classes, it should be stressed that generally the warm-season kinds are those from which the fruiting parts are eaten. Exceptions are many of the tropical and subtropical vegetables such as yams and sweet potatoes. Modern plant-breeding techniques have resulted in the development of cultivars of several vegetables (lettuce, tomatoes, snap beans) that could logically be placed in both of the above classes. Their genetic base has been so broadened that they respond to more extremes of environment than do their parent species.

Other means for modifying the restrictions imposed by the reaction of plants to their environment are the use of black plastic or paper mulches to improve soil heat, and the use of cloches or similar protectors (including greenhouses) to prolong the growing season, especially during the spring months. In tropical climates, cool-season crops are grown on mountainsides or on high-altitude plateaus.

Table 2
SOME CHARACTERISTICS OF SELECTED VEGETABLE KINDS

Life cycle	Crop	Edible portion(s)	Uses
Perennials[a]	Asparagus	Emerging shoots in spring	Fresh, frozen, canned
	Artichoke	Large, fleshy, flower bud	Fresh
	Cassava	Thickened root	Fresh, processed into flour
	Chive	Thick "sods" of hollow stems	Fresh in salads and soups; can be forced
	Potato	Tuber	Fresh, stored, processed
	Sweet potato	Fleshy, tuberous root	Fresh; can be forced
Biennials[b]	Beet	Root	Fresh; can be stored, pickled
	Broccoli	Masses of flower buds	Fresh, frozen
	Cabbage	Modified bud (head)	Fresh, processed; can be stored
	Carrot	Root	Fresh; can be stored
	Cauliflower	Fleshy, modified flower parts	Fresh, frozen, pickled
	Celery	Cluster of leaf bases	Fresh
	Leek	Thickened base of imbricated leaves	Fresh, can be stored
	Onion	Mature or immature bulb	Fresh, dried; can be stored, pickled
	Rutabaga	Root	Fresh; can be stored
Annuals	Bean	Immature pod; seed — immature or mature	Fresh, frozen, canned
	Cucumber	Fleshy fruit	Fresh, pickled
	Lettuce	Succulent leaves — loose or tightly wrapped "heads"	Fresh
	Muskmelon	Fleshy fruit	Fresh; short storage time
	Pea	Immature pods, seeds — immature or mature	Fresh, frozen, canned, stored
	Pepper	Fleshy fruit	Fresh, frozen, canned, processed
	Radish	Fleshy root	Fresh
	Spinach	Leaves and stems	Fresh, frozen
	Squash	Fleshy fruit	Fresh; can be stored
	Sweet corn	Immature seed	Fresh, frozen, canned
	Watermelon	Fleshy fruit	Fresh; short storage time

[a] Certain perennials are usually propagated vegetatively, e.g., potato, rhubarb.
[b] Biennials listed are grown from seed for annual harvest unless seed production is desired.

Table 3
ESTIMATES OF MAN-HOURS/ha, OUTPUT IN kg/MAN-HOUR, AND PERCENT INCREASE IN OUTPUT

	Man-hours/ha		Output in kg/man-hour		Increase in output
Crop	1959	1977	1959	1977	1959 to 1977(%)
Asparagus	71	29	5.9	15.8	269
Beans, snap	55	6	12.1	111.2	915
Carrots	43	31	81.4	161.1	198
Sweet corn	20	6	65.2	234.0	416
Onions	57	34	73.3	165.8	227
Peas	5	4	108.0	139.9	130
Potatoes	21	11	165.1	451.8	275
Tomatoes	76	34	24.8	83.3	336

From Ware, G. W. and McCollum, J. P., *Producing Vegetable Crops*, Interstate Printers & Publishers, Inc., Danville, Ill., 1980. With permission.

Note: U.S. data for 1959 and 1977.

Table 4
GENERAL CLASSIFICATION OF VEGETABLES BASED UPON TEMPERATURE REQUIREMENTS

Cool-season	Warm-season
Asparagus	Chick pea
Beet	Cucumber
Carrot	Lima bean
Cole crops[a]	Muskmelon
Garlic	Pepper
Lettuce	Pigeon pea
Onion	Snap bean
Parsley	Squash
Parsnip	Sweet corn
Pea	Sweet potato
Potato	Tomato
Rhubarb	Yam
Rutabaga	
Turnip	

[a] Cole crops include cabbage, cauliflower, broccoli, etc.

Table 5
SUMMARY OF TWO DECADES OF U.S. VEGETABLE PRODUCTION

Year	Area (1000 ha)	Production (10000 t)	Value ($ millions)
1957	2151	315	1679
1967	2092	372	2389
1977	1893	398	4674

Vegetable products are major food resources for people in all temperate and tropical countries. Most frequently they are used, raw or cooked, from a fresh state. However, increasing use is being made of freezing and canning to preserve the food parts and thus to make vegetables available from one producing season to the next. Improvements in storage and dehydration techniques also have expanded the utilization of produce, and in the case of dried vegetables, the cost of transport to distant consumers is much reduced.

PRODUCTION

Data for area devoted to vegetables, and for per capita consumption, are difficult to develop, mainly because a large volume of produce comes from gardens of nonfarm families, and such noncommercial units seldom enter agricultural statistics. A significant amount of vegetable produce is grown on town or city garden allotments. Shoemaker[6] estimates that during the early 1950s the U.S. and Canada had 6 million farms and 16 million nonfarm gardens. Of the former group, the farms listed as commercial vegetable growers numbered under 1 million. The ten leading vegetable kinds in North America are, in order of hectares producing, potatoes, tomatoes, sweet potatoes, peas, sweet corn, beans, melons (including muskmelons), cabbage, lettuce, and onions (Table 5).

It must be stressed that the huge number of home gardens and community market outlets in the developing countries makes it impossible to accumulate data that are fully representative. Thus the per capita production from the "other countries" is almost certainly somewhat more than the tabular data suggest (Table 6).

Information concerning the value of exports and imports is given in Figure 1. Data for per capita consumption of fresh and processed produce are in Figures 2 and 3. Although world vegetable data for production volume for export and import, and for per capita consumption, are not available, Burger has shown that for major European countries and for Japan, Australia, and the U.S., using data from the U.S. is sufficiently representative to indicate world trends.

Table 6
CHANGES IN AVERAGE ANNUAL PRODUCTION OF VEGETABLES IN THE PERIOD 1969—1980[7]

	Av annual production (thousands of tonnes)			Av annual production per capita (kg)		
Period	World	Developed market economies	Other countries	World	Developed market economies	Other countries
1969—1971	271,419	83,040	188,379	73.82	114.42	63.84
1978—1980	345,218	95,092	250,126	79.62	121.90	70.35

From Anon., FAO Production Yearbook, Vol. 34, Food and Agriculture Organization, Rome, 1981. With permission.

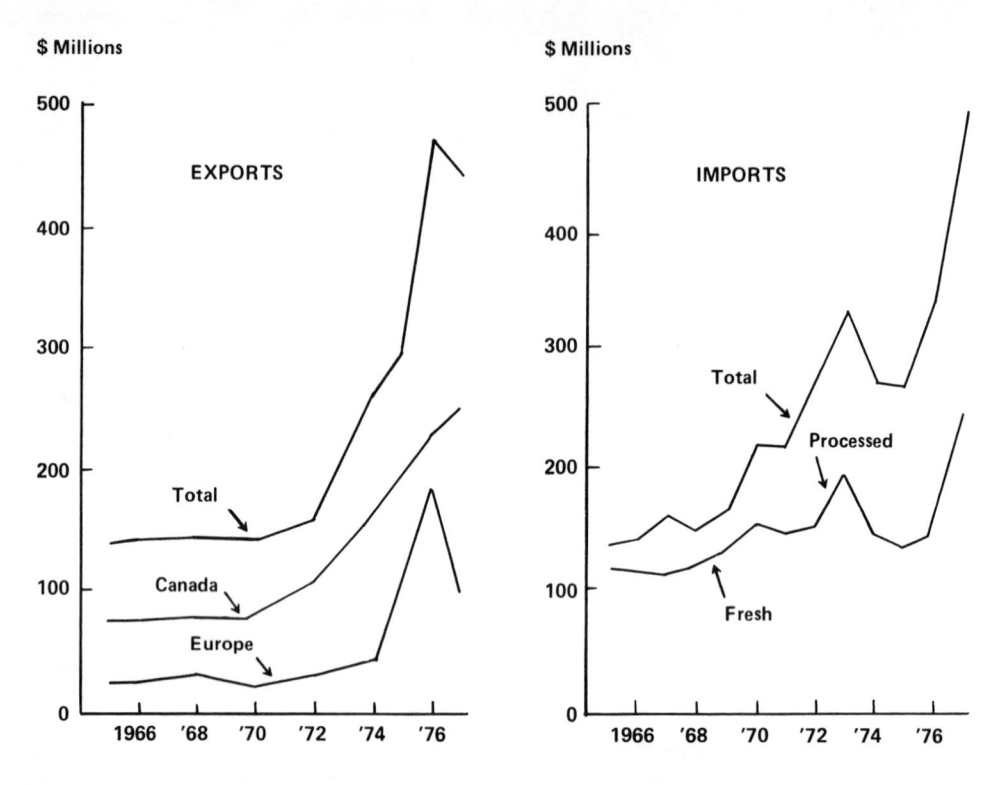

FIGURE 1. Value of vegetables exported from and imported into the U.S., 1965—1977.[1] (Data extrapolated from various U.S. Department of Agriculture reports.)

TYPES OF PRODUCTION UNITS

The ease with which many kinds of vegetables can be grown in relatively small areas, the great diversity of kinds of vegetables, and the widely disparate food products associated with vegetables all lead to a classification of production units into certain types. Six such classes are described here.

Home gardening — For many people, the practice of growing vegetables for home use is an enjoyable combination of physical exercise and a means for economizing in food costs. The crops grown are chosen partly as adapted to the soil and climate involved and partly on family preferences.

Market gardening — Usually this term is applied to production for local markets. Before train and truck transport became common, each community had peripheral farms that grew

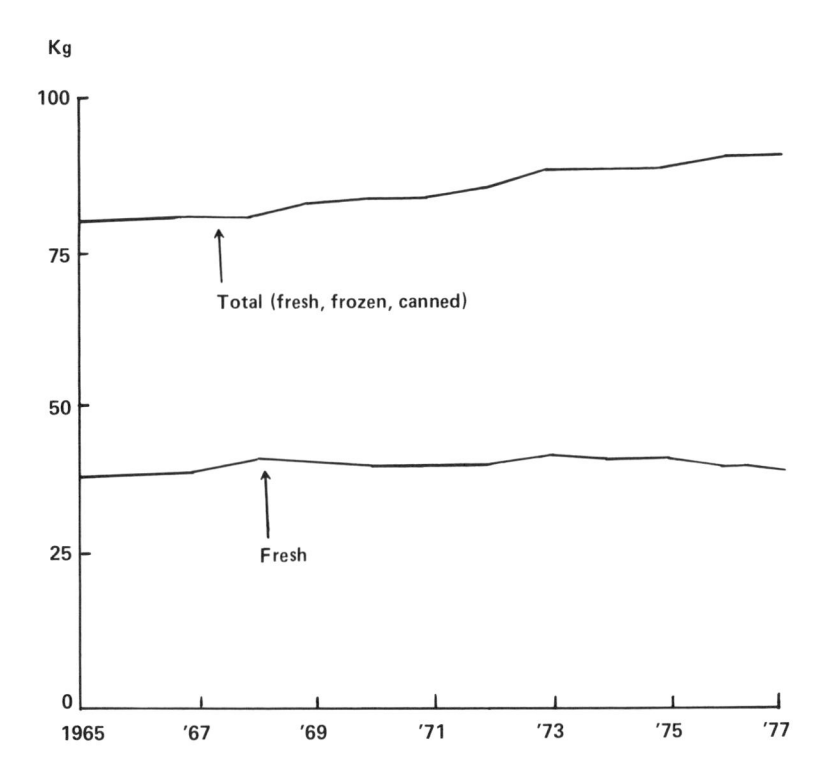

FIGURE 2. Vegetable consumption per capita (U.S.). (From Anon., FAO Production Yearbook, Vol. 33, Food and agriculture Organization, Rome, 1980.)

produce for local sale. The extent of this type of vegetable farming has been much curtailed in recent years, though there are signs of rejuvenation of many city produce markets.

Truck farming — The historical development of specialized farms to grow extensive amounts of a few kinds of vegetables required that they be located near large markets, or alternatively near rail or water transport. With the advent of motor trucks and refrigerated transport units, the "truck farms" are now sited more for suitability of soil and climate than for convenience of market location.

Production of crops for processing — These units are the largest-scale vegetable farms, often growing one or two kinds of vegetables, to supply the constantly expanding markets for frozen, canned, pickled, and dehydrated produce.

Pick-your-own producers — The growing of vegetables specifically for consumer harvesting is a modern version of the early practice in which townspeople "dropped by" a nearby farm for milk and eggs, and perhaps a peck of potatoes. The "Pick-Your-Own" or "U-Pick" farms are located so as to be accessible to the many families who live in apartments and condominiums. These are families whose urge to see and feel and take part in food harvesting can be satisfied by a short drive to the country.

Vegetable forcing — As noted elsewhere in this chapter, certain vegetable kinds are grown at other than normal times, by providing an "artificial" climate as in a greenhouse during winter, for example. Tomatoes, cucumbers, and lettuce are by far the most important such crops. For mushrooms, a stable, uniform climate in caves or special sheds will allow production on a year-round basis. Two other forced crops are rhubarb and endive. These are forced into winter growth in dark sheds or cellars, utilizing the stored energy of the roots that are lifted from the field in autumn, and provided with the period of low temperature necessary for growth to be resumed.

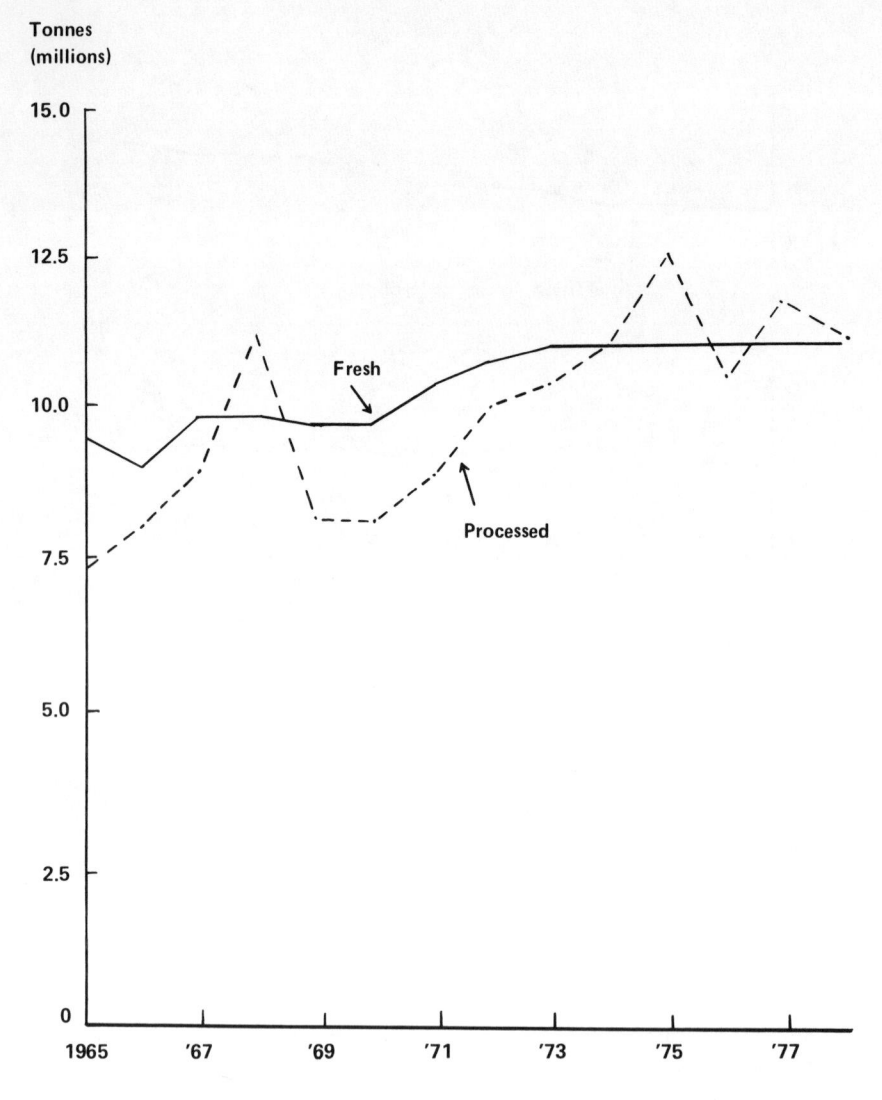

FIGURE 3. U.S. production of fresh vegetables relative to those processed.

RESEARCH TRENDS

Perhaps the most noteworthy advances in recent research involving vegetable crops has been that devoted to genetic manipulation within each of the many species making up this agricultural class. Tomato cultivars, for example, are now commonly available, either as hybrids or true-breeding cultivars, that are immune or highly resistant to the debilitating viral, fungal, and bacterial diseases that beset this crop. Some control of physiological disorders also has been achieved through breeding.

The genetic success story has been replicated for many crops; notably for parthenocarpy in cucumbers, male sterility in cole crops (cabbage family) so that hybrid seed production becomes commercially practicable, F_1 hybrids of sweet corn, disease resistance in cucumbers and melons, tip-burn resistance in lettuce, and resistance to Verticillium wilt in many vegetable crops.

The production of vegetable seed on a commercial basis is a highly specialized business and usually confined to relatively dry, long-season areas (Idaho, Utah, Montana, Alberta,

Table 7
AN APPROXIMATE RANGE OF SOIL REACTION TOLERANCE FOR SELECTED VEGETABLES

Slightly tolerant (pH 7.5—6.0)	Moderately tolerant (pH 7.5—5.5)	Very tolerant (pH 7.5—5.0)
Asparagus	Cabbage	Bean
Beet	Carrot	Eggplant
Cauliflower	Cucumber	Kohlrabi
Lettuce	Pepper	Pea
Muskmelon	Chard	Potato
Onion		Squash
Spinach		Sweet corn

California, British Columbia) where costs of drying the mature plants are relatively light. Modern facilities for maintaining good control of humidity and temperature have resulted in the seed industry being able to offer vegetable seed with a relatively high germinability.

Vegetable crop kinds, diverse as they are, nevertheless do not exhibit extremes in their reaction to soil acidity or alkalinity. This may be because they are not as dependent on mycorrhizal associations as are, for example, certain fruit and ornamental plant kinds such as blueberries, cranberries, and rhododendrons. Table 7 lists certain vegetables as they usually show response to soil acidity.

COMPOSITION OF VEGETABLE PRODUCE

With steady advances in understanding human nutritional requirements, there follows an equally steady interest in "natural foods". It is reasonable, therefore, to recognize a widespread concern to know just what is the makeup of the foods we eat. Table 8 provides a listing of the major food components of a selected group of vegetables.[8]

The vegetable portion of our daily diet is substantial and we should be at least as concerned about its health-giving properties as about the external appearance and texture of the produce. Hardh[10] states that researchers are investigating the use of organoleptic tests as quality determinants, to take the place of chemical analyses as a basis of classification.

Table 8
COMPOSITION OF 100 g OF FRESH VEGETABLE PRODUCT (EDIBLE PORTION) — SELECTED KINDS

Vegetable	Water (%)	Energy (cal)	Protein (g)	Fat (g)	Carbohydrates (g)	Ca (mg)	Fe (mg)	P (mg)	Vit. A (IU)	Thiamine (mg)	Riboflavin (mg)	Niacin (mg)	Vit. C (mg)
Beans, snap	90	30	2.1	0.2	6.5	56	0.8	44	425	0.90	0.11	0.5	19
Beets	87	43	1.6	0.1	9.9	16	0.7	33	20	0.03	0.5	0.4	10
Cabbage	92	24	2.0	0.2	7.2	52	0.6	38	130	0.05	0.05	0.3	47
Carrot	88	42	1.1	0.2	9.7	37	0.7	36	11,000	0.06	0.05	0.6	8
Lettuce	95	14	1.1	0.2	2.9	40	1.3	24	1,300	0.06	0.06	0.3	12
Muskmelon	91	30	0.7	0.1	7.7	14	0.4	16	3,400	0.04	0.03	0.6	33
Peas, green	78	84	6.3	0.4	14.4	26	1.9	116	640	0.35	0.14	2.9	27
Pepper, green	93	22	1.2	0.2	4.8	9	0.7	22	420	0.08	0.08	0.5	128
Potato	78	87	2.0	0.1	19.1	11	0.7	56	20	0.11	0.04	1.2	17
Rutabaga	87	46	1.1	0.1	11.0	66	0.4	39	580	0.07	0.07	1.1	43
Spinach	91	26	3.2	0.3	4.3	93	3.1	51	8,100	0.10	0.20	0.6	51
Squash, winter	85	50	1.4	0.3	12.4	22	0.6	38	3,700	0.05	0.09	0.6	3
Sweet corn	73	96	3.5	1.0	22.1	3	0.7	111	400	0.15	0.12	1.7	2
Sweet potato	69	126	1.8	0.7	28.0	30	0.7	49	7,700	0.10	0.06	0.7	22
Tomato	94	22	1.1	0.2	47.0	13	0.5	27	900	0.06	0.04	0.7	23

Adapted principally from USDA data, 1945—1963, as presented in Shoemaker, J. S., *General Horticulture*, J. P. Lippincott, New York, 1952); and Lorenz, O. A. and Maynard, D. M., *Knott's Handbook for Vegetable Growers*, 2nd ed., John Wiley & Sons, New York, 1980.

REFERENCES

1. **Anon.,** FAO Production Yearbook, Vol. 33, Food and Agriculture Organization, Rome, 1980.
2. **Everett, T. H.,** *New Illustrated Encyclopedia of Gardening,* Vol. 13, Graphic Press, New York, 1960.
3. **Ware, G. W. and McCollum, J. B.,** *Producing Vegetable Crops,* Interstate Printers and Publishers, Danville, Ill., 1980.
4. **Anon.,** Report of Economics, Statistics and Cooperative Service, U.S. Department of Agriculture, Washington, D.C., 1980.
5. **Burger, A.,** Intensive horticulture in the age of retarded economic growth, Proc. XXVI Int. Horticultural Congr., Hamburg, West Germany, 1982, 936.
6. **Shoemaker, J. S.,** *General Horticulture,* J. P. Lippincott, New York, 1952.
7. **Anon.,** FAO Production Yearbook, Vol. 34, Food and Agriculture Organization, Rome, 1981.
8. **Knott, J. E.,** *Vegetable Growing,* 4th ed., Lea & Febiger, Philadelphia, 1980.
9. **Lorenz, O. A. and Maynard, D. N.,** *Knott's Handbook for Vegetable Growers,* 2nd ed., John Wiley & Sons, New York, 1980.
10. **Hardh, J. E.,** Modern quality requirements for vegetables, Proc. XXVI Int. Horticultural Congr., Hamburg, West Germany, 1982, 701.
11. **Nath, P. and Swarup, U.,** Breeding tropical vegetable crops, Proc. XXVI Int. Horticultural Congress, Hamburg, West Germany, 1982, 563.
12. **Thompson, H. C. and Kelly, W. C.,** *Vegetable Crops,* 5th ed., McGraw-Hill, New York, 1957.

FRUIT CROPS

F. W. Liu

INTRODUCTION

Fruit is an important food item in our everyday diet, one for which there is no total substitute. It is indispensable to our health, supplying some essential nutrients, vitamins, and minerals, and it provides important intestinal regulatory agents, fibers, and pectic substances. No wonder there is a saying, "An apple a day keeps the doctor away."

Hundreds of plant species produce fruit which is used as "fruit", either in fresh or processed forms, by human beings. Many of these grow wild and are sometimes gathered and consumed by the local inhabitants. More important species are cultivated; and the most sophisticated agricultural technology is applied by the fruit industry in the production, processing, and marketing of fruit crops. The top ten species in terms of the quantity produced in the world are grape, banana, orange, apple, coconut, plantain, mango, pineapple, pear, and mandarin, in that order. (Large amounts of grapes are used for wine making and large amounts of coconuts are dried for copra instead of being consumed as "fruit", however.)

Americans consume 41 kg of fresh fruit and 62 kg equivalent of processed fruit per capita per year on average. The top ten fresh fruits consumed in the largest quantities by Americans are banana, apple, orange, grapefruit, peach, pear, grape, lemon, strawberry, and tangerine, in that order. About one half of the processed fruit consumed in the U.S. is frozen orange juice.

Nearly 100 important species of crops grown as "fruit" will be discussed in this chapter. Those species which are grown exclusively for dry nuts, stimulant beverages, and condiments, or which are generally used as vegetables, are not included. A table (Table 1) is used to list the fruit species and their common and scientific names, principal part consumed, and major uses. *Annona*, brambles, citrus, *Prunus*, and *Ribes* species are grouped to show the genetic closeness and characteristic similarities of the members in each group. Following Table 1, each species or group of species is given a brief description of the plant, climatic requirements, method(s) of propagation, characteristics or traits of fruit, and the utilization of the fruit.

Table 1
UTILIZATION OF FRUIT CROPS

Common name(s)	Scientific name(s)	Principal part consumed	Uses
Acerola (see Barbados cherry)			
Annona	*Annona* spp.	Flesh of fruit	Fresh fruit
Apple	*Malus domestica* Brokh.	Flesh of fruit	Fresh fruit, cooked (as for pie), applesauce, canned, dehydrated, frozen, juice, cider, vinegar
Apricot (see Prunus)			
Avocado	*Persea americana* Mill.	Flesh of fruit	Fresh fruit
Banana	*Musa* sp.	Flesh of fruit	Fresh fruit, puree, chips, dehydrated
Barbados cherry	*Malpighia glabra* L.	Flesh of fruit	Fresh fruit, juice, jelly, jam, preserves
Bitter orange (see Citrus)			
Blackberry (see Brambles)			
Blueberry	*Vaccinium* spp.	Total fruit	Fresh fruit, canned, frozen, juice, jam, jelly, preserves
Brambles	*Rubus* spp.	Total fruit	Fresh fruit, frozen, jam, jelly, preserves, canned, juice
Breadfruit	*Artocarpus altilis* (Patinson) Fosberg.	Flesh of fruit	Cooked
Bullock's heart (see Annona)			
Calamondin (see Citrus)			
Cantaloupe (see Melon)			
Carambola	*Averrhoa carambola* L.	Flesh of fruit	Fresh fruit, juice, canned, preserves
Cherimoya (see Annona)			
Cherry (see Prunus)			
Chinese gooseberry (see Kiwi)			
Citron (see Citrus)			
Coconut	*Cocos nucifera* L.	Kernel of nut, liquid endosperm	Fresh fruit, fresh drink, dried shredded, dried copra
Cooking banana (see Plantain)			
Citrus			
Calamondin	*Citrus madurensis* Loureiro	Flesh of fruit	Fresh fruit, sauce, juice ornamental
Citron	*C. medica* L.	Rind	Brined, candied
Grapefruit	*C. paradisi* Macf.	Flesh of fruit	Fresh fruit, canned, juice
Kumquat	*Fortunella* spp.	Total fruit	Fresh fruit, preserved, candied
Lemon	*C. limon* (L.) Burm. f.	Flesh of fruit	Fresh fruit, juice
Lime	*C. aurantifolia* Swing.	Flesh of fruit	Fresh fruit, juice
Mandarin, common	*C. reticulata* Blanco	Flesh of fruit	Fresh fruit, canned, juice
Mandarin, Satsuma	*C. unshiu* Marc.	Flesh of fruit	Fresh fruit, canned, juice
Orange	*C. sinensis* (L.) Osbeck	Flesh of fruit	Fresh fruit, juice
Pummelo	*C. grandis* (L.) Osbeck	Flesh of fruit	Fresh fruit
Tangerine (see Mandarin, common)			
Cranberry	*Vaccinium macrocarpon* Ait.	Whole fruit	Fresh fruit, sauce, juice, frozen
Currant (see Ribes)			
Date (date palm)	*Phoenix dactylifera* L.	Flesh of fruit	Fresh fruit, dried

Table 1 (continued)
UTILIZATION OF FRUIT CROPS

Common name(s)	Scientific name(s)	Principal part consumed	Uses
Dewberry (see Brambles)			
Durian	*Durio zibethinus* L.	Flesh of fruit	Fresh fruit
Fig	*Ficus carica* L.	Total fruit	Fresh fruit, dried, canned
Gooseberry (see Ribes)			
Grape			
European grape	*Vitis vinifera* L.	Flesh of fruit or total fruit	Fresh fruit, dried (raisin), juice, wine, canned, frozen, jelly, jam
Fox grape	*V. labrusca* L.	Flesh of fruit or total fruit	Fresh fruit, juice, wine, jelly, jam
Muscadine grape	*V. rotundifolia* Michx.	Flesh of fruit or total fruit	Fresh fruit, juice, wine, jelly, jam
Grapefruit (see Citrus)			
Guava	*Psidium guajava* L.	Flesh of fruit or total fruit	Fresh fruit, puree, juice, jelly
Ilama (see Annona)			
Jackfruit	*Artocarpus heterophyllus* Lam.	Flesh of fruit, seed	Fresh fruit, cooked, canned, dried, jam, cooked seed
Jaboticaba	*Myrciaria cauliflora* (DC.) Berg. (*Eugenia cauliflora* DC.)	Flesh of fruit	Fresh fruit
Japanese apricot (see Prunus)			
Jujube	*Ziziphus jujuba* Mill.	Flesh of fruit	Fresh fruit, dried, preserved
Kanran	*Canarium album* Raeusch	Flesh of fruit	Fresh fruit, preserved
Kiwi	*Actinidia chinensis* Planch.	Flesh of fruit	Fresh fruit
Kiwifruit (see Kiwi)			
Kumquat (see Citrus)			
Langsat	*Lansium domesticum* Corr.	Flesh of fruit	Fresh fruit
Lanzon (see Langsat)			
Leechee (see Litchi)			
Lemon (see Citrus)			
Lime (see Citrus)			
Litchi	*Litchi chinensis* Sonn.	Flesh of fruit	Fresh fruit, canned, dried
Longan	*Euphoria longana* Lam.	Flesh of fruit	Fresh fruit, canned, dried
Loquat	*Eriobotrya japonica* Lindl.	Flesh of fruit	Fresh fruit, canned
Lychee (see Litchi)			
Mamey sapote (see Sapote)			
Mandarin (see Citrus)			
Mango	*Mangifera indica* L.	Flesh of fruit	Fresh fruit, juice, canned, dried
Mangosteen	*Garcinia mangostana* L.	Flesh of fruit	Fresh fruit
Melon	*Cucumis melo* L.	Flesh of fruit	Fresh fruit
Mulberry	*Morus spp.*	Total fruit	Fresh fruit, juice, jam
Muskmelon (see Melon)			
Nectarine (see Prunus)			
Olive	*Olea europaea* L.	Flesh of fruit	Pickled, canned, oil
Orange (see Citrus)			
Papaw (see Papaya)			
Papaya	*Carica papaya* L.	Flesh of fruit, latex	Fresh fruit, dried, preserved, pickled (immature fruit), papain (latex)
Passion fruit	*Passiflora edulis* Sims.	Flesh of fruit	Fresh fruit, juice
Pawpaw (see Papaya)			
Peach (see Prunus)			

Table 1 (continued)
UTILIZATION OF FRUIT CROPS

Common name(s)	Scientific name(s)	Principal part consumed	Uses
Pear			
European pear	*Pyrus communis* L.	Flesh of fruit	Fresh fruit, canned, puree, jelly
Japanese pear	*P. serotina* Rehd.	Flesh of fruit	Fresh fruit, canned
Persimmon			
American persimmon	*Diospyros virginiana* L.	Flesh of fruit	Fresh fruit
Japanese persimmon	*D. kaki* L.	Flesh of fruit	Fresh fruit, dried
Pineapple	*Ananas comosus* (L.) Merr.	Flesh of fruit	Fresh fruit, canned, juice, dehydrated, bromellin
Plantain	*Musa paradisiaca* L.	Flesh of fruit	Cooked
Plum (see Prunus)			
Pomegranate	*Punica granatum* L.	Flesh of fruit	Fresh fruit, juice, drinks
Pomelo (see Citrus)			
Prune (see Prunus)			
Prunus			
Apricot	*Prunus armeniaca* L.	Flesh of fruit	Fresh fruit, canned, dried
Cherry, sweet	*P. avium* (L.) L.	Flesh of fruit	Fresh fruit, canned, frozen, brined
Cherry, sour (tart)	*P. cerasus* L.	Flesh of fruit	Fresh fruit, canned, frozen
Japanese apricot	*P. mume* Sieb. and Zucc.	Flesh of fruit	Preserves
Nectarine and peach	*P. persica* (L.) Batsch.	Flesh of fruit	Fresh fruit, canned, frozen, jelly
Peach (see Nectarine and Peach)			
Plum, European	*P. domestica* L.	Flesh of fruit	Fresh fruit, dried, canned
Plum, Japanese	*P. salicina* Lindl.	Flesh of fruit	Fresh fruit, preserved
Prune (see Plum, European)			
Pummelo (see Citrus)			
Quince	*Cydonia oblonga* Mill.	Flesh of fruit	Jelly, marmalades, preserves
Rambutan	*Nephelium lappaceum* L.	Flesh of fruit	Fresh fruit
Raspberry (see Brambles)			
Ribes	*Ribes* spp.	Total fruit	Fresh fruit, jelly, jam, juice, frozen
Rose-apple	*Syzygium jambos* (L.) Alst.	Flesh of fruit	Fresh fruit
Sapodilla	*Manilkara zapota* (L.) van Royen	Flesh of fruit	Fresh fruit
Sapote	*Calocarpum sapota* Merr. (*Lucuma mammosa* Gaertn.)	Flesh of fruit	Fresh fruit
Satsuma mandarin (see Citrus)			
Shaddock (see Citrus)			
Sour orange (see Citrus)			
Soursop (see Annona)			
Star-apple	*Chrysophyllum cainito* L.	Flesh of fruit	Fresh fruit
Strawberry	*Fragaria* spp.	Total fruit	Fresh fruit, frozen, jam, jelly, juice
Sugar-apple (see Annona)			
Sweet orange (see Citrus)			
Sweetsop (see Sugar apple)			
Tamarind	*Tamarindus indica* L.	Flesh of fruit	Fresh fruit, jam, juice, flavoring
Tangerine (see Citrus)			
Watermelon	*Citrullus vulgaris* Schrad.	Flesh of fruit	Fresh fruit, pickled (rind)
Wax-apple	*Syzygium samarangense* Merr. et Perry	Flesh of fruit	Fresh fruit
West Indian cherry (see Barbados cherry)			

ONE HUNDRED IMPORTANT FRUIT CROPS

Annona

At least five *Annona* species are grown for fruits: soursop (*A. muricata* L.), cherimoya (*A. cherimola* Mill.), sugar-apple or sweetsop (*A. squamosa* L.), ilama (*A. diversifolia* Saff.), and bullock's heart (*A. reticulata* L.). Some of the small to medium trees are evergreen, but others, such as cherimoya and sugar-apple, tend to shed all leaves before new shoots emerge in the spring. They are grown in tropical and warm subtropical regions; soursop requires a tropical climate and cherimoya and sugar-apple like a slightly cooler climate than pure tropical. *Annona* species are propagated by seeds or grafting. Fruit is a syncarp formed by the fusion of many carpels with the receptacle. Soursop fruit is oblong or heart shaped, 15 to 22 cm long, 1 to 2 kg in weight. The skin is green with many soft spines, white flesh, and has a sharp subacid and sweet flavor. Cherimoya fruit is heart shaped, 0.1 to 1.0 kg in weight, with rounded tubercles on the surface. The flesh is buttery or custard-like, with a pleasant blend of sweetness and mild acidity. Sugar-apple fruit is also heart shaped, but smaller and with more pronounced protuberances than cherimoya; the flesh is white and sweet. Cherimoya and sugar-apple fruit will not ripen properly below 15°C. Ilama and bullock's heart fruits are somewhat similar to cherimoya, but bullock's heart fruit has an insipidly sweet flavor, which is inferior to cherimoya. The fruit is consumed fresh, and is especially pleasant in ices and cool drinks.

Apple

The apple is a small- to medium-sized deciduous tree. Important varieties are grown in the temperate zone. Tropical apples are rare and poor in quality. The apple is propagated asexually, usually by grafting or budding. Fruit is a pome, various in shape (globular, oblate, conic), color (red, yellow, green), and size (5 to 12 cm in diameter). Fruit matures in summer or fall, and can be stored for many months in refrigerated (0 to 3°C) or controlled atmosphere storage for some varieties. Consumed fresh, cooked, or processed, although some varieties are only suitable for processing or cooking.

Avocado

The avocado is a medium to large evergreen tree grown in tropical and subtropical regions. It is propagated asexually, usually by grafting. The fruit is various in shape (spherical to pyriform), color (green, black, purple or reddish), and size (100 to 2000 g in weight). It matures in various seasons depending on varieties and does not ripen until it falls or is picked. The flesh contains little sugar, but has 3 to 30% oil. Fruit is consumed fresh as in salads, appetizers, dips, and "guacamole". Cooking impairs the flavor and appearance, but frozen products can be prepared.

Banana

The banana is a perennial herb which can be 2 to 10 m tall. It is grown in tropical region and subtropical region where there is no severe frost and propagated by suckers and corms or by tissue culture. The fruit is elongated, angled, and somewhat curved, 6 to 25 cm long, and is yellowish green, yellow, or red when ripe. Each plant can bear only one fruit bunch which consists of many fruits. Commercially grown bananas are seedless, but wild bananas are seedy and hardly worth eating. Bananas are harvested year-round in the tropics. They are normally harvested green and ripened before retailing or consumption. The fruit is susceptible to chilling injury if kept below 14°C. The flesh is very sweet and fragrant when fully ripe. It is mostly consumed fresh, but can be processed. Green fruit can be cooked and eaten under emergency conditions, but another group of bananas generally used for cooking is known as "cooking" banana and is discussed under plantain.

Barbados Cherry (Acerola, West Indian Cherry)

This large and densely branched shrub or small evergreen tree is grown in tropical and subtropical regions and propagated by air layering or hardwood cuttings. The fruit is small (2 to 3 cm in diameter), three-lobed, red when ripe, and very perishable; the flesh is yellow-orange and has very high vitamin C content (1000 to 2000 mg/100 g). It may be consumed fresh, and is excellent for juice by itself or in a mixture; can also be processed into other forms.

Blueberry

Although many blueberry species grow wild, only three have been bred and cultivated for fruit. These are the lowbush blueberry *(Vaccinium angustifolium)* from northern North America, the highbush blueberry *(V. corymbosum)* from the eastern U.S., and the rabbiteye blueberry *(V. ashei)* from the southern U.S. Wild populations of the lingonberry *(V. vitis-idaea)* are harvested for fruit in Canada and northern Europe. The lowbush blueberry is only 15 to 55 cm high, but the highbush blueberry is 3 to 5 m high. Blueberries are mainly grown in the temperate zone, and are propagated by cuttings or through tissue culture. The fruit is a small berry (<1.5 cm in diameter) with high vitamin C content. It may be consumed fresh or processed into various forms.

Brambles

Six major types of brambles are cultivated for fruit. These are the red raspberries *(Rubus strigosus* and *R. idaeus)*, the black raspberry *(R. occidentalis)*, the purple hybrid between the former species *(R. neglectus),* erect blackberries consisting of a heterogeneous mix of species and hybrids, trailing blackberries such as the Loganberry and Boysenberry, and dewberries. They are perennial plants with a biennial fruiting habit. Their slender canes may be erect, arched, or prostrate depending on types. They are mainly grown in the temperate zone, but some species can be grown in subtropical areas. Brambles are propagated by suckering, tip-layering, or tissue culture. The fruit is a small aggregate fruit. Blackberries are black, but raspberries can be black, red, yellow, or purple. The fruit is very perishable and should be picked carefully and cooled rapidly. It may be consumed fresh or processed into various forms.

Breadfruit

This large evergreen tree is grown in the tropical zone near the equator, and is propagated by seeds, root suckers, or root cuttings. The fruit is multiple fruit with a rough skin, and can reach 4 kg in weight. Ripe fruit is fairly sweet. Mature but unripe fruit is starchy and is consumed after cooking (boiling, steaming, or frying). Seeds can be roasted and taste like chestnuts.

Carambola

This small evergreen tree is grown in tropical and subtropical regions and propagated by grafting or air layering. The fruit is ovoid to ellipsoid, with five prominent longitudinal ribs and, therefore, star-shaped in cross-sections. It is yellowish-green to yellow and 7 to 14 cm long. The flesh is translucent, crisp, and juicy. The fruit has the best quality if ripened on tree; it is very susceptible to mechanical damage and must be handled with great care. Fruit may be consumed fresh or processed into other forms.

Coconut

This monocotyledonous large palm tree is grown in coastal regions of the tropics. The palm can grow in subtropical regions, but it bears fruit with a low yield of kernel and oil there. Propagated by seeds (whole nuts) or by tissue culture and harvested year-found in the tropics. Fruit is a large nut, three-sided, with a thick and tough husk and a hard interior shell, each weighing 1 to 2 kg or more when fresh. Immature fruit is harvested for the liquid endosperm (also known as coconut milk) for drinking. The mature nut contains a thick layer of coconut meat (solid endosperm) under the shell, besides some liquid endosperm. The coconut meat (kernel) is dried and sold as copra for oil and other uses. Some mature coconuts are eaten fresh (the meat is the edible portion) or dried shredded for culinary, bakery, and confectionary uses.

Citrus

Citrus fruits include many species of *Citrus, Fortunella*, and *Poncirus*. Trifoliate orange *(Poncirus trifoliata)* is grown for rootstock and other uses, but not for fruit for eating. Several species of kumquat *(Fortunella* spp.) are grown for fruit or for ornamental use. The nine species of *Citrus* listed in Table 1 are grown commercially for fruit. World production of citrus fruits approaches that of grapes. Trifoliate orange is a deciduous shrub or small tree; other species are small to medium evergreen trees. Citrus fruits are widely distributed in the world, from the equator to 41° latitude. Most of the high-quality citrus fruits are grown in the subtropical region between 23 to 35° latitude. The tolerance to cold temperature among various species is, in decreasing order, kumquat, mandarin, orange, grapefruit, lemon, pummelo, and lime. *Citrus* is propagated asexually, usually by grafting or budding. The fruit is a berry, various in size; green, yellowish-green, yellow, orange, or red in color; and with a high vitamin-C content. The calamondin is small (2 to 3 cm in diameter), rounded or oblate, with loose skin. It is orange colored, and normally too acid for fresh dessert, but is used for sauce or juice. The citron is medium- to large-sized (5 to 8 cm in diameter), oblong, with tight and thick skin (rind), and yellow colored; its rind is mostly brined for fruit cakes or candied. The grapefruit (also called pomelo) is a large fruit (8 to 12 cm in diameter), with thick and tight skin, greenish-yellow to yellow colored, and may be consumed fresh or canned or made into juice. The kumquat is a small (1.5 to 3 cm in diameter), rounded, oval or oblong fruit with thin, loose, and sweet-tasting skin; it is orange colored, and may be consumed fresh with the skin or preserved. The lemon is medium-sized (4 to 6 cm in diameter), oblong or oval, with tight skin, yellow colored, and very acid; it may be consumed fresh as flavoring or juice. The lime is small (3 to 4 cm in diameter), oval or oblong, with thin and tight skin; it is green colored at maturity and very acid, and may be consumed fresh as flavoring or juice. The common mandarin (including the tangerine) is a small to medium (3 to 8 cm in diameter) fruit, rounded or oblate, with thin and loose skin, yellow to orange colored. It is mostly consumed fresh, with some canned in segments or made into juice. The Satsuma mandarin is a medium-sized (5 to 6 cm in diameter) fruit with thin and loose skin; it is also orange colored, and may be consumed fresh or canned in segments or made into juice. The orange (sweet orange) is a medium-sized (5 to 7 cm in diameter) fruit, rounded or oval, with tight skin; it is orange or red colored, and may be consumed fresh or made into juice (Americans consume more orange juice in the form of frozen juice than any fruit in any other forms). The pummelo, also called the shaddock, is a large (8 to 20 cm in diameter) fruit with thick and tight skin; it is greenish-yellow to yellow colored, and may be consumed fresh after peeling off the skin and the segment membrane.

Cranberry

The cranberry is a perennial vine grown in acid bogs or swamp areas in cool temperate zones such as northern U.S. and southern Canada. Cranberries are propagated by cuttings. The fruit is small berries, which are best kept at 2°C after harvest. Cranberries are mostly processed into sauce or made into beverages; some are quick frozen.

Date

This monocotyledonous palm tree is a desert plant which requires a long hot summer for fruit growth and dry weather during flowering and fruit maturation. Leaves will be killed by $-7°C$. These dioecious plants are propagated by offshoots and require artificial pollination for fruit set. The fruit is a berry, oblong, 2 to 3 cm in diameter, and with a stony seed. It is harvested in the fall, and cured or dried for long-term storage. The fruit contains high sugar (50 to 85% after curing), and may be consumed fresh, dried, or used in baking.

Durian

The durian is a large evergreen tree grown in the real tropics and propagated by seeds or rarely by budding or air layering. Durian fruit is large (over 2 kg is common), with a spiny surface and about 1/3 edible flesh (the aril). The fruit has a strong distinct odor very much liked by tropical Asians, but frequently offensive to others who have not learned to enjoy it. It is known as the "king of fruits" and is mostly consumed fresh. Flesh of unripe fruit can be cooked or preserved. Seeds contain starch and can be cooked for consumption.

Fig

This small to medium deciduous tree is grown in subtropical and warm temperate regions where the summer is warm and dry and winter is mild but cool enough to satisfy the chilling requirement of the plant. Figs are propagated by cuttings. The fruit is a "synconium", which is a collective fleshy fruit in which the ovaries are borne upon an enlarged, more or less succulent, concave or hollow receptacle. It is pyriform, ovoid, or globose, 2 to 4 cm in diameter, and yellow, purple, or black. The fruit may be consumed fresh, dried or canned; some is made into fig paste and used in bakery.

Grape

The grape is a perennial deciduous vine, widely distributed in the world. Commercial vineyards are seen in temperate, subtropical, and tropical zones. The tonnage of grapes produced in the world is greater than for any other fruit. Three important species (listed in Table 1) and many hybrids are grown. Most grapes are propagated by cuttings, grafting, or tissue culture; muscadines are propagated by layering. The fruit is a berry, rounded or oval, 1 to 2.5 cm in diameter, various in color (green, white, red, or blue), seeded or seedless. Grapes ripen in summer and fall in temperate zone, but can ripen in winter and spring in tropical and subtropical regions. Some grapes are consumed fresh, but more are processed into wine or juice, or dried as raisins. Some varieties can be kept fresh for many months in refrigerated storage with preservatives such as sulfur dioxide.

Guava

The guava is a small evergreen tree grown in tropical and subtropical regions and propagated by seeds, grafting or inarching. The fruit is a berry, rounded or pyriform, generally with yellow skin and white, yellow, pink, or red flesh with no, few, or many small stony seeds. The fruit has very high vitamin-C content and may be consumed fresh or made into juice or puree which is then remanufactured into nectars, jam, jelly, and bakery products.

Jackfruit

This is a large evergreen tree, grown in tropical and warm subtropical regions. Jackfruit is propagated by seeds or rarely by grafting. The fruit itself is very large (up to 40 kg), with hard and prickly rind, and is borne on stalks which grow out of the trunk or large branches. The edible flesh, which is the aril surrounding each seed, has a distinct odor and sweet taste. It is mostly consumed fresh, but can be processed into various forms. Seeds contain starch and protein and can be consumed after cooking.

Jaboticaba

This is an evergreen small bushy tree or large shrub grown in tropical and warm subtropical regions and propagated by seeds. The fruit is a grape-like berry but with a thicker and tougher skin, produced directly upon the trunk and large branches singly or in a cluster, in several crops per year. It may be consumed fresh or made into jelly, jam, or wine.

Jujube

This deciduous shrub or small tree is grown in temperate and subtropical regions where there is enough summer heat and some winter chilling temperature. It is propagated by root suckers or grafting. The fruit is a drupe, oval or oblong, about 3 cm long, with a red parchment-like skin and a hard pit. Jujube is a popular fruit in China, and is consumed fresh, dried, canned, and also used as Chinese medicine.

Kanran

The kanran is a medium to large evergreen tree grown in subtropical regions and propagated by seeds. The fruit is a drupe, oval, about 3.5 cm long, with a large pit. It is usually too astringent to be consumed fresh as dessert, and thus is mostly preserved.

Kiwi (Kiwifruit, Chinese Gooseberry)

It is a deciduous perennial vine somewhat similar to the grape vine and is grown wild in central China. Kiwi is also commercially grown in warm temperate and cool subtropical regions where there is some chilling temperature in winter. It is propagated by grafting and is dioecious, needing male plants for pollination to set the fruit. The fruit is a berry about the size and shape of a hen's egg, with brownish skin covered with short and stiff hairs. The flesh has attractive green color, fine texture, and pleasant subacid flavor. Fruit may be consumed fresh, but can be made into jam, jelly, wine, or dried or powdered meat tenderizer.

Langsat

Also known as lanzon, this evergreen tree is grown in the tropical region of south Asia and is propagated by seeds or budding. The fruit is 2 to 4 cm long and 2 to 3 cm in diameter, with light-yellow skin covered with numerous short hairs. Edible flesh is the aril surrounding each of the four to six seeds per fruit. The fruit may be consumed fresh.

Litchi (Lychee, Leechee)

The litchi is a medium to large evergreen tree, a subtropical plant very demanding in climate requirements and is best adapted in areas where the summer is hot and wet and the winter is dry and cool but without severe frost. It is a popular fruit in southern China and Taiwan. Litchi is commonly propagated by air-layering, and less commonly by grafting and inarching. The fruit is borne in loose clusters, mature in summer, and rounded or more or less heart shaped, 2 to 4 cm long, with pink to red skin. The flesh (aril) is succulent, whitish, translucent, with excellent subacid flavor. Fruit is best consumed fresh, but can be dried or canned.

Longan

The longan is a medium to large evergreen tree grown in the subtropical zone. It is propagated by seeds, grafting, or air-layering. The fruit is borne in loose clusters, matures in late summer or early fall, and is round or nearly round, 1.5 to 2.5 cm in diameter, with brown skin. The flesh (aril) is succulent, whitish, translucent, and sweet, and a flavor somewhat inferior to litchi. It may be consumed fresh, dried, or canned.

Loquat

This small evergreen tree is grown in the subtropical zone. A mature dormant tree can withstand −12°C. Loquat is propagated by seeds, grafting, or air-layering. The fruit is borne in loose clusters, matures in spring and summer, and is oval, 3 to 4 cm long, with yellow to orange skin. The flesh is firm and juicy, white to orange, with sweet and subacid flavor. Loquat may be consumed fresh or canned, and can be frozen or made into jelly, jam, preserves, and pies.

Mango

The mango is a large evergreen tree grown in tropical and warm subtropical regions free from severe frost. It is one of the most popular tropical fruits and is propagated by seeds or grafting. The fruit is a large drupe, oval or kidney shaped, 100 to 700 g in weight, with yellow or partially red skin when ripe. Mango flesh is yellow to orange colored, soft, and juicy, with sweet and subacid flavor. The fruit contains a substance which may cause irritation and rash on lips and face of some people who are allergic to it. Mango fruit may be consumed fresh, or forzen, canned, dried, or made into juice, preserves, jelly, jam, etc.

Mangosteen

This medium-sized evergreen tree is purely tropical. It is propagated by seeds or by grafting, air-layering, and cuttings. The fruit is nearly rounded, 5 to 8 cm in diameter, with a thick and hard rind, and is purple colored when ripe. The flesh is a white and translucent aril, with a fine and mellow texture and excellent sweet and subacid flavor. It is known as the "queen of fruits". The fruit is very perishable and changes flavor in a few days, and is best when consumed fresh.

Melon

Other names for fruits of this annual vine are muskmelon and cantaloupe. The melon is generally classified as vegetable, but is commonly used as fresh dessert fruit. Melons may be grown in many areas where there is enough summer heat and it is not excessively wet, and are also grown in greenhouses. They are propagated by seeds, and there are many different types and varieties. The fruit is rounded, oval, or oblate, various in size (from a couple hundred grams to several kilograms), with smooth, netted, or rough or scaly skin. The flesh is soft, juicy, and sweet, with an orange, yellow, greenish, or white color. The fruit is mostly consumed fresh.

Mulberry

Many mulberry species bear edible fruit: *Morus nigra* (the black mulberry), *M. rubra*, *M. alba* var. *tatarica*, and *M. alba* var. *multicaulis*. More mulberries are grown for leaves as forage of silkworms than for fruit, however. This deciduous small tree or shrub is grown in warm temperate-zone and subtropical regions, and is propagated by cuttings or by seeds. The fruit is an aggregate fruit, cylindrical in shape, 1 to 3 cm long, red to dark purple in color, soft, juicy, and very perishable. It may be consumed fresh, or made into juice, jam, jelly, etc.

Olive

This large evergreen tree is grown in cool subtropical regions where there is a winter-period mean temperature below 10°C. It is propagated by cuttings or grafting. The fruit is a drupe, oval, 2 to 4 cm long, and turns color from straw green to pink, red, and black during maturation and ripening. It can be picked long before ripe and abscised, and may be pickled, canned, or used for oil.

Papaya (Pawpaw, Papaw)

This is a perennial herb with a succulent stem which can grow very tall (>10 m) and very large (>25 cm in diameter). Papayas are grown in tropical and warm subtropical regions free from severe frost and are propagated by seeds. They start to bear fruit several months after seed germination. There are male, female, and hermaphroditic plants. The fruit is a large berry, 7 to 20 cm in diameter, nearly spherical, oblong, or pyriform in shape, and is yellow colored when ripe. Papayas are harvested the year around. Tree-ripe fruit has the best quality, but mature green fruit is harvested for long-distance shipping and is ripened before consumption. The flesh is soft, yellow or pink in color, sweet, rich in vitamin A and C, and contains papain, a protein-digesting enzyme. It may be consumed fresh or dried or preserved. Immature fruit can be cooked or pickled and consumed as a vegetable. Latex can be collected from young fruit and made into meat tenderizer or used for pharmaceutical purposes.

Passion Fruit

Two types of passion fruit, the purple passion fruit *(Passiflora edulis)* and the yellow passion fruit (*P. edulis* f. *flavicarpa*), are grown for fruit. This is a woody perennial evergreen vine which is a robust climber under favorable conditions. It is grown in tropical and subtropical conditions and propagated by seeds or cuttings. The fruit is rounded or oval, 4 to 7 cm in diameter, with a tough rind, and is purple or yellow when ripe. Edible flesh is the aril surrounding each of the numerous seeds in a fruit. Fruit may be consumed fresh or made into juice.

Pear

This medium-sized deciduous tree is grown in temperate zones but some low-chilling varieties can also be grown in subtropical regions. Pears are propagated by grafting. The European pear is a pome, pyriform, 3 to 10 cm in diameter, yellow, red-shaded, or brown fruit when ripe, commonly harvested at maturity but unripe and ripened to a soft texture before consumption. The Japanese or oriental pear is a pome, rounded, or nearly rounded, or somewhat pyriformed, 5 to 8 cm in diameter, light yellow to brown in color, and is best consumed when it is firm and crisp rather than soft and mealy. Many varieties can tolerate several months of storage in refrigerated or controlled atmosphere. Pears are consumed fresh, canned, or dried. European pears can be made into puree.

Persimmon

This medium-sized to large deciduous tree is grown in temperate and subtropical regions. American persimmon grows wild in the U.S., bears small seedy fruit, and has little commercial value. Japanese or oriental persimmon bears large delicious fruit and is grown extensively in China and Japan. Persimmon is propagated by grafting. The fruit is oblate, heart shaped or nearly rounded, 3 to 8 cm in diameter, and yellow to red when ripe. Persimmon is normally harvested when the fruit has attained yellow to reddish color but is still firm. Some varieties can be eaten at this stage, but others are too astringent and are consumed after ripening or after astringency is removed. Fruit is consumed fresh or dried. Ripe fruit is extremely soft.

Pineapple

Pineapple is a monocotyledonous perennial herb with 50 to 70 thick leaves, each being 4 to 6 cm wide and 50 to 150 cm long, attached in a spiral to a short (<30 cm) stem. Pineapples are grown in warm subtropical to tropical regions, and are propagated by suckers, slips, and crowns which are buds emerged from stems, peduncles, and fruits, respectively. The fruit is an aggregate fruit, truncate to cylindrically shaped, 10 to 15 cm in diameter and

15 to 25 cm long. It is normally yellow when ripe but may remain greenish purple at high temperature. The flesh is whitish to golden yellow, juicy, rich in vitamin C, and contains bromelin, a protein-digesting enzyme. Pineapples may be consumed fresh, canned, frozen, dehydrated or made into juice.

Plantain

Also known as the cooking banana, this is a perennial herb similar to banana. Plantains are grown in the tropical zone and are an important diet of many tropical inhabitants. Plants are propagated by suckers or corms and bear fruit similar to bananas except that plantains contain more starch and less sugar than bananas. Green or partially ripe fruit is cooked (fried, mashed and fried, or boiled) before being consumed. Fruit is also made into banana flour or fermented and distilled for drink. Ripe fruit can be eaten fresh.

Pomegranate

This is a large shrub or small tree; usually deciduous, though there are varieties which are practically evergreen. It is grown in subtropical and warm temperate regions. Pomegranates grown in desert regions where there is a high sum of heat units are of particularly high quality. Plants are propagated by cuttings, layering, or seeds. The fruit is a large, 5 to 8 cm diameter, rounded, pome-like berry, many-celled, and many-seeded, with a tough and hard rind varying in color from pale yellow to deep purple red. Fruit may split before fully ripe and, therefore, is commonly harvested before ripe. Fruit can be stored for many months under refrigeration. The flesh (aril) is pink colored and juicy, with sprightly subacid flower. It may be consumed fresh or made into juice, drinks, and wine.

Prunus

Seven species of *Prunus* are listed in Table 1. These are small to medium-sized deciduous trees and are primarily temperate crops, but some varieties of Japanese apricot, peach, and Japanese plum can be grown in the subtropical region. Plants are propagated by grafting and cuttings. The fruit is a drupe. The apricot is roundish but usually somewhat flattened, 3 to 4 cm in diameter, and yellow overlaid with red. The flesh is firm and sweet, and may be consumed fresh, canned or dried. Sweet cherry is globular or somewhat depressed or in some forms heart shaped, 2 to 3 cm in diameter, yellow or red, with sweet flesh which may be consumed fresh, canned, frozen, or brined for remanufacturing. The sour cherry is globular or depressed, slightly smaller than the sweet cherry, red, with soft and sour flesh which is mostly canned or frozen for pies and other culinary uses. The Japanese apricot (called ''Mei'' in China) is similar in shape but smaller in size than the apricot and has yellow or greenish color. The fruit has dry and acid flesh and is mostly brined and preserved. The nectarine and peach are widely variable in shape and size, 2 to 8 cm in diameter, skin yellow or red, flesh white or yellow. The nectarine has no hairs and the peach has hairs at maturity. The fruit may be consumed fresh or canned. The European plum is usually ovoid to oblong, variable in size (usually 2 to 3 cm in diameter) and usually blue-purple in color. Some may be consumed fresh but more are dried and known as the prune. The Japanese plum is variable in shape (usually globular or somewhat depressed), 1 to 4 cm in diameter, yellow, red, or blue-purple in color, and has a flavor ranging from insipidly sweet to very acid. Fruit may be consumed fresh or preserved.

Quince

This small spreading-type deciduous tree is grown in the temperate zone and propagated by grafting or cuttings. The fruit is a pome, pyriform, 7 to 10 cm in diameter, yellow, with tart and gritty flesh. It is mostly made into jelly, marmalade, preserves, etc.

Ribes

Several species of *Ribes* are cultivated from the many found in northern temperate regions. The genus is divided into currants (red, black, golden, and ornamental) and gooseberries (red, green, yellow, blue, and white). Commonly grown species are *R. sativa* and *R. rubrum* (red currants), *R. americanum* (black currant), *R. odoratum* (golden currant), *R. cynosbati, R. reclinata,* and *R. uva-crispa* (gooseberries). These are deciduous shrubs or bushes, grown in cool temperate zones such as northeast and northwest regions of the U.S. and part of Europe. They are propagated by cuttings or layering. The fruit is a small berry (up to 60 g), seeded, globose to oblong, glabrous to prickly, and red, purple, or black at maturity. Commonly used for juice, jelly, jam, etc.

Rose-apple

This evergreen tree is distributed in tropical, subtropical, and warm temperate regions. It is generally propagated by seeds, but can also be propagated by cuttings, air layering, and budding. The fruit is globular, oval or oblate, 3 to 4 cm in diameter, and light yellow when ripe, with relatively thin edible flesh (pericarp) and hollow inside with one to three seeds. Fruit is consumed fresh or preserved.

Rambutan

The rambutan is a medium-sized evergreen tree grown in tropical regions and propagated by seeds, air layering, or budding. The fruit is borne in loose clusters, mature in summer, globular or oval, 3.5 to 4.5 cm in diameter, and is bright red or yellow with long soft protuberances looking like hairs. The flesh (aril) is similar to litchi but somewhat inferior in flavor. It is mostly consumed fresh. The peel contains tannin and saponine and has pharmaceutical use.

Sapodilla

This medium-sized to large evergreen tree is grown in tropical and warm subtropical regions and is propagated by seeds or grafting. The fruit is rounded to oval, 5 to 10 cm in diameter, with brown and scurfy skin. The flesh is light brown, with a smooth to granular texture, and is soft, juicy, and sweet when ripe. Fruit is usually picked when mature but firm, ripened off the tree before eating, and mainly consumed fresh.

Sapote (Mamey Sapote, Mamey Colorado)

This large evergreen tree is grown in tropical zones and propagated by seeds or inarching. The fruit is ovoid to ellipsoid, 7 to 15 cm in length, with thick, woody, russet-brown and scurfy-surfaced skin. The flesh is salmon pink to reddish brown, smooth to finely granular in texture, and sweet in flavor. It is usually consumed fresh, but can be used in ice cream, sherbert, jelly, drinks, and conserves.

Star-apple

This medium-sized to large evergreen tree is grown in tropical zones and commonly propagated by seeds, but can also be propagated by air-layering, inarching, or cleft grafting. The fruit is rounded or oblate, 5 to 10 cm in diameter, with smooth, waxy, green or purple skin. When cut across through the middle of fruit, the seeds spread out starlike. The flesh is white, with a smooth texture and mild sweet flavor. Fruit may be consumed fresh or preserved with sugar.

Strawberry

The cultivated strawberry *(Fragaria ananassa)* is a hybrid of two wild *Fragaria* species *(F. chiloensis × F. virginiana)*. This low perennial herb is grown in temperate zones as well as in the subtropics and high regions of the tropics. It is propagated by runner plants or tissue culture. In temperate zones, it is among the first of the fresh fruits harvested in the spring. The fruit is an aggregate fruit with a receptacle enlarged and becoming very fleshy (forming the "berry") and holding the seed-like fruitlets, or drupes, on the surface. It is red when ripe (white variety exists), hemispheric or somewhat elongated, and sometimes has a short neck, and is 2 to 4 cm in diameter. The flesh has a subacid and sweet pleasant flavor and a high vitamin C content. Strawberries may be consumed fresh or frozen or made into jam, jelly, ice cream flavoring, etc.

Tamarind

This medium-sized evergreen tree is grown in tropical and warm subtropical regions and is propagated by seeds, although budding is possible. The fruit is a succulent, edible pod, 7 to 20 cm in length and 2 to 3 cm in breadth, with brown pulp around and between the 2 to 6 seeds in a pod. It is rich in tartaric and other organic acids, sugar, protein (3.3%), and vitamin B (highest among all fruits). The fruit is edible fresh, but is chiefly used in acid drinks, flavoring preserves and chutneys, and to give zest to meat sauces.

Watermelon

Watermelon is an annual vine grown in various regions where there is a long and hot growing season. It is propagated by seeds. The fruit is rounded, oval or oblong, various in size (10 to 30 cm in diameter), with a thick rind. The flesh is crisp, juicy, and sweet; and is pink, red, or yellow in color when ripe. Watermelon is generally classified as a vegetable, but is mainly consumed fresh as a dessert fruit. The rind can be used in making conserves and pickles.

Wax-apple (Wax-Jambo)

This small to medium-sized evergreen tree is grown in tropical and subtropical regions and propagated by air-layering, cutting, or grafting. The fruit is pyriform to pyramid-shaped, 3 to 5 cm in diameter, with a waxy surface, and is white overlaid with pink. The flesh is the pericarp, crisp and juicy, with a gently sweet and subacid flavor. It may be eaten fresh without peeling; almost the entire fruit is edible.

REFERENCES

1. **Bailey, L. H.,** *Manual of Cultivated Plants Most Commonly Grown in the Continental United States and Canada,* Macmillan, New York, 1949.
2. **Chandler, W. H.,** *Deciduous Orchards,* Lea & Febiger, Philadelphia, 1942.
3. **Chandler, W. H.,** *Evergreen Orchards,* 2nd ed., Lea & Febiger, Philadelphia, 1958.
4. **Childers, N. F.,** *Modern Fruit Science,* 5th ed., Rutgers University, New Brunswick, N.J., 1973.
5. **Hackett, C. and Carolane, J.,** *Edible Horticultural Crops — a Compendium of Information on Fruit, Vegetable, Spice and Nut Species, Part I,* Academic Press, New York, 1982.
6. **Nagy, S. and Shaw, P. E.,**. *Tropical and Subtropical Fruits,* AVI, Westport, Conn., 1980.
7. **Reuther, W., Webber, H. J., and Batchelor, L. D.,** *The Citrus Industry,* Vol. 1, University of California Press, Berkeley, 1967.
8. **Samson, J. A.,** *Tropical Fruits,* Longman, New York, 1980.
9. **Simmonds, N. W.,** *Bananas,* 2nd ed., Longman, New York, 1982.
10. **Thompson, H. C. and Kelly, W. C.,** *Vegetable Crops,* 5th ed., McGraw-Hill, New York, 1957.
11. **Yang, C. F.,** *Manual of Fruit Trees in Taiwan* (in Chinese), Chiayi Agricultural Experiment Station, Taiwan, 1951.

DRUG CROPS

F. W. Cope

TEA

Camellia sinensis (L.) O. Kuntze
Camelliaceae Dum. = Theaceae Mirb.

The leaves of many plant species are used in the preparation of potable infusions, but the tea of commerce is derived from *Camellia sinensis* (synonyms: *Thea sinensis* L., *Camellia thea* Link).

When growing wild, the evergreen plant develops into a small tree, but under cultivation it is rendered bushy by the removal of two to three leaves of the most recently formed shoot tips, along with the terminal bud; pruning is also practiced. It is cultivated at high elevations in the tropics and at lower elevations in the subtropics.

Two principal groups of cultivars are usually recognized. The first group, known as China teas, constitute *C. sinensis* var. *sinensis;* the second group, known as Assam teas, constitute *C. sinensis* var. *assamica* (Mast.) Pierre. Visser[1] states, however, that these two varietal groups merit specific rank as *C. sinensis* and *C. assamica,* respectively. The China teas are slow growing, with small, erect, comparatively narrow, and markedly serrate leaves of a dark green color; the Assam teas are quick growing and have large drooping leaves.[2]

Harvesting consists of gathering soft vegetative shoot tips carrying two to three immature leaves, together with the apical bud. These young leaves have high caffeine and polyphenol contents, and yield the best tea quality.

Black tea is produced by an enzymic fermentation of the leaves. The harvested leafy shoots are first rapidly withered and then passed through rollers to release the sap containing the enzymes which break down the polyphenols by an oxidation process. The primary products of such enzymic oxidation are *o*-quinones which, on polymerization, produce numerous dark and astringent substances; some of these substances are soluble and are partially extracted in the process of infusion.[3]

Green tea (largely of China and Japan) is not subjected to a fermentation process, but is prepared by rapidly inactivating the enzymes by steam or heat treatment, so that dark condensation products are not formed.

Most existing tea plantations have been established from seed, where marked heterogeneity exists because of cross-pollination in the crop. Selection of elite plants, with clonal propagation as cuttings is now being practiced in many areas.

COFFEE

Coffea arabica L., *C. canephora* Pierre ex Froehner, *C. liberica* Bull ex Hiern
Rubiaceae

Coffee is used as the roasted product of the beans (endosperm of seeds) which when ground and brewed gives a stimulating beverage containing caffeine. It is thought that the roasting process, which gives the characteristic aroma, was discovered in Arabia in the 15th century.[4]

The three economically important *Coffea* species develop as small trees in the wild, but under cultivation they are kept pruned to produce new vegetative and flowering shoots. The species show dimorphic branching, associated with the presence of two buds in each leaf axil on the orthotropic vegetative growths. The true axillary buds give rise only to more orthotropic shoots, especially if the apical dominance of the shoot tip is removed, when, because of the opposite leaves, usually a pair of orthotropic growths results. Repeated removal of apical buds by pruning will increase the vegetative scaffolding of the plant; the orthotropic shoots are nonflowering. The extra axillary bud, standing above the true axillary bud, gives rise to plagiotropic flowering laterals, again usually in opposite pairs. The axils of leaves on the flowering branches have a linear series of usually six buds, but only three to four develop into inflorescences, each producing two to four white, fragrant flowers. Early pruning of the plagiotropic laterals can bring about the development of opposite pairs of flowering branches. Since coffee has no mechanism for shedding an overload of developing fruit, it can sometimes suffer from die-back due to exhaustion, so that cultivation of the crop demands a careful balance between vegetative and reproductive areas.

The fruit produced on the laterals is a small drupe from an inferior ovary, containing two endospermic seeds. It is the endosperm which develops the coffee aroma on roasting. The pulpy mesocarp lends itself to the making of alcoholic beverages.

Coffea arabica ("arabica coffee") provides some 80% of the world's coffee. It is considered the best of the three commercial species. The center of origin for this species is the Ethiopian highlands and the center of cultivation is Arabia. It is a tetraploid that no longer possesses the self-incompatibility of the diploid *Coffea* species and is largely self-pollinated.

Because of its origin in the highlands of Ethiopia, arabica coffee can only be successfully cultivated in the tropics at high elevations. At higher latitudes, it can be grown closer to sea level.

Two varieties are recognized. Variety *arabica* (= var. *typica* Cramer) is believed to have been taken from Yemen by the Dutch for planting in Java in 1699. One plant from Java was established in the Amsterdam Botanic Gardens some 17 years later, and it is from this one tree that much of the later planting of arabica coffee, principally in the New World, resulted — Brazil, Central America, and Jamaica ("Blue Mountain coffee").

The variety *bourbon* (B. Rodr.) Choussy is so called because it stems from material established by the French in the island called Bourbon (now Réunion) and taken from Arabia. It is possibly a mutant form of *arabica*. It gives higher yield than *arabica* and has replaced it in many parts, particularly in Brazil.

Coffea canephora (= *C. robusta* Linden, commonly called "robusta coffee") is found wild in the African equatorial forest belt, from sea level to higher elevations, and for this reason can be successfully grown in tropical lowlands. Tropical Africa and Asia are the principal producers of robusta coffee. Although it is more fruitful than *C. arabica,* it has not been extensively planted in the New World. The flavor developed from robusta beans is not as good as that from arabica beans, but it lends itself to the manufacture of instant coffee and is therefore experiencing increased demand. It is a diploid species with a strong incompatibility mechanism and is therefore cross-pollinated — largely by wind, since the pollen grains are light. Where clonal plantings are made, two or more cross-compatible clones must be interplanted to ensure the production of crop.

Coffea liberica ("liberica coffee") owes its specific name to its original discovery in Liberia, at low elevation. It requires heavy rainfall and high temperature for optimum bearing. It has never assumed commercial importance, and is used as a filler with other coffees; its bitter flavor is appreciated in some parts of the world (Far East). It is a self-incompatible diploid species, largely cross-pollinated by wind.

Both *C. arabica* and *C. liberica* are susceptible to the coffee leaf rust disease caused by *Hemileia vastatrix,* which exists in numerous races. *C. canephora* seem to be resistant to this serious disease of coffee.

COCOA, CACAO

Theobroma cacao L.
Sterculiaceae

Cocoa and chocolate are derived from the roasted cotyledons of the seeds of the highly variable species *T. cacao*. It seems that true chocolate flavor can be developed only after the seeds have been subjected to a process of fermentation in which the seeds are heat-killed and become permeated with acetic acid.

Cuatrecasas[5] recognizes two subspecies of *T. cacao;* they are subsp. *cacao* and subsp. *sphaerocarpum* (Chev.) Cuatr. He has erected four forms for *T. cacao* subsp. *cacao*. Two of these forms are characterized by the possession of elongated and tapering, warty-surfaced fruits enclosing plump seeds with white (recessive) cotyledons. A third from (= *T. leiocarpa* Bern.) possesses ovoid fruits with almost smooth walls, with plump seeds which may have white or pale violet cotyledons. Subsp. *sphaerocarpum* also has smooth-walled ovoid fruits but contains flat seeds with dark purple cotyledons. As far as is known, all subgroups of *T. cacao* are interfertile.

The cocoa trade distinguishes between criollo, forastero, and Trinitario cacao. Criollo types, now not commonly cultivated, are considered to be "fine" cacaos and are included in the three forms mentioned above. The forastero group now supplies the bulk of the world's cocoa and is identified with the *sphaerocarpum* subspecies. The trinitario cocoas resemble the heterogeneous population found in Trinidad and Tobago, resulting from centuries of hybridization between the originally introduced criollo and later introductions of forastero types.

Dimorphic branching is seen in cacao. Orthotropic shoots, with a 3/8 phyllotaxis and of determinate growth, give rise to three to five plagiotropic branches with 1/2 phyllotaxis and indeterminate growth, by the complete breaking up of the terminal meristem of the orthotropic growth into three to five smaller meristems. All parts of the tree can bear flowers ("cauliflory"). The species has an incompatibility mechanism which is unique in the angiosperms, depending on the failure of syngamy within the ovules of ovaries which have been incompatibly pollinated.

The center of origin of *T. cacao* is still a matter for debate. Cheesman[6] placed it on the lower eastern equatorial slopes of the Andes, but others claim Central America as the center. There is little dispute, however, about the center of cultivation, Central America, where criollo types have been under cultivation for possibly 2000 years.

Flowering is typically profuse, with the annual production of several thousand flowers per tree. The flowers are small and of an unusual construction which adapts them to pollination by small flying insects such as the Ceratopogonid midges. Pollination, however, is of such a level that only 1 flower out of 500 may develop into a fruit. The tree has a mechanism of shedding immature fruits, so that an annual harvest of 50 ripe fruits from a large tree would be uncommon in an unimproved population.

The cocoa "pod" contains 30 to 70 seeds, with smaller numbers being associated with criollo parentage. The seeds extracted from ripe pods are covered with a mass of white mucilagenous pulp material developed from hairs on the testa. The pulp is acid-sweet. Ripe harvested pods are usually cracked open in the field, the seeds being heaped up on banana leaves and, being exposed to the air, the mass becomes inoculated with yeast spores; bacteria are brought by fruit flies. This inoculation starts the fermentation process which may proceed for 1 to several days, according to the type of cacao; beans with purple cotyledons require longer fermentation times than those with white or pale-tinted cotyledons. The fermentation process on plantations is usually conducted in large wooden boxes with slatted bottoms for drainage; small-holders often allow the beans to "sweat" in the field, covered in some way to conserve the heat of fermentation. The process begins with the exothermic conversion of the sugars in the pulp to ethyl alcohol by the activity of yeasts, with the alcohol then being oxidized to acetic acid by the bacterial flora, with further evolution of heat in the mass. Temperatures within the fermenting beans may often reach 50°C. High temperature and penetration of alcohol and acetic acid kill the bean. After fermentation, the beans are dried to a moisture content of about 6%, and are often polished mechanically or by the working of feet into the mass of beans.

The dried, fermented beans contain about 50% by weight of a fat known as cocoa butter. Cocoa powder is manufactured from roasted beans in which the fat content has been reduced to about 25% under pressure; the extracted fat is then added to other roasted beans for the manufacture of eating chocolate, along with sugar and flavorings. Milk chocolate contains milk powder also.

Early attempts at crop improvement were by the production of clonal material, mainly as rooted cuttings, from selected mother trees. Modern efforts, however, show a reversion to seedlings of hybrid and synthetic cultivars.

COLA, KOLA

Cola acuminata (P. Beauv.) Schott & Endl., *C. anomala* K. Schum., *C. nitida* (Vent.) Schott & Endl., *C. verticillata* (Thonn.) Stapf ex A. Chev.
Sterculiaceae

The cola nut contains some 2% caffeine, a little theobromine, and kolanin (a glucoside), which is a heart stimulant.[7] The nuts (seeds with testas removed) are chewed for the stimulation derived from the purine alkaloids and kolanin, and they are greatly prized in West Africa. The seeds are sometimes used for the preparation of beverages.

The genus *Cola* contains about 60 species of small trees, all native to the West African evergreen forest regions. The nonendospermic seeds may measure up to 5 × 3 cm, and are produced in star-shaped follicles. *C. nitida* ("Gbanja kola") is the most important economic species, with *C. acuminata* ("Abata kola") holding minor importance; both species are tetraploids. *C. anomala* ("Bamenda kola") and *C. verticillata* ("Owe kola") are only rarely cultivated but are sufficiently prized to be left intact in the clearing of forests. *Cola* species show self-incompatibility, and cross-incompatibility is also common.

The evergreen tree has leathery leaves and demands continuously available water, for should dry conditions lead to defoliation, the tree usually dies.[8]

The important *C. nitida* is native to the forests of Sierra Leone, the Ivory Coast, and eastward into Ghana. It has been extensively planted in western Nigeria since 1912. It is also widely cultivated in Ghana. The apetalous flowers are produced in small axillary inflorescences; there are male flowers about 2 cm across in which the anthers are joined into a column, and rather larger hermaphrodite flowers, 3 to 4 cm across with two whorls of bilobed anthers. Pollination is by insects. The fruit may contain up to seven divergent follicles with four to ten seeds per carpel.[7] The seeds may reach a length of 5 cm, with a thin testa enclosing the two cotyledons. The seeds are skinned after keeping for a few days, and are then protected from drying by packing them between leaves. Some seeds of *C. nitica* have white cotyledons (preferred) and there are others with pink or red coloration.

As with the related *Theobroma cacao*, improvement in *Cola* was initiated by vegetative propagation of selected mother trees. Large seedling populations were produced in Nigeria in the 1960s by interpollination of outstanding yielders in *C. nitida;* seedlings selected from these populations will be used for the production of clones. Later, possibly, a second cycle of seedlings will be produced to continue the improvement.[8]

QUININE

Cinchona calisaya Wedd., *C. ledgeriana* Moens ex Tremen., *C. officinalis* L., *C. succirubra* Pav. ex Klotzsch, numerous hybrids between *Cinchona* spp.
Rubiaceae

The genus *Cinchona* comprises some 40 species, all of them evergreen shrubs or trees and indigenous to the Andes. The four species listed above are those which are grown for quinine production. Purseglove[9] states that *C. ledgeriana* is probably a variety of *C. calisaya*, and *C. succirubra*, a variety of *C. pubescens* Vahl.

The cortex of the stem, branches, and roots of the *Cinchona* trees is the source of several alkaloids, the most important being quinine. Quinidine, isomeric with quinine, is also present, as well as cinchonine and its isomer cinchonidine. Harvesting is by uprooting mature trees, which are then cut into lengths and beaten with clubs to loosen the bark; the cortex, or bark, of the stems and roots is then peeled off and dried to a 10% moisture content. This total destruction of trees for harvest in South America prompted the establishment of plantations in the Far East.

C. calisaya and *C. ledgeriana* are native to the southern Peruvian and Bolivian Andes, at elevations of 1200 to 1700 m. *C. officinalis* is a species found at higher elevations in northern Peru and Colombia. *C. succirubra* is found over a wide range of altitudes from Costa Rica to Bolivia.[9]

Flowers are borne in terminal panicles. The flowers are heterostylic and this promotes cross-pollination, which in turn ensures a high level of heterozygosity in the species. The inferior ovary develops into a small capsule containing 40 to 50 small winged seeds.

Although quinine has largely been replaced by synthetic antimalarials, it seems likely that a considerable market for the crop products will continue; new applications in medicine have been found.[10]

TOBACCO

Nicotiana tabacum L., *N. rustica* L.
Solanaceae

Both *Nicotiana* species are highly polymorphic amphidiploids and are self-compatible, in contrast to the diploid *N. alata* Link & Otto whose homomorphic gametophytic incompatibility system has given rise to the term "nicotiana type" of incompatibility.

Although *N. tabacum* is now by far the more important commercial species, *N. rustica* was once the source of smoking and chewing tobacco and material for rituals in pre-Columbian times in Mexico, the southwestern and -eastern U.S., and eastern Canada. It was the first tobacco to reach Europe and to be grown and exported by the American colonists.[11] Only a few areas of the world now have rustica tobacco under cultivation; it was formerly a good source of nicotine, but synthetic insecticides have greatly diminished the use of nicotine in plant protection.

N. tabacum was also used in pre-Columbian times in the New World, being cultivated in Central America and tropical South America. When first introduced into Europe, it was used as an ornamental and medicinally; only towards the end of the 16th century was it used for smoking purposes.

The plant is cultivated as an annual, with the leaves forming the crop. Inflorescences and lateral shoots are regularly removed from the growing plant to ensure the maximum carbohydrate (mainly starch) content of the leaves which at maturity become brittle and readily snap. Just before harvest, the plant is usually topped, whereby the topmost leaves are broken off. The leaves and stem are viscid to the touch from the presence of very numerous epidermal hairs; some of the multicellular hairs are simple, and others are glandular and secrete gummy substances. Nicotine is present throughout the plant, and grafting experiments have identified the root system as the site of nicotine synthesis.

Harvesting may be either by the periodic removal of mature leaves or by harvesting the whole plant when the optimum number of leaves have reached maturity.

The leaves have to undergo a carefully conducted curing process to develop acceptable smoking qualities. The simplest form of curing is air-curing where the leaves cure under cover, exposed to atmospheric conditions and controlled ventilation. The leaves should slowly lose their harvested green color and assume a yellow color, due to the breakdown of the chlorophyll pigments and the unmasking of the carotene and xanthophyll pigments of the leaf. Following the development of yellowing, the leaves are then dried rapidly by increasing the ventilation. This curing process often lasts for up to 8 weeks. Cigar tobacco is usually air-cured.

Sun-curing is a more rapid curing method. The leaves are usually allowed to undergo air-curing for a few days, then allowed to ferment for up to 36 hr, after which the materials are exposed to the warmth of the sun. Most Turkish tobacco in sun-cured.

Rapid curing is achieved by the application of artificial heat to the curing leaves. Fire-curing involves exposure of the leaves to heat and smoke in curing barns after a few days of yellowing with the fires being lit during the day on the floor of the barn. The leaves acquire a distinctive aroma under fire-curing, and are principally used in the manufacture of chewing tobaccos.

In flue-curing, the leaves are subjected to smokeless heat in the curing barn. Closed flues carry the products of combustion from an outside fire through the barn, warming the air within. Careful control of the temperature and humidity is required. The curing process starts with the temperature around 32 to 37°C (90 to 100°F) and relative humidity of the order of 85%, obtained by closing the ventilators of the barn. Over the next 2 days, the temperature is slowly raised to as high as 45°C (115°F) and humidity is decreased by partial opening of the ventilators. Further increase in temperature to about 50°C (125°F), with increased ventilation in the barn, occupies the 3rd day of curing, after which the temperature in the barn may be raised to 70°C (160°F) to ensure the drying out of the midrib of the leaf. A variant of flue-curing is to generate heat in the barn with thermostatically controlled propane gas burners. The end product of flue-curing or curing with smokeless gas burners is usually of a bright yellow color and is used in the manufacture of cigarettes. Mixtures of flue- and air-cured leaves are favored for pipe tobaccos.

The tobacco industry is a highly conservative one. Each manufacturer uses blends of tobacco leaf from diverse sources and is concerned that the ingredients of the blend shall be available year after year. Both grower and buyer of tobacco leaf therefore depend on one another, and neither will readily change from tradition.

After curing, the leaves are bulked up and usually allowed to undergo a mild fermentative process for a month or more, following which the leaves will be graded for quality. The moisture content of export leaf needs to be carefully adjusted to 10 to 12% to avoid payment of excise duty on excessive water content. The leaf is then baled, and it may be left to mature in the bale for a year or two before being subject to the manufacturing process.

Flavoring and conditioning materials — most of them trade secrets — are added to the leaf in the first stage of manufacture. Waste leaf material may be used to produce nicotine which, on oxidation, gives nicotinic acid, or niacin, which is vitamin B_3.

PYRETHRUM

Chrysanthemum cinerariaefolium (Trev.) Bocc.
Compositae (Asteraceae)

Pyrethrum is obtained from the dried capitula of *Chrysanthemum cinerariaefolium (C. cinerariifolium,* according to Contant[12]). It is a tufted, daisy-like perennial herb, often growing to a height of 60 cm. The leaves are pinnately dissected, with the segments themselves being pinnatifid. A rosette habit of growth is seen in the juvenile stages of development. The 3- to 4-cm diameter capitulum carries both disc and ray florets, and is borne at the end of a long, slender peduncle.

The insecticidal toxicity of pyrethrum comes from the two pyrethrins and two cinerins contained in the florets of the capitulum. These substances may be extracted with solvents from the dried, ground inflorescences, or the dry material may be used directly in the manufacture of mosquito coils. Insects are rapidly paralyzed by pyrethrum, but it is nontoxic to warm-blooded animals; despite the synthesis of large numbers of insecticides, pyrethrum has maintained its popularity. Pyrethrum assumed great importance in World War II, for it was found to be effective against flies, mosquitos, fleas, and lice, and was used to protect food against beetles and flow flies.[2]

The species is native to the Mediterranean Adriatic region. In countries with temperate climates the plant flowers only in summer for 1 to 2 months, but when cultivated in the tropics at high elevations, the production of inflorescences may continue for 9 to 10 months. Exposure to temperatures of 16°C or below for about 10 days seems to be needed for inflorescence induction.[13]

The highlands of Kenya produce most of the world supply of pyrethrum. The crop is often grown in rotation with wheat, maize, or grasses. The plants are established either from seed or from portions split off from mother plants. Only the capitula are harvested (by hand) when fully expanded; they are then spread out thinly on trays and dried in a current of warm air.

C. cinerariaefolium is one of the relatively few angiosperms which possess the sporophytic type of incompatibility system. Cross-pollination is therefore needed for seed development.

REFERENCES

1. **Visser, T.,** Tea, in *Evolution of Crop Plants,* Simmonds, N. W., Ed., Longman, London, 1976, 18.
2. **Purseglove, J. W.,** Theaceae, in *Tropical Crops. Dicotyledons 2,* Longman, Green & Co., London, 1968, 599.
3. **Eden, T.,** *Tea,* 2nd ed., Longman, Green & Co., London, 1965.
4. **Purseglove, J. W.,** *Tropical Crops. Dicotyledons 2,* Longman, Green & Co., London, 1968, 458.
5. **Cuatrecasas, J.,** Cacao and its allies: a taxonomic revision of the genus *Theobroma, Contrib. U.S. Natl. Herb.,* 35, 379, 1964.
6. **Cheesman, E. E.,** Notes on the nomenclature, classification and possible relationships of cacao populations, *Trop. Agric. (Trinidad),* 21, 144,
7. **Purseglove, J. W.,** Sterculiaceae, in *Tropical Crops. Dicotyledons 2,* Longman, Green & Co., London, 1968, 564.
8. **van Eijnatten, C. L. M.,** Kola, in *Evolution of Crop Plants,* Simmonds, N. W., Ed., Longman, London, 1976, 284.
9. **Purseglove, J. W.,** Rubiaceae, in *Tropical Crops. Dicotyledons 2,* Longman, Green & Co., London, 1968, 451.
10. **van Harten, A. M.,** Quinine, in *Evolution of Crop Plants,* Simmonds, N. W., Ed., Longman, London, 1976, 255.

11. **Purseglove, J. W.,** Solanaceae, in *Tropical Crops. Dicotyledons 2,* Longman, Green & Co., London, 1968, 523.
12. **Contant, R. B.,** Pyrethrum, in *Evolution of Crop Plants,* Simmonds, N. W., Ed., Longman, London, 1976, 33.
13. **Purseglove, J. W.,** Compositae, in *Tropical Crops. Dicotyledons 1,* Longman, Green & Co., London, 1968, 52.

FORAGE CROPS*

C. C. Sheaffer and D. K. Barnes

INTRODUCTION

Forages are crop species whose herbaceous plant parts are used for domestic animal feed. Forages can be utilized directly via animal grazing, harvested and fed fresh (green chop), or harvested and stored as hay, silage, or dehydrated products. Ruminant animal species are the primary forage consumers. They are able to digest complex forage carbohydrates (cellulose, hemicellulose) in the microbial-containing, rumen portion of their digestive tract.

FORAGE SPECIES

Most forage species are members of the Gramineae (grass) or Leguminosae (legume) plant families. There are over 100 species of grasses and legumes used for forage, but only about 50 of these are of major importance.[1] Most of the primary forage species are perennial, but annual and biennial species are also important (Table 1). Annual species such as crimson clover, common and hairy vetch, and annual ryegrass are often used as winter annuals in the southern U.S. to provide grazing, but can also be used as summer annuals in northern portions of the U.S. Many species are perennial under ideal environmental conditions, but are considered annual or biennial because they lack winter hardiness and disease resistance. Examples include red clover, alsike clover, and white clover when grown in certain areas of the U.S.

Many of the most successful perennial forage species spread by rhizomes (below-ground modified stems), by stolons (above-ground modified stems), by adventitious roots, or by vines in addition to being seed propagated. These features may increase persistence and forage yield but also may contribute to certain species such as kudzu and quackgrass becoming weeds. Bermuda grass and Bahia grass are routinely established from vegetative "sprigs". Kudzu and crownvetch are routinely established by planting either crowns or seeds.

ADAPTATION

Climate

Temperature and moisture are the primary climatological factors affecting forage-crop distribution and production.[7,48] Temperature influences crop survival as well as length of growing season. Generally, a longer growing season allows more harvests and greater forage yields. The distribution of forage species within the humid portion of North America is closely related to the minimum temperature within a region, while in nonhumid areas, rainfall is the most important factor (Table 1, Figure 1). Native species in low rainfall areas are generally adapted to a wider range of temperature than species in high rainfall areas. Native grasses such as bluestem, gramas, switchgrass and buffalo grass have considerable heat and cold tolerance.

Species are referred to as warm or cool season depending on the temperature or period of greatest growth. Warm-season species have greatest dry-matter accumulation in the summer and have limited growth in the spring and fall (Table 2). Conversely, the summer forage production of cool-season species is usually lower than in the spring and fall. For cool-season grasses and legumes, minimum, optimum, and maximum temperatures for dry-matter

---

* Tables follow text.

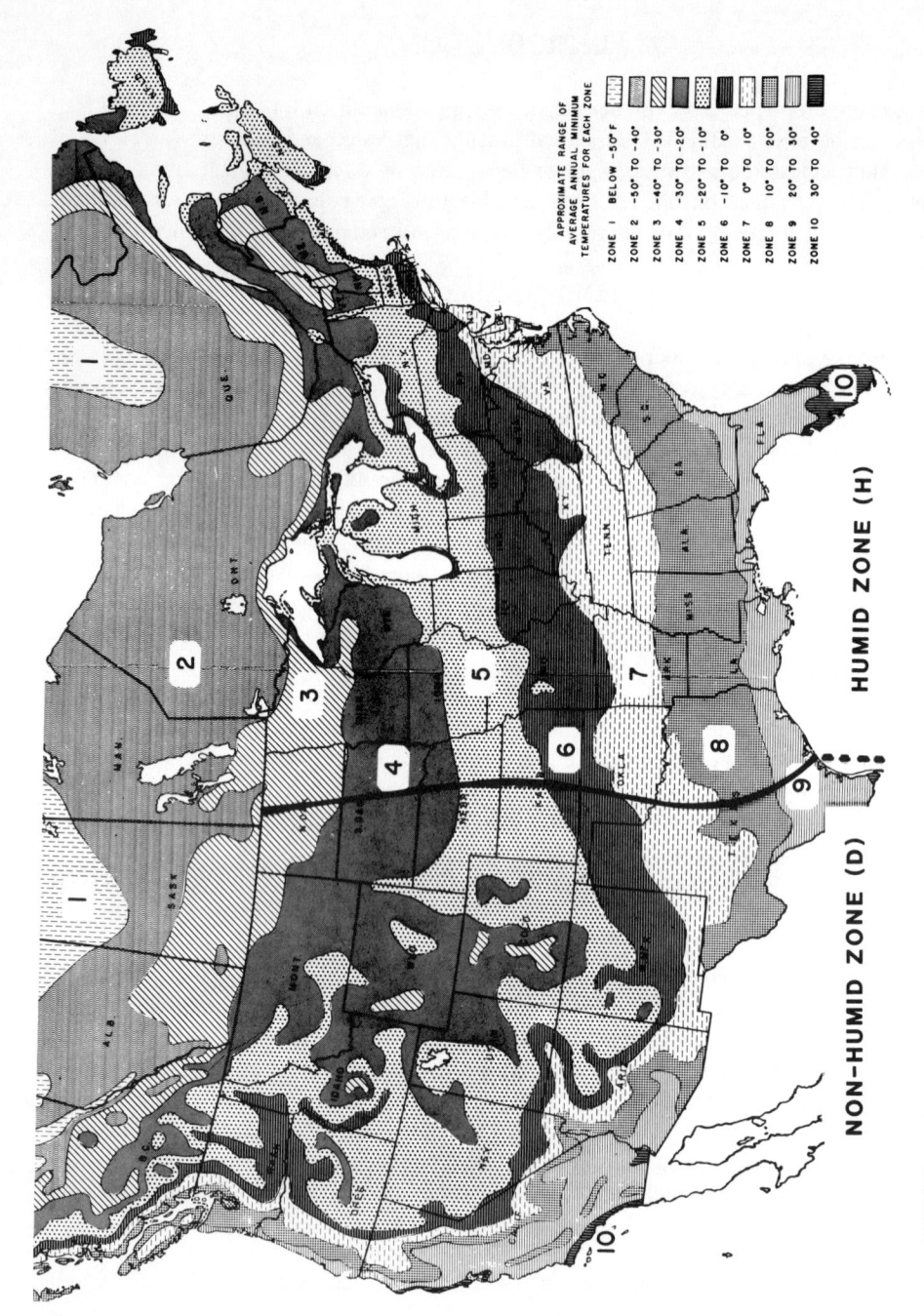

FIGURE 1. Plant hardiness zone map. (Miscellaneous publication 814, U.S. Department of Agriculture, Washington, D.C., 1960.)

accumulation are approximately 5, 20 to 25, and 30 to 35°C, respectively. Warm-season grasses and legumes (including tropical and subtropical species) have minimum, optimum, and maximum temperatures of approximately 15, 35, and 40 to 45°C, respectively.[58-60] Differences in growth response to temperature for cool- and warm-season species are presented in Tables 3, 4, and 5.

Although most subtropical, warm-season species are more susceptible to winter injury than cool-season species, variations in cold resistance exist among species and among cultivars within species. Stage of plant development also influences tolerance to low temperature (Tables 6 and 7). Among cool-season species such as alfalfa, variations in cultivar cold hardiness and disease resistance interact with environmental factors to influence winter hardiness and subsequent long-term yield and persistence (Table 8).

Species adapted to nonhumid and high-temperature regions usually have greater water-use efficiency and drought resistance than cool season, humid species (Table 9). Among the humid species, deeply taprooted legumes such as alfalfa are able to remain productive under limited rainfall by utilizing water deep within the soil profile. Most cool-season grasses are relatively shallow rooted and less effective foragers for available soil water than alfalfa. Bermuda grass naturally has greater drought resistance and water use efficiency than most warm-season grasses in the 6- to 10-H region and improved cultivars such as 'Coastal' have been developed with increased yield and water-use efficiency (Table 10).

Soils and Fertility

Although climatic factors have the greatest effect on general crop distribution, soil characteristics such as wetness, pH, and salinity influence forage-crop adaption and productivity within a region (Table 2). Acidity and salinity are soil characteristics limiting utilization of some forage species. Generally grasses are more tolerant than legumes of both problems. Low soil pH results from a loss of basic cations such as Ca, Mg, and K due to leaching and plant uptake with an increase in H and Al anions. Low soil pH decreases the availability of essential nutrients and increases the availability of toxic elements such as Al, Mn, and Zn. Low soil pH also usually decreases legume nodulation. Soil pH problems develop on highly eroded soils in the humid south and noncalcareous soils in the humid north. High-producing legumes such as alfalfa are less productive unless soil pH is corrected to desirable levels (6.0 to 7.0). Salt-affected soils contain high concentrations of soluble salts (saline soils) or adsorbed sodium (alkali or sodic soils). Salts originate from the crusts of the earth and are accumulated in arid and semiarid areas due to limited leaching, poor drainage, and high evapotranspiration and as a result of irrigation. Grasses which are native to these areas generally have a high salt tolerance.

Comprehensive information is lacking on differences in responses of forage grasses and legumes to most fertilizer nutrients. Information is available on the response of forage grasses to N. The extent of response is related to the yield potential of the grass species, harvest management, and the environment (Tables 11, 12, 13 and 14). For example, tall-growing and productive grasses such as reed canary grass have a greater yield response to N than Kentucky bluegrass which is a low-growing and less productive species (Table 13). Nitrogen fertilization rates should be based on economic returns rather than on maximizing yield. For example, warm-season range grasses are usually fertilized with only 45 kg/ha of N annually due to lack of soil moisture for production and to prevent invasion of undesirable species associated with higher rates of N fertilization.[70]

Biological Nitrogen Fixation

Forage legumes are able to utilize atmospheric nitrogen and reduce their dependence on soil N because of their symbiotic relationship with a specific bacterium (e.g., *Rhizobium meliloti* for alfalfa; (Table 15).[73] Nitrogen-fixation potential varies among legume species because of nodule type, nodule mass and host × *Rhizobium* strain interactions. Efforts are underway to breed for increased nitrogen fixation in alfalfa and some clover species. Nitrogen fixation has been reported for some tropical grass species (Table 16). This is due to association of grasses with *Azospirillum* spp. and *Azotobacter* spp.[75] Research on nitrogen fixation by cool-season grasses is limited.

FORAGE UTILIZATION

Grazing

Method of forage utilization depends on intensity of land utilization, forage species present, and climatic conditions in the region. Grazing of range and pasture is the cheapest method of forage utilization. Grazing results in forage wastage due to animal selectivity, trampling, and the inability to utilize forage at ideal stages of maturity; but it is the only practical method of utilizing rangeland, rocky areas, and hillsides. In the short-grass prairie region of the U.S. where forage production per hectare is low, grazing is the predominant form of utilization, although hay is often made from excess production.[76] Some forage species are more adapted to and persist better under grazing than other species. The best adapted are rhizominous and stoloniferous species that have a large basal-leaf area (Table 1). Many tall-growing legumes and grasses which are easily defoliated must be rotationally grazed to persist (Table 2). This is particularly true of legumes such as alfalfa which following defoliation require 30 to 40 days of regrowth to restore root and crown carbohydrates reserves.

In continuous grazing situations such as often occur with permanent pastures and rangelands, grazing is usually initiated when plants begin vegetative growth and continues until environmental conditions prohibit grazing. With rotational systems, grazing is initiated when plants are at less mature stages than advised for harvesting as hay or silage (Table 1).

Hay and Silage

Preservation of forage as hay or silage permits livestock feeding during periods when grazing is not possible. This occurs in much of the northern portion of the U.S. Mechanical harvesting of forages allows for confined feeding and for maximizing dry matter and nutrient yields per hectare.[77] Species with upright growth can be easily harvested and are preferred as hay or silage crops. Hay is aerobically stored forage which is dried to a moisture concentration of 20 to 25%. Significant forage dry-matter and quality losses frequently occur in field drying as a result of plant respiration and leaf shattering. Leaf shattering is particularly serious with legumes such as alfalfa. Silage is anaerobically stored forage with a moisture concentration of 40 to 85%. The higher moisture content at harvest reduces field and harvest losses compared to haymaking; however, storage losses are usually greater. In the northern states, mechanized feeding of low-moisture alfalfa silage (40 to 60% moisture) has become a widely accepted practice in dairy operations. Most forage grass and legume silage is low in energy and high in protein.

Forage yields in Table 1 represent average to maximum values for systems which maximize nutrient yields per hectare while ensuring stand persistance. This usually is at bud to early flowering stages of development of most forage legumes and at boot to early heading of grasses (Tables 17, 18, 19). Greatest dry matter yields can usually be obtained by harvesting at more mature stages of plant development or by making fewer harvests per growing season (Table 14).

Soil Conservation

Forages are important for soil conservation and soil fertility improvement. Most small-seeded forage species reduce wind and water erosion compared to row crops. Sod-forming species are most effective at reducing erosion due to their extensive soil coverage and soil-holding, fibrous root systems. When used in rotations, legumes supply biological fixed N to subsequent crops and increase soil tilth by organic matter addition (Table 20). Forage crops allow animal agriculture to exist in areas where cultivated grain crops cannot normally be grown. This is because they are adapted to diverse climatic and soil conditions. Examples include revegetated mining sites, steep slopes, badly eroded land, the short grass prairies, and high altitudes.[76,79] Certain species are adapted for growth on land used for disposal of wastewater effluent produced from sewage treatment (Table 21).

SEEDLING VIGOR

Seedling vigor is generally a limiting factor in the establishment of small-seeded forage legumes and grasses. Species with low vigor are particularly susceptible to competition for light and moisture by vigorous annual weeds and other forage species in mixtures.[51,80] Slow establishment also exposes new stands to wind and water erosion for long periods of time. Relative seedling vigor as shown in Table 2 consists of germination and seedling-development components. Generally, grasses have slower germination and less seedling vigor than the legumes.

Grasses have hypogeal emergence (cotyledons remain in the soil) while legumes have either hypogeal or epigeal emergence. There are two main types of grass seedling development that affect the establishment of some species. *Panicum, Andropogon,* and *Bouteloua* seedlings have an elongated subcoleoptile internode and a short coleoptile which results in crowns and adventitious roots near the soil surface.[81] Most cool-season grass species have a long coleoptile and no elongation of the subcoleoptile internode. Crowns and roots near the soil surface are more subject to extremes of temperature and moisture, and stand failures occur more frequently.

Seed size increases emergence pressures and ease of establishment of grasses and legumes. Grasses tend to have greater emergence pressure per plant than legumes. This is in part a function of their hypogeal emergence in which the cotyledons remain in the soil and the coleoptile is the first vegetative structure emerging. Most forage legumes have epigeal emergence in which the cotyledons are pulled through the soil by an elongated hypocotyl. Large-seeded legumes such as hairy and common vetch, which also have hypogeal emergence, are more easily established than smaller-seeded species such as alfalfa.

DISEASE, NEMATODES, AND INSECT PESTS

Forage crop species are exposed to many types of pests. This is because they are used over wide geographic areas and are exposed to wide ranges of summer-winter, wet-dry environments for periods of several years. Pest susceptibility can affect the area of adaptation, yield, and quality of a forage species. Forage legumes generally have more pest problems than forage grasses. As a group the forage legumes are relatively susceptible to crown and root rots and nematodes which affect persistence, and to foliar diseases which affect yield and forage quality (Table 22). The clovers are also seriously affected by viruses. Numerous insect species affect forage legumes, but they are more species-specific than plant diseases. Pests of forage grasses generally include damping-off fungi, foliar disease, crown- and root-feeding insects, and nematodes. It is necessary to be aware of the potential pest problems in an area and to select pest-resistant cultivars if the goal is to economically maximize forage yield and quality.

FORAGE QUALITY

Legumes and Grasses

Legumes generally have lower hemicellulose and cell-wall (neutral detergent fiber, NDF) concentrations than grasses. Conversely, the cell walls of legumes are more lignified and consequently less digestible (Tables 23 and 24). The extent of digestion of legumes and grasses is often similar, although legumes have a faster rate of digestion.[84] The greater rate of legume digestion allows faster passage through the digestive tract, less rumen fill, and greater intake than for grasses. By virtue of their ability to fix atmospheric N_2, legumes generally have greater crude protein concentrations than unfertilized grasses.

Average forage intake and digestibility are 12.5 and 12.8 units lower, respectively, for warm-season tropical than cool-season temperate grasses (Table 25).[86-88] Lower values for tropical grasses are due to differences in leaf anatomy, fiber levels, and sward structure. Tropical grass species have higher cell wall concentrations than temperate species, but similar lignin concentrations (Table 26). Increased temperature and light intensities which increase plant structural components may be partially responsible for lower quality in tropical species.[90] Average leaf/stem ratios and tiller densities are higher for temperate grass species; crude protein concentrations are similar.

Tropical grasses have C-4 photosynthesis and greater CO_2 fixation than C-3 temperate grasses.[85] Associated with the C-4 photosynthethic pathway in tropical grass leaves are greater vascular tissue and parenchyma bundle sheaths and fewer mesophyll cells than occur in temperate grasses (Table 27). Leaf lamina of the temperate grasses are more rapidly degraded than those of the tropical grasses. This is primarily due to the presence of greater amounts of easily degradable mesophyll cells in the leaves of temperate species and to the lower digestion of epidermal and parenchymal sheath cells in tropical species (Table 28).

Maturity

The forage quality of both grasses and legumes decreases as plant maturity increases (Tables 17, 18, 29, and 30). Declines in grass and legume digestibility of 0.1 to 0.7 units/day have been reported for tropical and temperate grasses and legumes, respectively.[87,93] The decline in forage quality is associated with a decrease in the leaf/stem ratio of the forage (Tables 31 and 32). The forage quality of stems declines at a faster rate than the forage quality of leaves. Harvest schedules with more frequent harvests during the growing season usually produce higher quality forage than schedules with less frequent harvests. This is because the forage is at less mature stages at initial and subsequent harvests (Tables 17 and 33).

Antiquality Components

Antiquality components of forages include allelochemicals and other plant constituents having detrimental effects on animal health or performance. Specific allelochemicals produced by plants include saponins, estrogens, tannins, glycosides, alkaloids, and coumarin. A partial listing is shown in Table 34. Other antiquality factors produced by microbial interaction with plants include mycotoxins and dicoumarol. Soluble leaf proteins are generally accepted as the principle foaming agents responsible for legume bloat, but the presence of tannins, which decrease protein solubility, and low rates of cell wall digestion may decrease the incidence of bloat by some legume species (Table 35).[96,108]

Grass tetany (hypomagnesemia) is a metabolic disorder of ruminants associated with a low Mg intake and low blood serum levels (10 ppm or less).[109] It is associated with perennial grasses having forage Mg levels of less than 0.20 to 0.25% or forage K/Ca + Mg ratios >2.2. Reed canary grass, orchard grass, and tall oat grass have been classified as species most likely to cause grass tetany in ruminants (Table 36).[111]

Table 1
CHARACTERISTICS OF FORAGE LEGUMES AND GRASSES

Common name	Scientific name	Principal use[a]	Life cycle[b]	Principal production area[c]	Seeds/kg (10³)	Seeding rate (kg/ha)	Growth habit	Maturity for hay harvest	Av maximum yield (Mg/ha)	Ref. (yield)
Legumes										
Alfalfa	*Medicago sativa* L.	H,S,G	P	2-8H	440	11—22	Bunch	10% bloom	6—22	8,9,10
Arrowleaf clover	*Trifolium vesiculosum* Savi	G,H	A	7-8H	880	6—9	Bunch	10% bloom	4—10	11
Alsike clover	*Trifolium hybridum* L.	H,G	P	2-5H	1544	7—9	Bunch	10% bloom	4—8	12,13
Birdsfoot trefoil	*Lotus corniculatus* L.	G,H,C	P	2-6H	827	9—13	Bunch	10% bloom	3—11	9,14,15
Crimson clover	*Trifolium incarnatum* L.	G,H	A	7-9H	308	22—34	Sod	10% bloom	3—5	12
Crownvetch	*Coronilla varia* L.	G,H,C	P	4-6H	243	6—11	Sod (rhizomes)	10% bloom	1—10	16
Cicer milkvetch	*Astragalus cicer* L.	G,H	P	3-5D	287	22—28	Sod (rhizomes)	10% bloom	4—8	17,18
Common lespedeza	*Lespedeza striata* Hook. & Arn.	G,H	A	6-10H	419	28—34	Bunch	10% bloom	2—4	19
Common vetch	*Vicia sativa* L.	G,H	A	7-10H	15	45—90	Bunch	10% bloom	2—6	20
Hairy vetch	*Vicia villosa* Roth	G,H	A	5-8H	44	45—90	Bunch	10% bloom	2—6	20
Korean lespedeza	*Lespedeza stipulacea* Maxim.	G,H	A	6-10H	496	22—28	Bunch	10% bloom	2—4	19
Kudzu	*Pueraria lobata* (Willd.) Ohwi	G,C,H	P	8-10H	88	7—11	Vine	Veg.	5—7	21
Red clover	*Trifolium pratense* L.	H,S,G	P	3-7H	606	17—22	Bunch	10% bloom	4—13	9,13,22—24
Sainfoin	*Onobrychis viciaefolia* Scop.	G,H	P	3-5D	66	34—39	Bunch	10% bloom	4—12	17,18,23
Sericea lespedeza	*Lespedeza cuneata* (Dumont) G. Don	G,H	P	6-10H	772	34—45	Bunch	Veg. (30—40 cm)	3—6	19
White clover	*Trifolium repens* L.	G	P	4-10H	1784	6—8	Sod (stolons)	10% bloom	1—8	8,12,13
White sweet clover	*Melilotus alba* Desr.	M,H,C	B,A	West 2-6H	573	13—17	Bunch	10% bloom	4—9	25,26

Table 1 (continued)
CHARACTERISTICS OF FORAGE LEGUMES AND GRASSES

Common name	Scientific name	Principal use[a]	Life cycle[b]	Principal production area[c]	Seeds/kg (10^3)	Seeding rate (kg/ha)	Growth habit	Maturity for hay harvest	Av maximum yield (Mg/ha)	Ref. (yield)
Yellow sweet clover	*Melilotus officinalis* Lam.	M,H,C	B	West 2-6H	573	13—17	Bunch	10% bloom	4—9	25,26
Grasses										
Annual ryegrass	*Lolium multiflorum* Lam.	G,H	A	7-9H	501	28—34	Bunch	Boot	5—10	14,27
Bahia grass	*Paspalum notatum* Flugge	G,H	P	8-10H	336	11—17	Sod (stolons and rhizomes)	Veg. (4- to 6-wk interval)	3—14	28—31
Bermuda grass	*Cynodon dactylon* (L.) Pers.	G,H	P	7-10H	3940	7—9	Sod (stolons and rhizomes)	Veg. (4- to 6-wk interval)	4—15	29,30,32
Big bluestem	*Andropogon gerardi* Vitman	G,H	P	West 3-6H	364	11—17	Bunch-sod (rhizomes)	Boot	4—7	33,34
Blue grama	*Bouteloua gracilis* Lag. ex Steud.	G	P	East 5-8D	1819	11—17	Bunch-sod (tillers)		2—6	33
Crested wheatgrass	*Agropyron desertorum* (Fisch. ex Link) Schult.	G,H,C	P	3-5D	386	7—13	Bunch	Early heading	2—4	35
Dallis grass	*Paspalum dilatatum* Poir.	G	P	3-10H	485	9—22	Bunch (rhizomes)	Veg. (4- to 6-wk interval <30 cm)	3—10	36
Kentucky bluegrass	*Poa pratensis* L.	G,H	P	2-6H	4800	17—28	Sod (rhizomes)	Early heading	2—7	37—40
Little bluestem	*Andropogon scoparius* Michx.	G	P	East 5-8D	573	13—22	Bunch-sod (rhizomes)	Veg.	34	34
Orchard grass	*Dactylis glomerata* L.	H,G	P	4-6H	1442	22—28	Bunch	Boot	4—12	12,14,39,41

Perennial ryegrass	*Lolium perenne* L.	G,H	P	7-8H	501	28—34	Bunch	Early heading	4—10	14
Quackgrass	*Agropyron repens* (L.) Beauv.	G,H	P	2-4H	243	7—13	Sod (rhizomes)	Boot	3—10	40,42
Reed canary grass	*Phalaris arundinacea* L.	G,H,C	P	2-5H	1175	9—13	Sod (rhizomes)	Boot	4—14	9,40,41
Smooth bromegrass	*Bromus inermis* Leyss.	G,H	P	2-5H	300	13—22	Sod (rhizomes)	Early heading	3—11	38,39,41
Sudan grass	*Sorghum bicolor* (L.) Moench	G,H	A	3-8H	121	17—84	Bunch	Veg. (90 cm)	6—10	43
Switchgrass	*Panicum virgatum* L.	G,H	P	West 3-6H	858	6—9	Bunch-sod	Early heading	2—8	34,44—46
Tall fescue	*Festuca arundinacea* Schreb.	G,H	P	5-7H	500	11—28	Bunch-sod	Boot	6—14	37,39,41,47
Timothy	*Phleum pratense* L.	H,G	P	2-5H	2712	9—13	Bunch	Early heading	2—11	37,39,41

[a] C = soil conservation, G = grazing, H = hay, M = green manure, S = silage.
[b] A = annual, B = biennial, P = perennial.
[c] See Figure 1 for location based on minimum temperatures.

Table 2
RELATIVE TOLERANCE OF ESTABLISHED LEGUMES TO ENVIRONMENTAL CONDITIONS AND FREQUENT HARVEST AND RELATIVE SEEDLING VIGOR[a2,3,5-7,50-57]

Species	Cold-frost	Drought	Wet	Soil acidity	Salinity	Frequent harvest	Seedling vigor
Legumes							
Alfalfa	G	G	P	P	P	F	F
Alsike clover	F	P	G	G	P	F	F
Arrowleaf clover	P	P	P	F	P	G	G
Birdsfoot trefoil	G	F	G	G	G	G	P
Crimson clover	F	P	P	G	P	P	G
Crownvetch	G	G	P	E	F	P	P
Cicer milkvetch	G	G	F	G	G	P	F
Common lespedeza	P	G	P	G	P	P	P
Common vetch	P	P	F	G	P	F	E
Hairy vetch	F	P	F	G	P	P	E
Korean lespedeza	P	G	P	F	P	P	P
Kudzu	P	G	P	E	F	F	F
Red clover	F	P	F	F	P	P	G
Sainfoin	F	G	P	G	G	P	G
Sericea lespedeza	P	G	F	E	E	P	P
White clover	F	P	G	F	P	E	F
White sweet clover	G	G	P	P	F	P	F
Yellow sweet clover	G	G	P	P	F	P	F
Grasses							
Annual ryegrass	P	P	G	F	F	E	E
Bahia grass	P	G	G	F	F	E	P
Bermuda grass	P	E	G	E	E	E	F
Big bluestem	E	E	P	G	P	P	P
Blue grama	E	E	F	F	G	E	P
Crested wheatgrass	E	G	F	G	G	G	G
Dallis grass	P	G	F	F	F	E	P
Kentucky bluegrass	E	P	G	F	P	E	P
Little bluestem	E	E	P	G	F	E	P
Orchard grass	F	F	F	F	F	E	G
Perennial ryegrass	P	P	G	F	F	E	E
Quack grass	E	G	G	G	G	G	F
Reed canary grass	E	G	E	G	F	G	P
Smooth bromegrass	E	G	F	F	G	P	G
Sudan grass	P	G	P	F	F	P	E
Switchgrass	E	E	F	G	G	P	P
Tall fescue	F	G	G	E	E	E	G
Timothy	G	F	F	G	P	P	F

[a] E = excellent, G = good, F = fair, P = poor.

Table 3
EFFECT OF TEMPERATURE REGIME ON RELATIVE GROWTH RATES OF WARM- AND COOL-SEASON SPECIES

Subfamily and species	Temp. regime (°C) 15/20	27/22	36/31
Festucoideae			
Pubescent wheatgrass	70[a]	98	61
Smooth bromegrass	75	98	65
Tall fescue	76	99	55
Kentucky bluegrass	68	98	62
Reed canary grass	66	93	61
Panicoideae			
Dallis grass	23	90	100
Guinea grass	23	89	100

[a] All values given indicate relative growth.

Adapted from Kawanabe, S., *Proc. Jpn. Soc. Plant Taxonomists,* 2, 17, 1968.

Table 4
GROWTH OF DALLIS GRASS AND PERENNIAL RYEGRASS OVER A RANGE OF CONSTANT TEMPERATURES

Species	Plant growth/day 5°C	10°C	15°C	20°C	25°C	30°C	35°C
Dallis grass	0	2	5	7	15	22	19
Perennial ryegrass	2	4	4	4	3	2	1

[a] Values are given in mg dry-weight increment/tiller.

Adapted from Mitchell, K. J., *N.Z.J. Sci. Tech.,* 38, 203, 1956.

Table 5
YIELD AND GROWTH RATES OF FIVE LEGUME SPECIES HARVESTED AT FIRST FLOWER FOLLOWING GROWTH IN FOUR TEMPERATURE REGIMES

Temp. regime and legume species	Days to first flower	Herbage yield (g/pot)	Root yield (g/pot)	Plant growth rate (mg/pot/day)
32/27°C				
Alsike clover	17	4.0	3.0	217
Red clover	18	3.8	2.2	146
Alfalfa	18	3.4	3.0	145
Sweet clover	31	9.3	5.6	292
Birdsfoot trefoil	52	7.4	4.3	190
LSD	5	0.8	0.8	47
27/21°C				
Alsike clover	19	5.1	2.9	237
Red clover	20	5.5	2.4	214
Alfalfa	20	5.6	3.3	242
Sweet clover	29	14.4	6.2	536
Birdsfoot trefoil	32	4.0	2.3	143
LSD	2	1.0	0.4	85
21/15°C				
Alsike clover	25	10.2	5.0	435
Red clover	29	7.4	3.7	256
Alfalfa	28	6.0	4.6	235
Sweet clover	32	18.0	6.5	540
Birdsfoot trefoil	39	4.8	2.7	149
LSD	1	1.2	0.8	68
15/10°C				
Alsike clover	43	15.8	6.6	408
Red clover	48	14.4	6.5	364
Alfalfa	40	7.1	5.6	190
Sweet clover	55	24.9	7.5	478
Birdsfoot trefoil	68	7.2	4.6	145
LSD	6	2.1	1.2	66

Note: LSD = least significant difference ($p = <0.05$).

Adapted from Smith, D., *Agron. J.*, 62, 520, 1970.

Table 6
SURVIVAL OF 7-WEEK-OLD LEGUME SEEDLINGS SUBJECT TO FREEZING TEMPERATURES

Temp. (°C)	Alfalfa	Red clover	Sericea lespedeza	Common lespedeza	Korean lespedeza
−4.1	100	100	15	0	0
−2.8	100	100	73	0	0
−2.5	100	100	84	95	59
−1.7	100	100	100	92	87

Survival (%)

Adapted from Tysdal, H. M. and Pieters, A. J., *Agron. J.*, 26, 923, 1934.

Table 7
SURVIVAL OF FORAGE GRASSES AND LEGUMES WHEN SUBJECT TO FREEZING AT DIFFERENT STAGES OF DEVELOPMENT

Survival of forage grasses (%)

Stage of development	Timothy	Reed canary	Kentucky blue	Smooth brome	Meadow fescue
Seed	97	79	60	110	95
Plumule out	45	48	37	95	57
One leaf	118	62	90	110	86
Two leaves	98	31	97	65	89
Three to four leaves	93	61	100	86	71
Four to six leaves	100	98	98	96	100

Survival of forage legumes (%)

	Alfalfa	Sweet clover	Alsike clover	Red clover	White clover
Seed	131	163	102	130	188
Lifting up soil	87	72	99	69	79
Cotyledons open	1	2	5	3	1
One true leaf	6	2	8	5	16
Second true leaf	6	1	2	3	0
Third true leaf	23	0	9	1	1
Three to five leaves	48	5	41	39	56
Four to six leaves	62	6	55	38	62
Six to eight leaves	49	2	41	27	46
Seven to nine leaves	98	62	79	82	84

Adapted from Arakeri, H. R., and Schmid, A. R., *Agron. J.*, 41, 182, 1949.

Table 8
FORAGE YIELDS ASSOCIATED WITH DIFFERENCES IN WINTER HARDINESS AND BACTERIAL WILT RESISTANCE FOR EIGHT ALFALFA VARIETIES ESTABLISHED IN 1967 AT ROSEMOUNT, MINN.

Cultivar	Winter hardiness[a]	Bacterial wilt resistance[b]	1969	1971	1973	1975	1977	10-year total	Stand 10—77 (%)
Ramsey	VW	R	10.5	9.5	7.2	11.3	8.7	89.6	41
Iroquois	W	VR	10.5	9.6	7.3	10.3	7.4	87.5	23
Vernal	W	R	10.0	8.9	6.3	10.5	7.6	84.1	20
Saranac	MW	R	10.6	9.4	6.5	10.0	5.4	81.2	10
Ranger	W	MR	9.4	8.5	6.0	9.6	6.6	77.2	10
Arnim	MW	S	10.2	7.6	0.0	0.0	0.0	37.6	1
Franks Langmeiler	W	S	10.1	6.9	0.0	0.0	0.0	37.4	1
Alt Franken-Schmidt	W	S	9.8	6.8	0.0	0.0	0.0	35.9	1
LSD (0.05)			0.1	0.9	0.9	0.9	0.9		

Forage yield (Mg dry matter/ha)

[a] VW = very winter hardy, W = winter hardy, MW = moderately winter hardy.
[b] VR = very resistant, R = resistant, MR = moderately resistant, S = susceptible.

Adapted from Barnes, D. K., Frosheiser, F. I., and Martin, N. P., *Proc. 4th Minnesota Forage and Grassl. Council Symp. Rochester, Minn.*, Minnesota Forage and Grassland Council, St. Paul, 1979, 34.

Table 9
TRANSPIRATION AND WATER-USE EFFICIENCY (WUE) OF FORAGE SPECIES GROWN IN THE GREENHOUSE

Cultivar/species	Amount (cm)	As a fraction of evapotranspiration (%)	WUE (kg/ha/cm H_2O)
Legumes			
Alsike clover	11.3	59	141
Lutana cicer milkvetch	7.2	56	119
Remont sainfoin	12.4	64	137
Travois alfalfa	12.7	63	147
Vernal alfalfa	12.1	74	138
Grasses			
Tall fescue	13.8	67	167
Creeping foxtail	13.1	67	130
Intermediate wheatgrass	11.9	63	136
Orchard grass	12.3	67	169
Smooth bromegrass	15.5	69	135
Blue grama	8.6	59	263
Crested wheatgrass	8.9	62	165
Slender wheatgrass	8.8	53	154

Adapted from Fairbourn, M. L., *Agron. J.*, 74, 62, 1982.

Table 10
WATER USE BY FIVE GRASSES FERTILIZED WITH TWO N RATES

	Water use/N rate (kg/ha)	
Species	56	224
Common Bermuda grass	3753[a]	1334
Coastal Bermuda grass	913	327
Suwannee Bermuda grass	687	519
Pensacola Bahia grass	1202	957
Pangola grass	1155	1192

[a] Values given indicate kg H_2O/kg day matter.

Adapted from Burton, G. W., Prine, G. M., and Jackson, J. E., *Agron. J.*, 49, 498, 1957.

Table 11
HARVEST FREQUENCY AND N FERTILIZATION EFFECTS ON BERMUDA GRASS YIELDS AT TIFTON, GA.

Harvest			Yield/cultivar and year (1 or 2)[a]							
			Callie		Coastal		Coast-cross 1		Midland	
Interval (weeks)	No.	N level (kg/ha)	1[b]	2	1	2	1	2	1	2
2	12	336	11.7	9.3	12.6	9.8	13.3	8.8	11.8	9.0
		672	16.1	12.6	15.1	11.6	16.3	10.9	15.4	11.0
4	6	336	15.4	9.7	14.4	9.5	15.1	8.8	13.5	8.3
		672	19.7	10.6	16.4	12.4	17.6	10.7	16.4	11.7
8	3	336	27.4	15.3	14.5	15.4	19.6	11.6	21.5	14.3
		672	29.5	17.5	23.6	18.8	22.1	9.9	21.7	16.6

[a] Values given indicate megagrams (Mg) of dry matter per hectare.
[b] Yield reductions from year 1 to year 2 were associated with winter injury.

Adapted from Monson, W. G. and Burton, G. W., *Agron. J.*, 74, 371, 1982.

Table 12
HARVEST FREQUENCY AND N FERTILIZATION EFFECTS ON BERMUDA GRASS YIELDS AT MARYLAND

	Yield/harvest no. (May—Oct.)		
kg/ha	3	4	7
0	3.3[a]	3.0	2.0
112	10.7	8.7	4.7
224	15.3	12.9	8.9
448	18.0	15.7	12.9
672	18.5	16.0	13.4
896	18.1	16.4	13.3

[a] All values given indicate megagrams (Mg) of dry matter per hectare.

Adapted from Decker, A. M., Hemkin, R. W., Miller, J. R., Clark, N. A., and Okorie, A. V., *Md. Agric. Exp. Stn. Bull.*, No. 487, 1971.

Table 13
RESPONSE OF COOL-SEASON GRASS SPECIES TO N FERTILIZATION[a]

Species	Yield/annual N rate (kg/ha)[b]					
	0	112	224	224/2	448	448/2
Kentucky bluegrass	5.3	7.3	9.6	10.4	11.9	11.0
Tall fescue	5.0	8.7	10.1	11.4	11.8	12.9
Orchard grass	5.4	7.5	9.7	10.3	11.7	11.5
Reed canary grass	5.0	9.3	11.2	11.4	13.9	15.1
Smooth bromegrass	5.8	8.6	9.2	10.7	13.7	14.2
Timothy	8.5	10.9	11.7	12.7	12.4	13.2

[a] All species harvested three times.
[b] All values given indicate megagrams of dry matter per hectare.

Adapted from Colyer, D., Alt, F. L., Balasko, J. A., Henderlong, P. R., Jung, G. A., and Thang, V., *Agron. J.*, 69, 514, 1977.

Table 14
EFFECT OF N FERTILIZATION AND HARVEST FREQUENCY ON YIELD OF COOL-SEASON GRASSES

	\multicolumn{6}{c}{Yield/N rate and harvest frequency}
	168 (kg N/ha)			336 (kg N/ha)		
Species	3	5	8	3	5	8
Kentucky bluegrass	5.0[a]	5.1	5.0	2.5[b]	6.0	7.0
Tall fescue	12.0	9.0	7.5	7.5[b]	12.0	11.0
Orchard grass	13.5	9.5	7.0	14.0	13.0	11.0
Reed canary grass	7.5	4.0	1.5	6.0[b]	4.0	2.0
Smooth bromegrass	10.0	5.5	3.0	15.5	9.0	5.0
Timothy	6.0	6.5	4.5	5.0[b]	7.5	7.0

[a] All values given indicate megagrams of dry matter per hectare.
[b] Yield reduction for high N rate associated with invasion of undesirable species.

Adapted from Jung, G. A., Balasko, J. A., Alt, F. L., and Stevens, L. P., *Agron. J.*, 66, 517, 1974.

Table 15
ESTIMATES OF NITROGEN FIXATION BY FORAGE LEGUMES

Species	N fixation (kg N/ha)
Alfalfa	148—290
White clover	128—268
Ladino clover	165—189
Red clover	154
Sweet clover	140
Birdsfoot trefoil	54
Alsike clover	133
Vetch	110—184
Korean lespedeza	193

Adapted from References 71 and 72.

Table 16
NITROGENASE ACTIVITY AS MEASURED BY ACETYLENE REDUCTION IN TROPICAL FORAGE GRASSES

	\multicolumn{2}{c}{Nitrogenase activity (nmol C_2H_4/[hr × g dry roots])}
Species	Minimum	Maximum
Bermuda grass	17	269
Pangola grass	21	404
Guinea grass	20	299
Bahia grass	2	283
Napier grass	5	954

Adapted from von Bulow, J. F. W., in Dobereiner, J., Burris, R. H., and Hollaender, A., Eds., *Limitations and Potentials for Biological Nitrogen Fixation in the Tropics*, Plenum Press, New York, 1978, 75.

Table 17
INFLUENCE OF HARVEST FREQUENCY ON GROWTH STAGE AT FIRST HARVEST AND ON IN VITRO DIGESTIBILITY (IVDDM), NEUTRAL DETERGENT FIBER (NDF), AND CRUDE PROTEIN (CP) CONCENTRATIONS OF FOUR COOL-SEASON PERENNIAL GRASSES

		Grass species			
Quality and yield factors	Cuttings per year[a]	Orchard grass	Tall fescue	Reed canary grass	Smooth bromegrass
		% (Dry matter basis)			
IVDDM	2	61.4	61.1	55.7	58.9
	3	66.9	65.4	61.7	66.5
	4	69.6	69.1	68.7	69.8
NDF	2	58.2	57.9	56.8	56.4
	3	55.1	55.2	54.8	54.9
	4	52.1	52.5	49.1	50.1
CP	2	12.7	12.9	13.2	14.0
	3	14.8	13.8	16.9	16.6
	4	16.5	15.5	21.0	20.2
		Megagrams per hectare			
Dry matter yield	2	10.6	10.4	12.3	9.9
	3	10.4	10.9	11.2	10.0
	4	8.5	9.2	7.9	6.7
IVDDM yield	2	6.5	6.4	6.9	6.4
	3	7.0	7.1	6.9	6.9
	4	5.9	6.4	5.4	5.6
CP yield	2	1.3	1.4	1.6	1.4
	3	1.5	1.5	1.9	1.7
	4	1.4	1.4	1.6	1.3

[a] Growth stages at first harvest were seed, heading, and vegetative for two, three, and four cuttings per year, respectively.

Adapted from Marten, G. C. and Hovin, A. W., *Agron. J.*, 72, 378, 1980.

Table 18
EFFECT OF STUBBLE HEIGHT AND STAGE OF HARVEST ON YIELD AND QUALITY OF SWITCHGRASS

Stubble height (cm)	Development stage	Total season yield (Mg dry matter/ha)	NDF	IVDDM	CP
8	Early jointing	7.6	69.4	59.1	14.6
	Late jointing	7.8	70.9	57.2	13.1
	Early boot	8.2	72.3	55.1	11.4
	Boot	9.0	73.7	53.1	9.8
	Heading	9.1	75.2	51.5	8.2
23	Early jointing	5.1	71.3	60.2	15.2
	Late jointing	5.2	71.6	58.6	13.7
	Early boot	5.2	72.0	57.2	12.2
	Boot	5.3	72.4	55.6	10.7
	Heading	5.4	72.8	54.0	9.2

Forage quality 1st harvest (% dry matter)

Adapted from Anderson, B. and Matches, A. G., *Agron. J.*, 75, 119, 1983.

Table 19
EFFECT OF STAGE OF MATURITY AT HARVEST ON CUTTINGS PER YEAR AND DRY MATTER AND CRUDE PROTEIN (CP) YIELDS OF ALFALFA

Stage of maturity	Cuttings/year	Dry matter	CP
Prebud	8	16.2	4.2
Bud	7	19.3	4.5
10% bloom	6	22.1	4.7
50% bloom	5	22.5	4.4

Yield (Mg/ha)

Adapted from Weir, W. C., Jones, L. G., and Meyer, J. H., *J. Anim. Sci.*, 19, 5, 1960.

Table 20
LEGUME N IN HERBAGE, CROWNS, AND ROOTS OF FOUR SPECIES SUBJECTED TO HAY (ALL FORAGE REMOVED) AND GREEN MANURE (NO FORAGE REMOVED) HARVEST MANAGEMENTS IN THE SEEDING YEAR

		Legume N (kg N/ha)			
Management	Species	Total	Herbage	Crowns	Roots
Hay harvest	Nondormant alfalfa	44		26	18
	Dormant alfalfa	71		34	37
	Red clover	97		60	37
	Sweet clover	46		18	28
Green manure	Nondormant alfalfa	184	103	35	46
	Dormant alfalfa	207	94	49	64
	Red clover	226	149	43	34
	Sweet clover	287	104	26	157

Adapted from Groya, F. L. and Sheaffer, C. C., in Proc. 28th National Alfalfa Improvement Conf., Davis, Calif., ARS-USDA, North Central Region, Peoria, 1982.

Table 21
FORAGE DRY MATTER AND CRUDE PROTEIN YIELD AND LONG-TERM PERSISTENCE OF 8 FORAGE SPECIES HARVESTED 3 TIMES PER YEAR AND IRRIGATED WITH 5 OR 10 CM OF WASTEWATER EFFLUENT PER WEEK

	Forage dry matter/ effluent (cm/wk)		Crude protein/ effluent (cm/wk)		Stand persistence/ effluent (cm/wk)	
Species	5	10	5	10	5	10
Alfalfa	7.5[a]	7.3[a]	2.1[a]	1.7[a]	28[b]	2[b]
Smooth bromegrass	8.2	6.9	1.6	1.6	9	8
Timothy	7.2	7.1	1.3	1.5	39	6
Kentucky bluegrass	6.8	7.2	1.3	1.6	96	77
Orchard grass	8.5	8.1	1.8	1.6	61	55
Tall fescue	12.8	11.5	1.7	2.2	94	12
Quackgrass	8.5	9.6	1.6	2.2	19	80
Reed canary grass	10.5	11.2	2.2	2.6	50	46

[a] Values in this column indicate megagrams per hectare.
[b] Values in this column indicate a percentage.

Adapted from Marten, G. C., Clapp, C. E., and Larson, W. E., *Agron. J.*, 71, 650, 1979.

Table 22
RELATIVE IMPORTANCE OF TYPES OF INSECT AND DISEASE PESTS OF FORAGE LEGUMES[3,5-7,56]

	Insect pests						Diseases				
Species	Aphids	Leaf-hoppers	Plant bugs	Leaf feeders	Crown and root feeders	Seed production	Damping off	Crown and root rots	Vascular wilts	Foliar diseases	Viruses	Nematodes
Alfalfa	2[a]	3	2	2	2	3	2	3	3	2	1	2
Alsike clover	2	2	1	1	3	2	2	3	1	2	3	2
Arrowleaf clover	1	1	1	1	1	2	1	2	1	1	3	2
Birdsfoot trefoil	1	1	2	1	2	2	2	3	1	2	1	2
Crimson clover	1	1	1	2	1	2	1	2	1	3	3	2
Crownvetch	1	1	1	1	1	1	1	2	1	2	1	1
Cicer milkvetch	1	1	1	1	1	1	1	1	1	1	1	1
Common and Korean lespedeza	1	1	1	1	2	1	2	2	1	2	1	3
Common and hairy vetch	2	2	2	1	2	1	2	3	2	2	1	2
Kudzu	1	1	1	2	1	1	1	2	1	2	1	2
Red clover	2	1	1	1	3	2	2	3	1	2	2	2
Sainfoin	1	1	1	1	2	2	2	3	2	2	1	2
Sericea lespedeza	1	1	1	2	2	1	1	2	1	2	1	2
Sweet clover	2	1	1	2	3	1	2	2	1	1	1	1
White clover	2	2	1	1	3	3	2	3	1	2	3	3

[a] 1 = infrequent problem, 2 = occasional problem, 3 = frequent problem.

Table 23
LEGUME AND GRASS CELL-WALL COMPOSITION, INDIGESTIBILITY, AND RATE CONSTANTS FOR DIGESTION

Species and cultivar	Development	Soluble dry matter (%)	Cellulose (%)	Hemicellulose (%)	Lignin (%)	CWIVI (%)[a] 48 hr	CWIVI (%)[a] 72 hr	CW digestion rate constant (hr^{-1})[b]
Legumes								
Alfalfa	Prebloom	66	21	6	7	46	45	0.191
Saranac	Early pod, some leaf loss	47	30	8	15	60	58	0.073
Birdsfoot trefoil	Prebloom	65	19	9	6	51	51	0.174
Viking	Early pod, some leaf loss	56	25	7	12	63	61	0.060
Ladino clover	Prebloom	82	11	4	3	19	19	0.309
Common	Dry heads	62	21	4	13	44	40	0.063
Red clover	Prebloom	53	23	21	2	22	22	0.091
Kenland	Dry heads, 90%	54	25	8	13	57	54	0.063
Crown vetch	Budding	72	15	8	4	53	50	0.103
Chemung	Podded, some leaf loss	54	26	7	14	73	73	0.097
Common Vetch	Early pod	60	23	1	16	51	54	0.118
	Podded, some leaf loss	55	23	9	11	61	59	0.057
Grasses								
Kentucky bluegrass	Vegetative	60	18	20	2	12	12	0.153
Common	Mature	34	30	26	8	45	38	0.048
Smooth bromegrass	Vegetative	50	24	23	3	19	15	0.183
Blair	Mature	32	33	24	10	50	50	0.073
Tall fescue	Vegetative	50	24	23	3	16	16	0.131
Ky-31	Mature	39	29	23	9	51	48	0.058
Orchard grass	Vegetative	64	16	17	2	15	12	0.128
Potomac	Mature	35	31	23	10	64	61	0.050
Orchard grass	Boot	38	31	24	5	29	26	0.077
Pennlate	Mature	29	35	25	11	54	46	0.046
Reed canary grass	Vegetative	62	18	18	1	13	10	0.183
Common	Mature	33	31	26	9	50	45	0.053

[a] In vitro cell-wall indigestibilities.
[b] Rate constants for in vitro disappearances of digestible cell walls.

Adapted from Smith, L. W., Goering, H. K., and Gordon, C. H., *J. Dairy Sci.*, 55, 1140, 1972.

Table 24
PERCENTAGE OF CRUDE PROTEIN (CP), ACID DETERGENT FIBER (ADF), AND NEUTRAL DETERGENT FIBER (NDF) IN ALFALFA, TEMPERATE GRASSES, AND SUBTROPICAL GRASSES

Species and maturity	Northern U.S.[a] CP	ADF	NDF	Southern U.S. CP	ADF	NDF
Alfalfa						
Bud to first flower	>19	<31	<40	25—30	30—32	33—41
First flower midbloom	17—19	31—35	40—46	19—27	34—37	40—47
Mid to full bloom	13—16	36—41	46—51	22	35	42
Postbloom	<13	>41	>51	17—18	37—41	51
Grasses[b]						
Vegetative to boot[c]	>18	<33	<55	18—19	32—33	64—69
Boot to early head[d]	13—18	34—38	55—60	8—18	34—40	64—79
Head to milk[e]	8—12	39—41	61—65	6—11	39—43	70—80
Dough[f]	<8	>41	>65	4—9	39—47	71—81

[a] See Figure 1; north 4—5H; south 6—8H.
[b] North — Smooth bromegrass, orchard grass, reed canary grass, and tall fescue.
[c] South — Pangola digit grass and Bermuda grass (2 to 3 weeks regrowth).
[d] South — Bahia grass (2 weeks); Pangola digit grass and Bermuda grass (4 to 6 weeks regrowth).
[e] South — Bahia grass (4 to 6 weeks); Pangola digit grass and Bermuda grass (8 weeks regrowth).
[f] South — All grasses (10 weeks regrowth).

Adapted from Rohweder, D., Jorgensen, N., and Barnes, R. F., *Feedstuffs*, 48, 22, 1976.

Table 25
RANGES IN VOLUNTARY INTAKE BY SHEEP AND NUTRIENT DIGESTIBILITY OF TEMPERATE AND TROPICAL GRASSES

Grass species	Intake (g/day/W)	Digestibility (%)
Temperate		
Ryegrass	48—89	53—88
Tall fescue	58—91	52—83
Orchard grass	30—94	45—83
Smooth bromegrass	62—64	53—68
Timothy	52—92	48—77
Tropical		
Paspalum spp. (e.g., Bahia grass)	41—62	38—65
Bermuda grass	46—88	45—71
Digit grass	40—73	42—72
Napier grass	61—66	60—65
Guinea grass	40—81	45—66

Adapted from Mott, G. O., in *Proc. 14th Int. Grassl. Congr., Lexington, Ky.*, Westview Press, Boulder, 1981, 35.

Table 26
FORAGE QUALITY PARAMETERS OF WARM- AND COOL-SEASON GRASS SPECIES (% DRY MATTER)

Grass species	IVDMD[a]	Protein	Ether extract	Ash	NDF[b]	ADF[c]	Hemicellulose	Holocellulose	PML[d]
Warm-season									
Bermuda grass (4-week)	66.1	18.7	3.4	7.2	60.0	31.9	28.1	53.6	3.5
Bermuda grass (8-week)	54.9	13.9	2.8	7.2	62.9	39.0	23.9	55.6	5.5
Bahia grass (4-week)	59.6	15.7	3.9	6.2	71.0	35.7	35.3	60.8	3.4
Bahia grass (8-week)	53.2	9.2	1.6	5.9	67.5	35.0	32.5	76.0	5.3
Pangola grass (4-week)	54.5	7.0	3.1	4.7	69.4	41.7	22.7	57.3	6.3
Pangola grass (8-week)	48.6	5.9	3.1	5.1	67.0	29.5	37.6	47.4	4.8
Cool-season									
Tall fescue (4 week)	65.6	13.2	5.6	8.3	58.2	33.6	24.7	41.6	3.2
Kentucky bluegrass (4-week)	58.1	15.5	4.0	7.1	54.0	30.6	23.5	43.5	4.3
Bromegrass (4-week)	64.2	14.3	4.5	8.8	56.2	34.3	21.9	51.1	4.8
Orchard grass (4-week)	62.8	14.8	4.4	8.2	57.9	33.3	24.6	43.9	4.1
Kentucky 31 tall fescue (4-week)	62.7	14.2	3.9	8.8	58.4	31.4	25.5	45.2	3.4
Timothy (4-week)	66.8	13.4	4.3	8.8	55.6	34.7	20.9	42.0	4.1

[a] IVDMD = in vitro dry-matter digestibility.
[b] NDF = neutral detergent fiber.
[c] ADF = acid detergent fiber.
[d] PML = permanganate lignin.

Adapted from Barton, F. E., Amos, H. E., Burdick, D., and Wilson, R. L., *J. Anim. Sci.*, 43, 504, 1976.

Table 27
PERCENTAGE OF TISSUE TYPES IN LEAF BLADES OF TROPICAL AND TEMPERATE GRASS SPECIES

Grass species	Total vascular tissue	Parenchyma bundle sheath	Mesophyll	Total epidermis
Warm-season				
Coastal Bermuda grass	37.0 ± 6.2	28.5 ± 6.3	27.0 ± 14.2	13.1 ± 3.5
Coastcross-1 Bermuda grass	35.5 ± 8.1	26.5 ± 6.2	23.5 ± 14.2	14.9 ± 3.5
Pensacola Bahia grass	17.9 ± 2.8	11.1 ± 2.4	52.4 ± 7.6	12.7 ± 3.7
Pangola digitgrass	13.4 ± 2.3	9.5 ± 2.3	32.4 ± 5.3	24.2 ± 3.4
Dallis grass	14.1 ± 3.2	9.8 ± 2.5	43.8 ± 6.1	17.9 ± 3.2
Cool-season				
Kentucky 31 tall fescue	12.9 ± 2.9	5.4 ± 1.2	60.6 ± 4.9	10.8 ± 2.0
Kenhy tall fescue	11.3 ± 2.1	4.3 ± 1.7	62.3 ± 5.0	9.6 ± 2.0
Timothy	13.0 ± 2.1	4.9 ± 1.3	54.8 ± 5.1	13.2 ± 3.0
Orchard grass	22.1 ± 4.9	9.1 ± 4.2	53.5 ± 4.5	10.2 ± 3.3
Smooth bromegrass	19.7 ± 2.2	6.6 ± 1.0	52.8 ± 7.6	9.5 ± 3.5
Kentucky bluegrass	11.0 ± 2.4	4.1 ± 0.9	64.7 ± 5.8	8.8 ± 2.0

Adapted from Akin, D. E. and Burdick, D., *Crop Sci.*, 15, 661, 1975.

Table 28
PERCENTAGE OF TISSUE TYPES GROUPED ACCORDING TO RELATIVE EASE OF IN VITRO DEGRADATION BY MICROORGANISMS

Grass species	Rapidly degraded, mesophyll and phloem	Slowly degraded, epidermis and parenchymal sheath	Nondegraded, sclerenchyma and lignified vascular tissue
Warm-season			
Coastal Bermuda grass	31.0	54.7	14.3
Coastcross-1 Bermuda grass	27.0	56.2	16.8
Pensacola Bahia grass	54.7	36.4	8.9
Dallis grass	45.9	45.0	9.1
Pangola digit grass	33.8	57.9	8.2
Cool-season			
Smooth bromegrass	55.6	25.6	19.0
Orchard grass	56.0	29.4	15.7
Timothy	56.2	31.4	12.4
Kentucky bluegrass	66.0	21.8	12.3
Ky 31 tall fescue	62.0	27.0	12.2
Kenhy tall fescue	63.8	23.5	12.7

Adapted from Akin, D. E. and Burdick, D., *Crop Sci.* 15, 661, 1975.

Table 29
FORAGE QUALITY OF ALFALFA AND SMOOTH BROMEGRASS AT VARIOUS STAGES OF MATURITY

Species/stage	Crude protein	Crude fiber	Acid detergent fiber	Neutral detergent fiber	Lignin
Alfalfa					
Early vegetative	22[a]	24	31	41	8
Late vegetative	20	27	34	44	9
Early bloom	17	31	38	48	10
Mid bloom	16	33	40	50	11
Full bloom	15	35	42	52	12
Mature	14	37	44	55	12
Smooth bromegrass					
Early vegetative	15	24	31	60	3
Late vegetative	11	33	40	68	5
Boot	9	33	38	67	5
Late bloom	4	40	44	72	8

[a] All valves given indicate percentage of dry matter.

Adapted from National Research Council, Nutrient Requirements of Dairy Cattle, National Academy of Sciences, Washington, D.C. 1971.

Table 30
YIELD AND QUALITY OF SUDAN GRASS

Growth Stage	Dry matter yield (mg/ha)	Dry matter (%)	Crude protein (%)	Cellulose (%)	Cell-wall components (%)	Acid detergent fiber (%)	Lignin (%)
First cutting							
Vegetative	2.0	16.8	15.5	27.0	51.8	28.6	2.1
Boot	3.6	18.5	10.5	28.5	59.1	34.7	2.6
Early bloom	6.2	23.5	7.3	37.8	66.2	41.6	4.5
Full bloom	6.7	28.8	7.5	36.9	65.1	42.1	5.3
Seed forming	9.4	33.1	4.8	38.7	66.0	43.3	5.7
Seed mature	10.1	39.9	4.2	40.3	68.6	45.3	6.2
Regrowth							
Vegetative	3.4	18.0	11.9	28.4	59.1	38.3	2.9
Vegetative	3.4	20.2	10.0	31.7	58.6	33.8	2.7
Vegetative	2.4	19.8	9.0	33.9	60.4	35.1	2.7
Boot	3.1	22.7	7.2	35.5	65.5	38.8	2.4

Adapted from Jung, G. A. and Reid, R. L., *W. Va. Univ. Agric. Exp. Stn. Bull.*, No. 524T, 1966.

Table 31
IN VITRO DIGESTIBILITY (IVD) OF TIMOTHY AT FIRST CUTTING AND PROPORTIONS (P) OF THESE FRACTIONS IN THE WHOLE PLANT DRY WEIGHT (%)

Growth stage	Whole plant (calculated) IVD	Leaf blade IVD	Leaf blade P	Leaf sheath IVD	Leaf sheath P	Stem (including inflorescence) IVD	Stem (including inflorescence) P
Vegetative with little sheath elongation	85	85	88	83	12	—	—
Considerable sheath elongation	83	84	74	80	23	—	3
Some internodal elongation of stem	80	82	59	74	24	84	17
Considerable internodal elongation	78	81	55	70	25	81	20
Flag leaf	73	77	44	69	22	71	34
Early emergence	72	76	37	71	25	70	38
Full emergence, with slight elongation of head	62	76	21	54	20	59	59

Adapted from Terry, R. A. and Tilley, J. M. A., *J. Br. Grassl. Soc.*, 19, 363, 1964.

Table 32
IN VITRO DIGESTIBILITY (IVD) OF LEAF AND STEM FRACTIONS OF YOUNG AND MATURE ALFALFA

	IVD (%)	Fraction of whole plant drymatter (%)
Immature alfalfa		
Whole plant	75	100
Leaf	81	53
Stem (mean)[a]	69	47
0—15 cm	79	19
15—30 cm	65	19
30—45 cm	55	9
Mature alfalfa		
Whole plant	66	100
Leaf	78	39
Stem (mean)	55	61
0—15 cm	77	11
15—30 cm	63	12
30—45 cm	50	15
45—60 cm	46	15
60—75 cm	40	7

[a] Stem fractions were ranged in height from the tops of the plants.

Adapted from Terry, R. A. and Tilley, J. M. A., *J. Br. Grassl. Soc.*, 19, 363, 1964.

Table 33
EFFECTS OF CUTTING INTERVAL ON YIELD, CHEMICAL COMPOSITION AND ON LEAF AND STEM COMPONENTS OF BERMUDA GRASS

Cutting interval (weeks)	Stem length (cm)	DM yield (Mg/ha)	Crude protein (%)	Crude fiber (%)	IVDDM (%)	Leaf (%)
3	7.6	15.2	18.5	27.0	65.2	85.2
4	11.7	16.2	16.4	29.1	63.7	81.4
6	20.5	19.9	13.3	31.6	59.7	67.2
8	27.1	19.9	10.7	32.9	56.6	52.4
12	36.1	20.0	9.0	33.4	52.5	47.6
24	61.9	14.6	8.4	33.9	43.2	36.0

Adapted from Burton, G. W., Jackson, J. E., and Hart, R. H., *Agron. J.*, 55,500, 1963.

Table 34
ALLELOCHEMICALS AND ANTIQUALITY CONSITUENTS OF SELECTED FORAGES

Species	Constituent	Effect	Ref.
Alfalfa	Soluble protein	Bloat in ruminants	96
	Saponins	Growth, palatability	97
	Estrogenic flavonoid	Animal fertility	98
Clovers (general)	Soluble protein	Bloat in ruminants	96
Red clover	Slaframine *(Rhizoctonia)*	Palatability, slobbering	99
White clover	Estrogenic flavonoid	Animal fertility	98
	Cyanogenic glucosides	HCN poisoning	98
Crownvetch	Glycosides (coronilline and cathertine)	Palatability	100—102
	β-Nitropropionic acid	Nonruminant poisoning	
Sweet clover	Coumarin, dicoumarol	Palatability, anticlotting factor	98, 101
Sericia lespedeza	Tannins	Palatability, digestibility	98, 101, 103
Tall fescue	Mycotoxins	Vasoconstrictor, fescue foot	104—106
	Alkaloid (perloline)	Palatability	
Reed canary grass	Alkaloids (gramine, tryptamine, carboline)	Palatability, diarrhea, gain	101, 105
Bahia grass	Mycotoxins	Vasoconstrictor	107
Sudan grass	Cyanogenic glycoside (dhurrin)	HCN poisoning	98

Table 35
CHARACTERISTICS OF BLOAT-CAUSING AND BLOAT-SAFE LEGUMES
(RESULTS EXPRESSED AS A PERCENTAGE OF RESULTS WITH ALFALFA)

	Bloat	In vitro DMD	Gas production in vitro	Sheep rumen fluid Soluble protein N	Chlorophyll	Tanning
Alfalfa	Yes	100	100	100	100	No
Red clover	Yes	128	83	—	—	No
White clover	Yes	130	88	—	—	No
Birdsfoot trefoil	No	80	62	12	16	Yes
Cicer milkvetch	No	80	83	53	51	No
Sainfoin	No	47	58	23	28	Yes

Adapted from Howarth, R. E., Cheng, K. J., Fay, J. P., Majak, W., Lees, G. L., Goplen, B. P., and Costerton, J. W., in *Proc. 14th Int. Grassl. Congr., Lexington, Ky.*, Westview Press, Boulder, 1981, 719.

Table 36
AVERAGE FORAGE CATION CONCENTRATION AND CATION RATIOS OF FORAGE GRASSES

Species	K(%)	Mg (%)	(Ca + Mg) (meq)
Orchard grass	3.96	0.27	2.47
Tall oatgrass	3.97	0.28	2.25
Reed canary grass	3.60	0.27	2.23
Crested wheatgrass	2.68	0.19	2.10
Tall wheatgrass	2.98	0.21	2.03
Meadow foxtail	3.28	0.25	1.94
Kentucky bluegrass	2.03	0.28	1.72
Smooth bromegrass	2.50	0.19	1.71

Adapted from Thill, J. L. and George, J. R., *Agron. J.*, 67, 89, 1975.

REFERENCES

1. **Hodgson, H. J.**, Forage crops, *Sci. Am.*, 234, 61, 1976.
2. **Ahlgren, G. H.**, *Forage Crops*, 2nd ed., McGraw-Hill, New York, 1956.
3. **Bogdan, A. V.**, *Tropical Pasture and Fodder Plants*, Longman, New York, 1977.
4. **Chessmore, R. A.**, *Profitable Pasture Management*, Interstate Printers and Publishers, Danville, Ill., 1979.
5. **Heath, M. E., Metcalfe, D. S., and Barnes, R. F.**, Eds., *Forages*, 3rd ed., The Iowa State University Press, Ames, 1973.
6. **Stefferud, A.**, Ed., *Grass — the Yearbook of Agriculture, 1948* U.S. Department of Agriculture, Washington, D.C., 1948.
7. **Wheeler, W. A.**, *Forage and Pasture Crops*, D Van Nostrand, New York, 1950.
8. **Decker, A. M., MacDonald, H. A., Wakefield, R. C., Jung, G. A., and Burger, O. J.**, Cutting management of alfalfa and ladino clover in the Northeast, *Rh. I. Agric. Exp. Stn. Bull.*, 356, 1960.
9. Minnesota Agricultural Experiment Station, Varietal trials of farm crops, Minnesota Rep. 24, St. Paul, 1983.

10. **Tesar, M. B.,** Fertility and management practices for a 10-ton yield in Michigan, in Proc. 27th Alfalfa Improvement Conf., Madison, Wis., ARS-USDA, North Central Region, Peoria, Ill., 1980, 43.
11. **Hoveland, C. S., Carden, E. L., Buchanan, G. A., Evans, E. M., Anthony, W. B., Mayton, E. L., and Burgess, H. E.,** Yuchi arrowleaf clover, *Ala. Agric. Exp. Stn. Bull.,* 396, 1969.
12. **Sigafus, R. E. and Taylor, N. L.,** Kentucky forage variety trails, Ky Agric. Exp. Stn. Prog. Rep., Lexington, 253.
13. **Dunavin, L. S.,** Legume investigations in West Florida, in Smith, R. R., Ed., Clovers and Special Purpose Legume Research, Progress Report 13, U.S. Department of Agriculture and the University of Wisconsin, Madison, 1980, 4.
14. **Van Keuren, R. W. and Myers, D. K.,** Ohio Forage Report — 1979, Ohio Agricultural Research and Development Center, Wooster, 1979.
15. **Collins, M.,** Yield and quality of birdsfoot trefoil stockpiled for summer utilization, *Agron. J.,* 74, 1036, 1982.
16. **Brann, D. E. and Jung, G. A.,** Influence of cutting management and environmental variation on the yield, bud activity and autumn carbohydrate reserve levels of crown vetch, *Agron. J.,* 66, 767, 1974.
17. **Cooper, C. S.,** Evaluation of sainfoin, birdsfoot trefoil, and cicer milkvetch seeded alone and in mixtures for pasture, in Clovers and Special Purpose Legume Research, Smith, R. R., Ed., Progress Report 8, U.S. Department of Agriculture and the University of Wisconsin, Madison, 1975, 20.
18. **Meyer, D. W.,** Yield, regrowth and persistence of sainfoin under fertilization, *Agron. J.,* 67, 439, 1975.
19. **Offutt, M. S. and Baldridge, J. D.,** The lespedezas, in *Forages,* 3rd ed., Heath, M. E., Metcalfe, D. S., and Barnes, R. F., Eds., Iowa State University Press, Ames, 1973, 189.
20. **Taylor, R. W., Griffin, J. L., and Meche, G. A.,** Potential of vetch (*Vicia* spp.) for forage and cover crop production in the rice area of Louisiana, in Clovers and Special Purpose Legumes Research, Smith, R. R., Ed., Progress Report 14, U.S. Department of Agriculture and the University of Wisconsin, Madison, 1981, 34.
21. **Miles, I. E. and Gross, E. E.,** A compilation of information on kudzu, *Miss. Agric. Exp. Stn. Bull.,* No. 326, 1939.
22. **Taylor, R. W., Griffin, J. L., and Meche, G. A.,** Evaluation of annual clovers, white clovers and red clovers for forage production in Southwest Louisiana, in Clovers and Special Purpose Legumes Research, Smith, R. R., Ed., Progress Report 14, U.S. Department of Agriculture and the University of Wisconsin, Madison, 1981, 29.
23. **Carleton, A. E., Cooper, C. S., Roath, C. W., and Krall, J. L.,** Evaluation of sainfoin for irrigated hay in Montana, in Cooper, C. S. and Carleton, A. E., Eds., Sainfoin Symposium, *Mont. Agric. Exp. Stn. Bull.,* No. 627, 1968, 44.
24. **Smith, R. R. and Rohweder, D.,** Arlington Red Clover, Res. Rep. R2762, University of Wisconsin, Madison, 1975.
25. **Groya, F. L. and Sheaffer, C. C.,** Effect of seeding year harvest management on legume nitrogen availability for succeeding crops, in Proc. 28th National Alfalfa Improvement Conf., Davis, Calif., ARS-USDA, North Central Region, Peoria, 1982.
26. **Sanderson, M. A., Meyer, D. W., and Casper, H. H.,** Sweetclover variety trials, in Clovers and Special Purpose Legumes Research, Smith, R. R., Ed., Progress Report 15, U.S. Department of Agriculture and the University of Wisconsin, Madison, 1982, 72.
27. **Frakes, R. V.,** The ryegrasses, in *Forages,* 3rd ed., Heath, M. E., Metcalfe, D. S., and Barnes, R. F., Eds., Iowa State University Press, Ames, 1973, 307.
28. **Beaty, E. R., Stanley, R. L., and Powell, J. D.,** Effect of height of cut on yield of Pensacola bahiagrass, *Agron. J.,* 60, 356, 1968.
29. **Burton, G. W., Prine, G. M., and Jackson, J. E.,** Studies of drought tolerance and water use of several southern grasses, *Agron. J.,* 49, 498, 1957.
30. **Brown, W. E., Spiers, J. M., and Thurman, C. W.,** Performance of five warm-season perennial grasses grown in southern Mississippi, *Agron. J.,* 68, 821, 1976.
31. **Stanley, R. L., Beaty, E. R., and Powell, J. D.,** Forage yield and percent cell wall constituents of Pensacola bahiagrass as related to N fertilization and clipping height, *Agron. J.,* 69, 501, 1977.
32. **Monson, W. G. and Burton, G. W.,** Harvest frequency and fertilizer effects on yield, quality, and persistence of eight bermudagrasses, *Agron. J.,* 74, 371, 1982.
33. **Newell, L. C. and Keim, F. D.,** Effects of mowing frequency on the yield and protein content of several grasses grown in pure stands, *Neb. Agric. Exp. Stn. Res. Bull.,* No. 150, 1947.
34. **Warnes, D. D., Newell, L. C., and Moline, W. J.,** Performance evaluation of some warm season prairie grasses in Nebraska environments, *Neb. Agric. Exp. Stn. Bull.,* No. 241, 1971.
35. **Lorenz, R. J. and Rogles, G. A.,** Irrigated forage crops for North Dakota, *N. D. Agric. Exp. Stn. Bull.,* No. 437, 1961.
36. **Hart, R. H. and Burton, G. W.,** Prostrate vs. common dallisgrass under different clipping frequencies and fertility levels, *Agron. J.,* 58, 521, 1966.

37. **Jung, G. A., Balasko, J. A., Alt, F. L., and Stevens, L. P.**, Persistence and yield of 10 grasses in response to clipping frequency and applied nitrogen in the Allegheny highlands, *Agron. J.*, 66, 517, 1974.
38. **Jung, G. A. and Kocher, R. E.**, Influence of applied nitrogen and clipping treatments in winter survival of perennial cool season grasses, *Agron. J.*, 66, 64, 1974.
39. **Schmid, A. R., Elling, L. J., Hovin, A. W., Lueschen, W. E., Nelson, W. W., Rabas, D. L., and Warnes, D. D.**, Pasture species clipping trials in Minnesota, *Minn. Agric. Exp. Stn. Tech. Bull.*, No. 317, 1979.
40. **Marten, G. C., Clapp, C. E., and Larson, W. E.**, Effects of municipal wastewater effluent and cutting management on persistence and yield of eight perennial forages, *Agron. J.*, 71, 650, 1979.
41. **Marten, G. C. and Hovin, A. W.**, Harvest schedule, persistence, yield and quality interactions among four perennial grasses, *Agron. J.*, 72, 378, 1980.
42. **Schoper, R. P., Malzer, G. L., and Simkins, C. A.**, Influence of N and P on the yield and chemical composition of quack grass growing on an organic soil, *Agron. J.*, 71, 1034, 1979.
43. **Jung, G. A. and Reid, R. L.**, Sudangrass: studies on its yield management, chemical composition and nutritive value, *W. Va. Agric. Exp. Stn. Bull.*, No. 524T, 1966.
44. **Anderson, B. and Matches, A. G.**, Forage yield, quality, and persistence of switchgrass and caucasian bluestem, *Agron. J.*, 75, 119, 1983.
45. **Berg, C. C.**, Forage yield of switchgrass (*Panicum virgatum* L.) in Pennsylvania, *Agron. J.*, 63, 785, 1971.
46. **Koshi, P. T., Stubbendieck, J., Eck, H. V., and McCully, W. G.**, Switchgrasses: forage yield, forage quality, and water-use efficiency, *J. Range Manage.*, 35, 623, 1982.
47. **Matches, A. G.**, Management, in *Tall Fescue*, Vol. 20, Buckner, R. C. and Bush, L. P., Eds., American Society of Agronomy, Madison, 1979, 171.
48. **Wilsie, C. P.**, *Crop Adaptation and Distribution*, W. H. Freeman, San Francisco, 1962.
49. Plant Hardiness Zone Map Misc. Publ. 814, U.S. Department of Agriculture, Washington, D.C., 1960.
50. **Beard, J. B.**, *Turfgrass Science and Culture*, Prentice-Hall, Englewood Cliffs, N.J., 1973, 658 pp.
51. **Blaser, R. E., Skrdla, W. H., and Taylor, T. H.**, Ecological and physiological factors in compounding forage seed mixtures, *Adv. Agron.*, 4, 179, 1954.
52. **Bolton, J. L. and McKenzie, R. E.**, The effects of early spring flooding on certain forage crops, *Sci. Agric.*, 26, 3, 1946.
53. **Carleton, A. E., Austin, R. D., Stroh, J. R., Wiesner, L. E., and Scheetz, J. G.**, Cicer milkvetch seed germination, scarification and field emergence studies, *Mont. Agric. Exp. Stn. Bull.*, No. 627, Bozeman, 1971.
54. **Dodds, D. and Vasey, E. H.**, Forages for Salt-Affected and Wet Soils, Circular R-584, North Dakota State University Extension Service, Fargo, 1979.
55. **Elliott, C. R., Hoyt, P. B., Nyborg, M., and Siemens, B.**, Sensitivity of several species of grasses and legumes to soil acidity, *Can. J. Plant Sci.*, 53, 113, 1973.
56. **Duke, J. A.**, *Handbook of Legumes of World Economic Importance*, Plenum Press, New York, 1981.
57. **McElgunn, J. D. and Lawrence, T.**, Salinity tolerance of altai wild ryegrass and other forage grasses, *Can. J. Plant Sci.*, 53, 303, 1973.
58. **McWilliam, J. R.**, Response of pasture plants to temperature, in Wilson, J. R., Ed., *Plant Relations in Pastures*, CSIRO, Melbourne, Australia, 1978, 17.
59. **Evans, L. T., Wardlow, I. F., and Williams, C. M.**, Environmental control of growth, in Barnard, C., Ed., *Grasses and Grasslands*, St. Martin's Press, New York, 1966, 102.
60. **Smith, D.**, Physiological considerations in forage management, in *Forages*, 3rd ed., Heath, M. E., Metcalf, D. S., and Barnes, R. F., Eds., Iowa State University Press, Ames, 1973, 403.
61. **Kawanabe, S.**, Temperature responses and systemics of the Gramineae, *Proc. Jpn. Soc. Plant Taxonomists*, 2, 17, 1968.
62. **Mitchell, K. J.**, Growth of pasture species under controlled environment. I. Growth at various levels of constant temperature, *N.Z.J. Sci. Tech.*, 38, 203, 1956.
63. **Smith, D.**, Influence of temperature on the yield and chemical composition of five forage legume species, *Agron. J.*, 62, 520, 1970.
64. **Tysdal, H. M. and Pieters, A. J.**, Cold resistance of three species of lespedeza compared to that of alfalfa, red clover and crown vetch, *Agron. J.*, 26, 923, 1934.
65. **Arakeri, H. R. and Schmid, A. R.**, Cold resistance of various legumes and grasses in early stages of growth, *Agron. J.*, 41, 182, 1949.
66. **Barnes, D. K., Frosheiser, F. I., and Martin, N. P.**, Choosing the best alfalfa varieties, in *Proc. 4th Minnesota Forage and Grassl. Council Symp., Rochester, Minn.*, Minnesota Forage and Grassland Council, St. Paul, 1979, 34.
67. **Fairbourn, M. L.**, Water use by forage species, *Agron. J.*, 74, 62, 1982.

68. **Decker, A. M., Hemkin, R. W., Miller, J. R., Clark, N. A., and Okorie, A. V.,** Nitrogen fertilization, harvest management and utilization of 'Midland' bermudagrass (*Cynodon Dactylon* (L.) Pers., *Md. Agric. Exp. Stn. Bull.,* No. 487, 1971.
69. **Colyer, D., Alt, F. L., Balasko, J. A., Henderlong, P. R., Jung, G. A., and Thang, V.,** Economic optima and price sensitivity of N fertilization for six perennial grasses, *Agron. J.,* 69, 514, 1977.
70. **Launchbaugh, J. L. and Owensby, C. E.,** Kansas rangelands: their management based on a half century of research, *Kans. Agric. Exp. Stn. Bull.,* No. 622, 1978.
71. **LaRue, T. A. and Patterson, T. G.,** How much nitrogen do legumes fix, *Adv. Agron.,* 34, 15, 1982.
72. **Heichel, G. H., Barnes, D. K., and Vance, C. P.,** Nitrogen fixation by forage legumes, and benefits to the cropping system, in *Proc. 6th Minnesota Forage and Grassl. Council, Symp., St. Paul,* Minnesota Forage and Grassl. Council, St. Paul, 1981, 1.
73. **Vance, C. P.,** Nitrogen fixation in alfalfa: an overview, in *Proc. 8th Alfalfa Symp., Bloomington, Minn.,* Certified Alfalfa Seed Council, Woodland, Cal. and Minnesota Forage and Grassland Council, St. Paul, 1978, 34.
74. **von Bülow, J. F. W.,** Plant influence in symbiotic nitrogen fixation, in Döbereiner, J., Burris, R. H., and Hallaender, A., Eds., *Limitations and Potentials for Biological Nitrogen Fixation in the Tropics,* Plenum Press, New York, 1978, 75.
75. **Döbereiner, J.,** Potential for nitrogen fixation in tropical legumes and grasses, in Dobereiner, J., Burris, R. H., and Hallaender, A. Eds., *Limitations and Potentials for Biological Nitrogen Fixation in the Tropics,* Plenum Press, New York, 1978, 13.
76. **Harlan, J. R.,** Nature range, *Okla. Exp. Stn. Bull.,* 547, 1960.
77. **Burdick, D., Spencer, R. R., and Kohler, G. O.,** Processing and utilization of forage crops, in *CRC Handbook of Processing and Utilization in Agriculture,* Wolff, I. A., Ed., CRC Press, Boca Raton, Fla., 1982, 361.
78. **Weir, W. C., Jones, L. G., and Meyer, J. H.,** Effect of cutting interval and stage of maturity on the digestibility and yield of alfalfa, *J. Anim. Sci.,* 19, 5, 1960.
79. **Ruffner, J. D.,** The comparative performance and adaption of crownvetch on strip mine spoil and other critical areas in West Virginia and Western Pennsylvania, in *Proc. Crown Vetch Symp.,* The Pennsylvania State University, University Park, 1964, 17.
80. **McWilliams, J. R., Clements, R. J., and Dowling, P. M.,** Some factors influencing the germination and early seedling development of pasture plants, *Aust. J. Agric. Res.,* 21, 19, 1970.
81. **Hyder, D. N.,** Morphogenesis and management of perennial grasses in the United States, in *Plant Morphogenesis as the Basis for Scientific Management of Range Resources,* Kreitlow, K. W., and Hunt, R. H., Eds., U.S. Department of Agriculture, Washington, D.C., 1971, 89.
82. **Smith, L. W., Goering, H. K., and Gordon, C. H.,** Relationships of forage compositions with rates of cell wall digestion and indigestibility of cell walls, *J. Dairy Sci.,* 55, 1140, 1972.
83. **Rohweder, D., Jorgensen, N., and Barnes, R. F.,** Using chemical analyses to provide guidelines in evaluating forages and establishing hay standards, *Feedstuffs,* 48, 22, 1976.
84. **Goodrich, R. D. and Meiske, J. C.,** Rate and extend of digestion-aids for explaining some nutritional observations, in *Proc. 40th Minnesota Nutrition Conf.,* University of Minnesota, St. Paul, 1979, 51.
85. **Mott, G. O.,** Potential productivity of temperature and tropical grassland systems, in *Proc. 14th Int. Grassld. Congr., Lexington, Ky.,* Westview Press, Boulder, 1981, 35.
86. **Minson, D. J. and McLeod, M. N.,** The digestibility of temperate and tropical grasses, in *Proc. 11th Int. Grassld. Congr., Surfers Paradise, Australia,* University of Queensland Press, St. Lucia, 1970, 719.
87. **Moore, J. E. and Mott, G. O.,** Structural inhibitors of quality in tropical grasses, in Matches, A. G., Ed., *Anti-quality Components of Forages,* Crop Science Society of America, Madison, 1973, 53.
88. **Minson, D. J.,** The nutritive value of tropical pastures, *J. Aust. Inst. Agric. Sci.,* 37, 255, 1971.
89. **Barton, F. E., Amos, H. E., Burdick, D., and Wilson, R. L.,** Relationship of chemical analysis to in vitro digestibility for selected tropical and temperate grasses, *J. Anim. Sci.,* 43, 504, 1976.
90. **Henderson, M. S. and Robinson, D. L.,** Environmental influences on fiber component concentrations of warm-season perennial grasses, *Agron. J.,* 74, 573, 1982.
91. **Akin, D. E. and Burdick, D.,** Percentage of tissue types in tropical and temperate grass leaf blades and degradation of tissues by rumen microorganisms, *Crop Sci.,* 15, 661, 1975.
92. **National Research Council,** Nutrient Requirements of Dairy Cattle, National Academy of Sciences, Washington, D.C., 1971.
93. **Barnes, R. F. and Gordon, C. H.,** Feeding value and on-farm feeding, in Hanson, C. H., Ed., *Alfalfa Science and Technology,* Vol. 15, American Society of Agronomy, Madison, 1972, 601.
94. **Terry, R. A. and Tilley, J. M. A.,** The digestibility of the leaves and stems of perennial ryegrass, cocksfoot, timothy, tall fescue, lucerne and sainfoin as measured by an in vitro procedure, *J. Br. Grassl. Soc.,* 19, 363, 1964.

95. **Burton, G. W., Jackson, J. E., and Hart, R. N.,** Effects of cutting frequency and nitrogen on yield, *in vitro* digestibility, and protein, fiber and carotene content of coastal Bermudagrass, *Agron. J.,* 55, 500, 1963.
96. **Howarth, R. E., Cheng, K. J., Fay, J. P., Majak, W., Lees, G. L., Goplen, B. P., and Costerton, J. W.,** Initial rate of digestion and legume pasture bloat, in *Proc. 14th Int. Grassl. Congr., Lexington, Ky.,* Westview Press, Boulder, 1981, 719.
97. **Hanson, C. H., Pedersen, M. W., Berrang, B., Wall, M. E., and Davis, K. H.,** The saponins in alfalfa cultivars, in Matches, A. G., Ed., *Anti-quality Components of Forages,* Crop Science Society of America, Madison, 1973, 33.
98. **Barnes, R. F. and Gustine, D. L.,** Allelochemistry and forage crops, in Matches, A. G., Ed., *Anti-quality Components of Forages,* Crop Science Society of America, Madison, 1973, 1.
99. **Taylor, N. L.,** Red clover and alsike cover, in *Forages,* 3rd ed., Heath, M. E., Metcalf, D. S., and Barnes, R. F., Eds., Iowa State University Press, Ames, 1973, 148.
100. **Reynolds, P. J., Jackson, C., Lindahl, I. L., and Henson, P. R.,** Consumption and digestibility of crown vetch *(Cornilla varia)* forage by sheep, *Agron. J.,* 59, 589, 1967.
101. **Marten, G. C.,** Effect of deleterious compounds on animal preference for forage and on animal performance, in Wheeler, J. L. and Mochrie, R. D., Eds., *Forage Evaluation: Concepts and Techniques,* American Forage and Grassl. Council and CSIRO, Lexington, Ky., 1981, 39.
102. **Bustine, D. L. and Moyer, B. G.,** Review of mechanisms of toxicity of 3-hetropropanoic acid in non-ruminant animals, in *Proc. 14th Int. Grassl. Congr., Lexington, Ky.,* Westview Press, Boulder, 1981, 736.
103. **Wilkins, H. L., Bates, R. P., Henson, P. R., Lindahl, I. L., and Davis, R. E.,** Tannin and palatability in sericea lespedeza, *Agron. J.,* 45, 335, 1953.
104. **Boling, J. A., Hemken, R. W., Bush, L. P., Buckner, R. C., Jackson, J. A., and Yates, S. G.,** Role of alkaloids and toxic compound(s) in the utilization of tall fescue by ruminants, in *Proc. 14th Int. Grassl. Congr., Lexington, Ky.,* Westview Press, Boulder, 1981, 722.
105. **Kendall, W. A. and Sherwood, R. T.,** Palatability of leaves of tall fescue and reed canarygrass and of some of their alkaloids to meadow voles, *Agron. J.,* 67, 667, 1975.
106. **Hoveland, C. S., Haaland, R. L., King, C. C., Anthony, W. B., Clark, E. M., McGuire, J. A., Smith, L. A., Grimes, N. W., and Holliman, J. L.,** Association of *Epichloe typhina* fungus and steer performance on tall fescue pasture, *Agron. J.,* 72, 1046, 1980.
107. **Ward, C. Y. and Watson, V. H.,** Bahiagrass and carpetgrass, in *Forages,* 3rd ed., Heath, M. E., Metcalfe, D. S., and Barnes, R. F., Eds., Iowa State University Press, Ames, 1973, 314.
108. **Marten, G. C. and Walgenbach, R. P.,** Overcoming the bloat potential of alfalfa, in *Proc. 8th Alfalfa Symp., Bloomington, Minn.,* Certified Alfalfa Seed Council, Woodland, Cal. and Minnesota Forage and Grassland Council, St. Paul, 1978, 13.
109. **Grunes, D. L.,** Grass tetany of cattle and sheep, in Matches, A. G., Ed., *Anti-quality Components of Forages,* Crop Science Society of America, Madison, 1973, 113.
110. **Thill, J. L. and George, J. R.,** Cation concentrations and K to Ca + Mg ratio of nine cool season grasses and implications with hypomagnesaemia, *Agron. J.,* 67, 89, 1975.
111. **Odom, J. W., Haaland, R. L., Hoveland, C. S., and Anthony, W. B.,** Forage quality response of tall fescue, orchardgrass and phalaris to soil fertility level, *Agron. J.,* 72, 401, 1980.

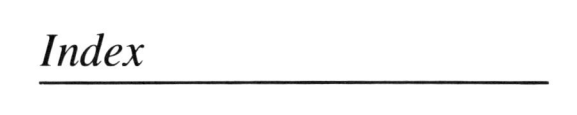

Index

INDEX

A

Abaca, 179
Abela × grandiflora, 66, 73
Abscissic acid, 30
Abutilon theophrasti, 102—104, 178
Acerola, 196
Acid detergent fiber, 239
Actinidia chinensis, 197, 203
Adsorption, 23
Adzuki bean, 169
African yam bean, 138, 169
Agave fibers, 178—179
Agropyron repens, see Quackgrass
Agropyron species, see Wheatgrass
Agrostemma githago, 104
Air classification, 168, 171
Air pollution, and salt tolerance, 58
Akee, 184
Akund, 181
Aleppo pine, 66
Aleuritus fordii, see Tung
Alfalfa
  bacterial wilt resistance, 230
  and bloat, 245
  characteristics, 223
  composition, 239
  digestibility, 238, 243
  dry-matter production, 93, 236
  greenhouse growth, 230
  harvesting, 235, 236
  insect and disease pests, 237
  nitrogen content, 236
  nitrogen fixation, 219, 233
  quality, forage, 242
  salt tolerance, 58, 60, 67, 71
  seedling vigor, 221, 226
  as silage, 220
  temperature effects, 228—230
  toxins and antiquality constituents, 244
  water relations
    moisture stress, symptoms of, 43
    roots, 14, 17
    seasonal water use, 44
    stomates, 35
    transpiration, 47—48
  weeds and, 104, 105
  world production, 77, 87
Algerian ivy, 66
Alkaligrass, 60
Alkali sacaton, 60
Alkaloids, 171, 244
Allelochemicals, 244
Allium species, see also Garlic; Leek; Onion, 184, 186
Almond
  salt tolerance, 64, 65, 67, 72
  world production, 86

Alocasia macrorrhiza, 138, 143
Alsike clover, see Clover; Clover, alsike
Aluminum, 219
Amaranth/amarinth, 102, 104, 118, 122
Amaranthus hybridus, 104
Amaranthus retroflexus, 102
American elm, 72
Amino acids
  oilseed crops, 156—158
  protein crops, 167
Amoracia rusticana, see Horseradish
Amorphophallus, 138
Ananas comosus, see Pineapple
Andropogon, see Bluestem
Annona, 65, 195, 196, 199, 200
Annual ryegrass, see Ryegrass
Annual species
  forage crops, 217
  vegetables, 186
  weeds, 100
Antioxidants, oats as, 117
Antiquality components, forage crop, 222, 244, 245
Anu, 138
Apium graveolens, see Celery
Apomixis, 108
Apple, 195
  horticultural features and uses, 199
  salt tolerance, 65
  utilization, 196
  water relations, 5, 34
  world production, 82
Apricot, see also *Prunus* species, 42
Arachis hypogaea, see Peanut
Arbignya speciosa, see Babussa
Arborvitae, Oriental, 66, 73
Arbutus unedo, 66
Arid, defined, 6
Arracacha, 138
Arracacia xanthorrhiza, 138
Arrowhead, 138
Arrowleaf clover, 223
Arrowroot, 138
Artemisia biennis, 102
Artichoke
  characteristics, 185—186
  salt tolerance, 62, 70
  world production, 82
Artichoke, Jerusalem, 70, 138, 185
Artocarpus altilis, 196
Artocarpus heterophyllus, 197, 202
Asclepias syriaca, 181
Asparagus
  botanical classification, 184
  characteristics, 186
  economics of production, 186
  salt tolerance, 62, 71
  soil pH limits, 191
  temperature requirements, 187

Asparagus bean, 184
Aspartame, 128
Aspergillus fumigatus, 140
Aster, China, 73
Astragalus cicer, see Cicer milkvetch
Atrazine, 108
Attalea, 180
Avena fatuis, 104
Avena sativa, see Oats
Averrhoa carambola, 196
Avocado
  horticultural features and uses, 199
  salt tolerance, 64, 65, 68—70
  water relations, 47
  world production, 82
Azospirillum, 220
Azotobacter, 220

B

Babussa, 148, 150
  characteristics of oil, 151, 154
  composition of farm feeds, 156
Bacterial wilt, 230
Bahia grass, 217
  characteristics, 224
  composition, 239
  nitrogenase activity, 233
  quality parameters, 240
  seedling vigor, 226
  tissue types, 241
  toxins and antiquality constituents, 244
  water use, fertilization and, 231
Bambara groundnut, 169
Banana, 195
  horticultural features and uses, 199
  utilization, 196
  water relations, 17, 42, 47
  world production, 82
Barbados cherry, 196, 200
Barley
  as cereal crop, 118
    uses, 117—118, 121, 122
    world production, 79, 117, 119—120
  dry-matter accumulation, 94, 95
  salt tolerance, 58, 60, 63, 67, 70
  water relations, 4, 42
  weeds and, 104, 105
Barley, foxtail, 102, 108
Barnyard grass, 102, 104, 108
Bauhinia purpurea, 66
Bean, asparagus, 189
Bean, broad/faba
  as protein crops, 168
  salt tolerance, 60
  toxins, 170
  water relations, 42
  world production, 85
Bean, castor, see Castor bean
Bean, Jack, 169, 170

Bean, lima, see also Bean crops, 70, 169—170, 187
Bean, mung, 70, 169
Bean, scarlet runner, 184
Bean, snap, see also Bean crops, 185
  composition, 192
  temperature requirements, 187
  water relations, 43
  weeds and, 99
Bean, yam, 139
Bean crops, 183, 184
  economics of production, 186
  production levels, 187
  as protein crops, 167—169
  salt tolerance, 58, 60, 70
  soil pH limits, 191
  water relations
    growth, 37, 38
    leaf water potential-stomatal conductance relationship, 31
    moisture stress, 42, 43
    roots, 17
    seasonal evapotranspiration, 47
    stomates, 34, 35
  weeds and, 99, 104
  world production, 82, 85
Bearded sprangletop, 100, 104
Beet (*Beta vulgaris*), 184, 186
  composition, 192
  dry-matter accumulation, 94, 96
  dry-matter production, 92, 93
  salt tolerance, 58, 60, 62, 67, 71
  soil pH limits, 191
  as sugar crop, 128—129
  temperature requirements, 187
  water relations, 4, 18, 42, 47
  weeds and, 100, 104
  world production, 77, 83
Bentgrass, 60, 63
Bermuda grass, 102, 217
  composition, 239, 244
  digestibility, 239
  dry-matter production, 93
  fertilization effects, 231, 232
  nitrogenase activity, 233
  quality parameters, 240
  salt tolerance, 60, 63
  seedling vigor, 226
Beta vulgaris, see Beet
Beverage and drug crops, see Drug crops
Biennial vegetables, 186
Bindweed, field, 102
Biological weed control, 106—107
Birdsfoot trefoil
  bloat and, 245
  characteristics, 223
  digestibility, 238
  insect and disease pests, 237
  nitrogen fixation, 233
  salt tolerance, 61, 63
  seedling vigor, 226

temperature effects on growth, 228
Bitter yam, 139
Blackberry, see *Rubus* species
Black pine, Japanese, 66
Black raspberry, see *Rubus* species
Blighia sapida, 184
Blueberry, 196, 200
Blue dracaena, 66, 73
Blue grama, see Grama, blue
Bluegrass, Kentucky, see Kentucky bluegrass
Bluestem, 217
    characteristics, 224
    seedling vigor, 221, 226
    salt tolerance, 60
Boehemeria nivea, see Ramie
Bombax malabaricum, 181
Borassus, 180
Boron tolerance, 69—73
Bottlebrush, 73
Bougainvillea spectabilis, 66
Bouteloua gracilis, see Grama, blue
Boxwood, Japanese, 66, 73
Boysenberry, see *Rubus* species
Brachiaria platyphylla, 104
Brambles, see *Rubus* species
Brassica campestris var. *rapa*, see Turnip
Brassica cauloraps, 184
Brassica chinensis, 184
Brassica napus, see Rape/rapeseed
Brassica napus var. *napobrassica*, see Rutabaga
Brassica oleracea varieties, 170, 184
    characteristics, 186
    salt tolerance, 62, 70
    temperature requirements, 187
    var. *acephala*, see Kale
    var. *botrytis*, see Cauliflower
    var. *capitata*, see Cabbage
    var. *gemmifera*, see Brussels sprouts
    var. *italica*, see Broccoli
    water relations, 42
Brassica pekinensis, 184
Brassica rapa, see Turnip
Breadfruit, 196, 200
Broadleaf signalgrass, 104
Broccoli, see also *Brassica oleracea* varieties, 42, 70, 186
Bromegrass
    characteristics, 224
    quality parameters, 240
    salt tolerance, 63
    tissue types, 241
Bromegrass, smooth
    cations in, 245
    composition, 239, 245
    digestibility, 238, 239
    dry matter and protein yields, 236
    fertilization effects, 232, 233
    greenhouse growth, 230
    harvest frequency effects, 234
    quality, forage, 242
    seedling vigor, 226

    temperature effects, 227, 229
    tissue types, 241
Bromus species, see Bromegrass
Broomcorn, 180
Brush cherry, 66, 73
Brussels sprouts, see *Brassica oleracea* varieties
Buckwheat, 105, 117, 118, 122
Buffalo grass, 217
Buffel grass, 60
Bullock's heart, see also Annona, 200
Burford holly, 64
Burnet, 60
Butter bean, see Bean, lima
Buxus microphylla, 66, 73

C

Cabbage, see also *Brassica oleracea* varieties
    characteristics, 186
    composition, 192
    production levels, 187
    salt tolerance, 58, 62, 70
    soil pH limits, 191
    water relations, 5, 34, 42
    world production, 77, 81
Cacao, see also Cocoa, 211—212
Cajanus cajan, see Pigeon pea
Calathea allouia, 138
Calcium, 348
Calendula officinalis, 73
California poppy, 73
Callistemon citrinus, 73
Callistemon viminalis, 66
Callistephus chinensis, 73
Calocarpum sapota, 198, 207
Calotropis, 181
Camellia sinensis, see Tea
Canarium album, 197, 203
Canary grass, reed, see Reed canary grass
Canavalia, 169, 170
Cannabis sativa, see Hemp
Canna edulis, 138
Cantaloupe, 14, 82, 185, 197
Capillary phenomena, 23
Capsicum species, see Pepper
Carambola, 196, 200
Carbohydrates, unavailable, 171
Carbon fixation, see Dry matter accumulation
Carcia papaya, see Papaya
Carissa grandiflora, 66, 73
Carnation, 73
Carrot
    botanical classification, 184
    characteristics, 186
    salt tolerance, 62, 70
    soil pH limits, 191
    temperature requirements, 187
    water relations, 17
    world production, 81
Carthamus tinctorius, see Safflower

Caryota, 180
Casaba melon, 185
Cashew, 86
Cassava, 183
   botanical classification, 184
   characteristics, 186
   as starch crop, 138, 139—141
   world production, 80
Cassia obtusifolia, 104
Castor bean, 147, 148
   characteristics of oil, 151, 154
   composition of farm feeds, 156
   salt tolerance, 65
   uses, 150
   water relations, 42
   world production, 84
Cations, forage, 245
Cattail, 92
Cauliflower, see also *Brassica oleracea* varieties
   characteristics, 186
   salt tolerance, 67, 70
   soil pH limits, 191
   world production, 82
Ceiba pentandra, 181
Celeriac, 184
Celery
   botanical classification, 184
   characteristics, 186
   salt tolerance, 62, 70
Ceniza, 66, 73
Cereal crops, see also Protein crops
   production statistics, 77—79, 119, 120
   proteins of, 167
   salt tolerance, 70
   species used as, 117—118
   uses, 119—123
   water relations, 42
Chamaerops humilis, 66, 181
Chard, 184, 191
Chavar, 138
Chemicals
   and drought resistance, 50
   herbicides, 107—109
Chenopodium album, 100, 104
Cherimoya, 65, 195, 196, 199, 200
Cherry, see also *Prunus* species, 34, 42
Cherry, Barbados, 196, 200
Cherry, brush, 66, 73
Cherry plum, 66
Cherry tomato, 185
Chestnut, 86
Chestnut, water, 138
Chickpea, 168, 184, 187
China aster, 73
Chinese cabbage, 184
Chinese gooseberry, see also Kiwi, 203
Chinese hibiscus, 66, 73
Chinese holly, 66, 72
Chinese water chestnut, 138
Chinese yam, 139
Chive, 184, 186

Chloride tolerance, 67—68
Chrysanthemum cinerariaefolium, 215
Chrysophyllum cainito, 198, 207
Chufa, 138
Chuno, 138, 139
Cicer arietinum, 168, 184, 187
Cicer milkvetch
   and bloat, 245
   characteristics, 223
   greenhouse growth, 230
   insect and disease pests, 237
   seedling vigor, 226
Cichorum indica, 185
Cinchona, 213
Citrullus vulgaris, see Watermelon
Citrus species, 195
   horticultural features and uses, 201
   salt tolerance, 64, 65, 67—70, 72
   utilization, 197
   water relations, 17, 42, 47
   world production, 77, 82
Climate, see also Temperature
   dry-matter production, 92
   forage crop adaptability, 217—219, 227—230
   and water use, 44, 46—48
   and water-use efficiency, 47, 48
Clover, alsike, see also Clover crops, 60, 230
Clover, arrowleaf, 223
Clover, Berseem, see also Clover crops, 61, 63
Clover, Hubam, 61
Clover, ladino, 61
Clover, red, see also Clover crops, 61
   and bloat, 245
   temperature sensitivity, cold hardiness, 228
   toxins and antiquality constituents, 244
Clover, sweet, 61
   characteristics, 223—224
   insect and disease pests, 237
   nitrogen content, 236
   nitrogen fixation, 233
   salt tolerance, 70
   seedling vigor, 226
   temperature effects, 228—229
   toxins and antiquality constituents, 244
Clover, white, see also Clover crops, 61, 244, 245
Clover crops
   and bloat, 245
   characteristics, 223—224
   classification, 217
   digestibility, 238
   insect and disease pests, 237
   nitrogen content, 236
   nitrogen fixation, 233
   salt tolerance, 60—61
   seedling vigor, 226
   temperature effects on growth, 228
   temperature sensitivity, cold hardiness, 229
   toxins and antiquality constituents, 244
   water relations, 17, 46
   weeds and, 105
   world production, 87

Cluster bean, 60, 169
Cocklebur, common, 104
Cocoa, 211—212
    water relations, 47
    world production, 77, 78, 86
Coconut
    as fiber crop, 179—180
    horticultural features and uses, 200—201
    as oilseed crop, 148, 150—151, 154
    utilization, 197
    world production, 84
Cocos nucifera, see Coconut
Cocoyam, 138, 142—143
Coffee, 207—208
    water relations, 42, 47
    world production, 78, 86
Coir, 179—180
Cola, 212
Cola auminata, see Cola
Cole, see *Brassica oleracea* varieties
Coleus, 34
Colloidal hydration, 23
Colocasia esculenta, 138, 183, 184
Common groundsel, 108
Common vetch, see Vetch
Competition, defined, 103
Conductance, 28, 32—34
Contratoxicants, 109
Convolvulus arvensis, 102
Copra, 147
    protein and nitrogen, 155, 156
    world production, 84
Corchorus capsularis, 85, 177
Cordelauxia edulis, 169
Cordyline indivisa, 66, 73
Corn, 183
    as cereal crop, 118
        uses, 121, 122
        world production, 119—120
    as C_4 plant, 41
    characteristics, 186
    composition, 192
    dry-matter production, 92, 94, 95
    economics of production, 186
    as oilseed crop, 148
        characteristics of oil, 151
        composition, 162
        farm feeds, 156
        oil composition, 153
        uses, 150
    production levels, 187
    as protein crop, 167
    salt tolerance, 58, 60—62, 67, 70
    soil pH limits, 191
    as sugar crop, 127, 128, 130—132
    temperature requirements, 187
    as vegetable crop, 184
    water relations
        dessication effects, 38, 39
        growth, 41
        leaf water potential, 30—31
        moisture deficit, effects of, 50, 51
        moisture stress, 42, 43
        roots, 10, 18
        seasonal water use, 44
        season and temperature effects, 19, 24, 25
        stomates, 34, 35
        tissue content, 5
        transpiration ratio, 48
        yield losses, causes of, 4
    weeds and, 99, 100, 104, 105
    world production, 77, 79, 117
Corn cockle, 92, 104
Corn sweeteners, 127
Coronilla varia, see Crown vetch
Cotoneaster, 64, 66
Cotoneaster congestus, 66
Cotton/cottonseed
    as fiber crop, 174—176
    as oilseed crop, 145, 147—149
        characteristics of oil, 151, 153
        protein and nitrogen, 155, 156
        uses, 150
    salt tolerance, 58, 60, 67, 71
    water relations, 5, 17, 31, 47
    weeds and, 99, 100, 104
    world production, 77, 84, 85
Coumarin, 244
Cowpea, 168
    salt tolerance, 58, 60, 61, 70
    weeds and, 105
Crabgrass, large, 104
Crambe, 146, 151, 154, 163
Cranberry, 197, 201
Crape myrtle, 66
Creeping bent grass, 63
Creeping foxtail, 230
Crimson clover, see Clover, crimson; Clover crops
Crin vegetal, 181
Critical moisture, 11
Croceum ice plant, 66
Crop rotation, 105—106
Crotalaria juncea, 178
Crown vetch, 217, 223
    insect and disease pests, 237
    seedling vigor, 226
    toxins and antiquality constituents, 244
Cucumber, 170
    botanical classification, 185
    characteristics, 186
    as forced crop, 189
    salt tolerance, 62, 67, 70
    soil pH limits, 191
    temperature requirements, 187
    world production, 81
Cucumis melo, see Melon
Cucumis sativus, see Cucumber
Cucurbita pepo, see Pumpkin; Squash
Cultural practices, weed control, 105—106
Curcuma zedoaria, 138
Current, see also *Ribes* species, 82
Cush-cush yam, 139, 141, 184

Cyamopsis tetragonobola, 60, 169
Cyanogenetic glycosides, 170, 244
Cydonia oblonga, 198, 206
Cynara scolymus, see Artichoke
Cynodon dactylon, see Bermuda grass
Cyperus esculentus, 102, 138
Cyperus rotundus, 102
Cyrtosperma chamissomis, 138, 143

D

Dactylis glomerata, see Orchardgrass
Dalapon, 108
Dallis grass
  characteristics, 224
  salt tolerance, 61
  seedling vigor, 226
  temperature effects on growth, 227
  tissue types, 241
Dandelion, 185
Dark respiration, see Respiration
Dasheen, see Taro
Date
  environmental factors, 47, 65
  horticultural features and uses, 202
  utilization, 197
  world production, 82
Daucus carota, see Carrot
Delosperma alba, 66
Delphinium, 73
Delphinium barbeyi, 102
Dewberry, 196
Dianthus caryophyllus, 73
Dicoumarol, 244
Dictyosperma, 180
Digitaria sanguinalis, 104
Digit grass, 239, 241
Dihydroxyphenylalanine (DOPA), 171
Dioscorea, see Yam
Diospyros, see Persimmon
Disease, see Pathogens
Dodonaea viscosa, 66
Dolichos uniflorus, 169
DOPA (dihydroxyphenylalanine), 171
Dracaena, blue, 66, 73
Drosanthemum hispidum, 66
Drought, 3, 5
Drought stress
  and photosynthesis, 36
  resistance to, 48, 50—51
  responses to, 28, 30—37
Drug crops
  cocoa/cacao, 211—212
  coffee, 207—208
  cola, 212
  pyrethrum, 215
  quinine, 213
  tea, 207
  tobacco, 213—214
  world production, 77, 78, 86

Dry-matter production
  carbon dioxide assimilation, 91
  forages, 234—236
  growth rates, 92
  maximum productivity, 93
  photosynthesis, 89—90
  world areas, 95—96
  yields, 93—94
Durian, 197, 202
Durio zibethinus, 197

E

East Indian arrowroot, 138
Echinochloa crusgalli, 102, 104
Eddo, see Taro
Eggplant, 185
  salt tolerance, 62
  soil pH limits, 191
  world production, 82
Elaeagnus pungens, 66, 72
Elaeis guineensis (oil palm), see Palm kernel; Palm oil
Eleocharis dulcis, 138
Elephant yam, 138
Elm, American, 72
Elymus species, 62
Endive, 185
Energy system, water relations, 21, 23, 25—29
English ivy, 35
Eriobotrya japonica, 65, 197, 204
Eschscholtzia californica, 73
Estrogenic flavonoids, 244
Eugenia cauliflora, 197, 203
Euonymus japonica, 66, 72
Euphorbia pulcherrima, 73
Euphorbia longana, 197, 203
European fan palm, 66
Evaporation, see Water relations
Evergreen pear, 66

F

Faba bean, see Bean, broad
False hellebore, Western, 102
False yam, 138
Fatty acids, see Oilseed crops
Favism, 170
Feed, see also Forage crops
  cassava as, 140
  cereal crops used as, 117, 121—123
Feijoa sellowiana, 66, 72
Fenugreek, 169
Fertilizer
  forage crops, 219, 231—233
  and water-use efficiency, 47, 49
Fescue, 61, 229
Fescue, tall
  characteristics, 225

composition, 239
digestibility, 238, 239
dry matter and protein yields, 236
fertilization effects, 232, 233
greenhouse growth, 230
harvest frequency effects, 234
quality parameters, 240
seedling vigor, 226
temperature effects on growth, 227
tissue types, 241
toxins and antiquality constituents, 244
Festuca species, see Fescue
Fiber crops
    brush and braiding, 180
    classification, 173
    cordage, 178—180
    filling, 180—189
    future trends, 180—181
    malvaceous, 178
    salt tolerance, 60
    textile and fabric, 174—178
        cotton, 174—175
        flax, 175—176
        hemp, 176—177
        jute, 177
        ramie, 177
    world production, 77, 78, 85
Ficus carica, see Fig
Field bean, see Bean; Bean crops
Field bindweed, 102
Field capacity, 8—9, 11
    evapotranspiration at, 19, 23, 25, 26
    and growth, 40
Fig, 65, 70, 202
Flat pea, 169
Flax
    as fiber crop, 175—176
    as oilseed crop, 147, 148
        area, yield, and production, 149
        characteristics of oil, 151
        composition, 154, 156, 163
        protein and nitrogen, 155, 156
        uses, 150, 156
        world production, 145
    salt tolerance, 60
    water relations, 47
    world production, 85
Flax, New Zealand, 180
Flowering, moisture stress and, 37
Forage crops
    adaptation, 217—220, 228—233
        climate, 217, 219, 227—230
        nitrogen fixation, 219—220, 233
        soils and fertility, 219, 231—233
    characteristics, 223—225
    pathology and insect pests, 221, 237
    quality, 221—222, 234—245
        antiquality components, 222, 244, 245
        legumes and grasses, 221—222, 238—243
        maturity, 222, 234—236
    salt tolerance, 60—62

seedling vigor, 221, 226
species, 217
utilization, 220—221
water relations, 41
weeds and, 101, 105
world production, 77, 87
Fortunella, 201
Foxtail, creeping, 230
Foxtail, giant, 104
Foxtail, meadow, 61
Foxtail barley, 102, 108
Fragaria, see Strawberry
French bean, see Bean; Bean crops
Fruit crops, see also specific crops
    salt tolerance, 62—65, 67—70
    species of economic importance, 199—208
    utilization, 196—198
    water relations, 17, 42, 43
    world production, 77, 78, 82
Fufu, 141
Fungal diseases
    resistance to, 170
    weed control, 106—107
Furcraea foetida, 180

G

Garcinia mangostana, 197, 204
Gardening units, 188—189
Garlic, 70, 82, 184, 187
Geography, see also World production
    classification according to rainfall, 6
    and water use, 44, 46—48
Geranium, 34, 35, 73
Giant foxtail, 100, 104
Giant taro, 138, 142—143
Gladiolus, 73
Glossy abelia, 66, 73
Glossy privet, 66
Gluten, 117
Glycine max, see Soybean
Glycosides, 139, 170, 244
Goa bean, 169
Gooseberry, see *Ribes* species
Gooseberry, Chinese, 197, 203
Gossypium, see Cotton
Grain crops, see also Cereal crops; specific crops
    salt tolerance, 60, 70
    water relations
        moisture stress and, 42—43
        roots, 17, 18
        seasonal evapotranspiration, 47
    weeds and, 99, 100
    world production, 87
Gram, black, 169
Gram, green, 169
Grama, 217
Grama, blue, 61, 230
    seedling vigor, 221, 226
Grape, 195

horticultural features and uses, 202
   salt tolerance, 65, 67, 68, 70
   utilization, 197
   water relations, 5, 47
   world production, 82
Grapefruit, see also *Citrus* species, 47
Grasses
   C_3 plants, 41
   forage, see also Forage crops, 230, 238—239
   salt tolerance, 60—62
   seedling vigor, 226
   temperature sensitivity, cold hardiness, 228, 229
   vegetable species, see Corn
   water relations, 18, 43, 46
Grass pea, 169, 170
Grass tetany, 222
Grazing, weed control, 106
Greater yam, 139, 184
Green gram, 169
Greenhouse growth, forage crops, 230
Green manure, 236
Groundnuts, see also Peanut, 18, 169
Groundsel, common, 108
Guar, 60, 169
Guava, 197, 202
Guava, pineapple, 72
Guayale, 65
Guinea grass, 227, 233, 239

H

Hairs, leaf, 40
Hairy nightshade, 104
Hairy vetch, see Vetch
Hansa potato, 138
Harding grass, 61
Harvest frequency, forages, 234—236
Harvest index, see also Yield, 93, 94
Harvest time, forages, 235
Hawthorn, Indian, 66, 73
Hay, see also Forage crops, 87
Hazelnut, 86
Heavenly bamboo, 66
Hedera canariensis, 66
Hedera helix, 35
Helianthus annuus, see Sunflower
Helianthus tuberosum, 79, 138, 185
Hellebore, western false, 102
Hemicellulose, 171
Hemp
   as fiber crop, 176—177
   world production, 85
Hemp, Mauritius, 179, 180
Hemp, sunn, 178
Henquen, 178—179
Herbicides, 106—109
Hibiscus cannabinium, 178
Hibiscus esculentus, 62, 184
Hibiscus rosa-sinensis, 66, 73
Hibiscus sabdariffa, 178

Hibiscus trionum, 103
Hitchenia canlina, 138
Holly, American, 35
Holly, Chinese, 66, 72
Holly, Burford, 64
Hops, 83
Hordeum vulgare, see Barley
Horsegram, 169
Horseradish, 184
Humid, defined, 6
Hyacinth bean, 169
Hydraulic resistance, 28
Hydrologic cycle, see also Water relations, 6
Hymenocyclus croceus, 66

I

Icacina senegalensis, 138
Ice plants, 66
Ilex cornuta, 66, 72
Ilex opaca, 35
Indian hawthorn, 66, 73
Insects, 4, 100, 105—106
Integrated weed management, 105
Interference, defined, 103
Intoxicating yam, 139
Ipomoea batatas, see also Potato, sweet
Ipomoea purpurea, 104
Irrigation, see also Salt tolerance, 22
   forages, 236
   and moisture stress, 42, 43
   strategies for, 50—51
   weed dispersal via, 101, 102
Italian stone pine, 66
Ivy, 35
Ivy, Algerian, 66

J

Jaboticaba, 197, 203
Jack bean, 169, 170
Jackfruit, 197, 202
Japanese apricot, 198
Japanese black pine, 66
Japanese boxwood, 66, 73
Japanese pittosporum, 66, 72
Jerusalem artichoke, 138, 185
Johnson grass, 102, 106
Jojoba
   as oilseed crop, 146, 151, 154, 164
   salt tolerance, 65
Jujube, 65, 197, 203
Juniperus chinensis, 66, 72
Jute, 85, 177

K

Kale, see *Brassica oleracea* varieties

Kallar grass, 61
Kanran, 197, 203
Kapok, 181
Kenaf, 178
Kentucky bluegrass
    cation composition, 245
    digestibility, 238
    dry matter and protein yields, 236
    fertilization effects, 219, 232, 233
    quality parameters, 240
    salt tolerance, 70
    seedling vigor, 226
    temperature effects on growth, 227
    temperature sensitivity, cold hardiness, 229
    tissue types, 241
Kersting's groundnut, 169
Kidney bean, see Bean crops
Kiwi, 197, 203
Kochia, 102, 104
Kohlrabi, see also *Brassica oleracea* varieties, 191
Kola, see Cola
Kudzu, 217, 223, 226
Kumquat, 196
Kwashiorkor, 140

L

Lablab niger, 169
Lacuma mammosa, 198, 207
Lagerstroemia indica, 66
Lambsquarters, common, 100, 104
Lampranthus productus, 66
Langsat, 197, 203
Lansium domesticum, 197, 203
Lantana camara, 66, 72
Lanzon, 203
Large crabgrass, 104
Larkspur, 73
Larkspur, tall, 102
Lathrogen, 170
Lathyrus odoratus, 73
Lathyrus sativus, 169, 170
Latuca sativa, see Lettuce
Laurustinus, 66, 72
Leaf orientation, 38, 40
Leaf water potential, 26—27, 29, 30, 52
Lectins, 170
Leechee, 197, 203
Leek, 184, 186
Legume crops, see also Protein crops; Vegetable crops
    forage, see also Forage crops
        characteristics, 223—224
        digestibility, 238
        greenhouse growth, 230
        seedling vigor, 226
        temperature sensitivity, cold hardiness, 228, 229
    proteins of, 167
    water relations, 42

Lemon, see *Citrus* species
Lens culinaris, see Lentil
Lentil, 85, 167, 168
Leopoldinia, 180
Lepidium meyenni, 138
Leptochloa fascicularis, 100, 104
Lespedeza
    characteristics, 223
    insect and disease pests, 237
    nitrogen fixation, 233
    seedling vigor, 226
    temperature sensitivity, cold hardiness, 228
    toxins and antiquality constituents, 244
    world production, 87
Lesser yam, 139
Lettuce, 185
    characteristics, 186
    composition, 192
    as forced crop, 189
    production levels, 187
    salt tolerance, 58, 62, 70
    soil pH limits, 191
    temperature requirements, 187
    water relations, 5
    weeds and, 99
Leucophyllum frutescens, 66, 73
Licania rigida, 148, 151, 154
Ligustrum japonicum, 72
Ligustrum lucidum, 66
Lima bean, 70, 169, 170, 187
Lime, see *Citrus* species
Linseed, see also Flax/flaxseed, 84, 145
Linum usitatissimum, see Flax/flaxseed
Lipids, see Oilseed crops
Liquidambar styraciflua, 66
Liriodendron tulipifera, 66
Litchii, 197, 203
Litchi sinensis, see Litchii
Llama heart fruit, 200
Lolium species, see Ryegrass
Longan, 197, 203
Loquat, 65, 197, 204
Lotus corniculatus, see Birdsfoot trefoil
Lotus root, 138
Louvana, 169
Love grass, 61
Lupine, white, 169
Lupinus species, 70, 169, 171
Lychee, 197, 203
Lycopersicon, see Tomato

M

Maco, 138
Macuna pruriens, 169
Magnesium, 69, 222
Magnolia grandiflora, 66
Mahonia aquifolium, 66, 72
Maize, see Corn
Malt, 117

Malus domestica, see Apple
Mamey sapote, 198, 207
Manganese, 219
Mangifera indica, see Mango
Mango, 197, 204
    salt tolerance, 65
    water relations, 42
    world production, 82
Mangosteen, 197, 204
Manihot esculenta, see Cassava
Manilkara zapota, 198, 207
Manioc, see Cassava
Maple, Norway, 5
Maple, sugar, 128, 132—133
Marama bean, 169
Maranta araundinacea, 138
Marigold, 73
Market gardening, 188—189
Mat bean, 169
Matrix potential, 23
Mauritius hemp, 179, 180
Meadow fescue, see also Fescue, 229
Meadow foxtail, 245
Meals
    oilseed, 155—157
    protein crop processing, 168, 170
Medicago sativa, see Alfalfa
Melilotus species, see also Clover, sweet
    characteristics, 223—224
    salt tolerance, 61
Melon, 170, 185
    horticultural features and uses, 204
    production levels, 187
    utilization, 197
    water relations, 18, 43
    world production, 78, 81, 82
Metroxylon sagus, see Sago palm
Milk vetch, 61
Millet, see also Sorghum
    as cereal crop, 118, 119, 122
    uses, 117
    weeds and, 105
    world production, 79
Millet, foxtail, 60
Millet, pearl, 42
Milling, protein crop processing, 168, 170
Moisture stress, see Water relations
Morning glory, tall, 104
Mulberry, 197, 204
Mung bean, 169
Musa, see Banana; Plantain
Musa textilis, 179
Mushroom, 189
Muskmelon, 185—186
    composition, 192
    salt tolerance, 62, 70
    soil pH limits, 191
    temperature requirements, 187
    utilization, 198
Mustard, 70
Mycoherbicide, 107

Mycotoxins, 222, 244
Myrciaria cauliflora, 197, 203

N

Nandina domestica, 66
Napier grass
    digestibility, 239
    dry-matter production, 92, 93
    nitrogenase activity, 233
Nasturtium, 5
Nasturtium officinale, 184
Natal plum, 66, 73
Nectarine, see *Prunus* species
Nelumbo nucifera, 138
Nephelium lappaceum, 198
Nerium oleander, 66, 73
Net photosynthesis, 34, 36
Netted melon, 185
New Zealand spinach, 184
Nicotiana tabacum, see Tobacco
Nightshade, hairy, 104
Nitrogenase, 233
Nitrogen content, see also Protein
    forages, 236
    oilseed crops, 157, 158
Nitrogen fertilizer
    and forage crops, 219, 231—233
    and water-use efficiency, 49
Nitrogen fixation, see also *Rhizobium*, 219—220, 233
Norway maple, 5
Nut crops, 70, 77, 78, 86

O

Oak, 5, 35
Oat
    as cereal crop, 118
        uses, 117, 121, 122
        world production, 119—120
    salt tolerance, 60, 61, 70
    water relations
        moisture stress, critical periods, 42
        stomates, 34, 35
        transpiration ratio, 48
        yield losses, causes of, 4
    world production, 79, 119—120
Oat, wild, 104
Oatgrass, tall, 61, 222, 245
Oca, 138
Oilseed crops, see also Protein crops
    area, yield and producers, 149
    characteristics of oils, 152
    corn, 162
    crambe seed, 163
    flax, 163
    jojoba, 164
    lipid and fatty acid content, 153—154

peanut, 159—161
  protein and amino acid content, 156
  protein meal equivalent, 155
  rapeseed, 159
  salt tolerance, 70
  soybeans, 157, 158
  sunflower seeds, 158
  uses, 145, 150—152
  water relations, 42, 47
  world production, 77, 78, 83—84, 145, 147, 148
Oiticica, 148, 151, 154
Okra, 62, 184
Olea europaea, see Olive
Oleander, 66, 73
Olive
  horticultural features and uses, 204
  as oilseed crop, 148, 150, 151, 153
  salt tolerance, 65
  utilization, 150, 197, 204
  water relations, 42
  world production, 84
Onion, 184, 186
  economics of production, 186
  production levels, 187
  salt tolerance, 58, 62, 70
  soil pH limits, 191
  temperature requirements, 187
  water relations, 14, 47
  weeds and, 99
  world production, 81
Onobrychis viciaefolia, see Sainfoin
Orange, see also *Citrus* species, 47, 77
Orchard grass
  composition, 239, 245
  digestibility, 238, 239
  dry matter and protein yields, 236
  fertilization effects, 232, 233
  and grass tetany, 222
  greenhouse growth, 230
  harvest frequency effects, 234
  quality parameters, 240
  salt tolerance, 61
  seedling vigor, 226
  tissue types, 241
Orchid tree, 66
Oregon grape, 66, 72
Oriental arborvitae, 66, 73
Ornamentals, salt tolerance, 64, 66—67, 72—73
Oryza sativa, see Rice
Osmotic adjustment, 32—33
Osmotic potential, 27, 32
Osmotic pressure, 23
Oxalis, 73
Oxalis tuberosa, 138
Ozone, 58

P

Pachyrrhizus, 169
Pak-choi, 184

Palm, European fan, 66
Palm, Panama hat, 180
Palm, roots of, 18
Palm, sago, 138, 143
Palm kernel, see also Palm oil, 147, 148
  protein and nitrogen, 155, 156
  world production, 84
Palm oil, 84, 148
  characteristics of oil, 151
  composition of oil, 154
  uses, 150
  world production, 84
Panama hat palm, 180
Pangola digit grass, 239, 241
Pangola grass, 231, 233, 240
Panic grass, blue, 61
Panicum, seedling vigor, 221
Panicum virgatum, see Switchgrass
Pansy, 73
Papaw, see Papaya
Papaya, 65, 82, 197, 205
Parsley, 71, 184, 187
Parsnip, 62, 184, 187
Paspalum dilatatum, see Dallis grass
Paspalum notatum, see Bahia grass
Paspalum species, digestibility, 239
Passiflora edulis, see Passionfruit
Passionfruit, 65, 197, 205
Pastinaca sativa, 62, 184, 187
Pasture crops, 43, 101
Pathogens
  alfalfa sensitivity, 230
  disease resistance, 170
  forage crop, 221, 237
  weed control, 106—107
  weeds and, 100
  yield loss and, 4
Pawpaw, see Papaya
Pea, 183—184
  characteristics, 186
  composition, 192
  economics of production, 186
  production levels, 187
  as protein crops, 167, 168, 170
  salt tolerance, 62, 70
  soil pH limits, 191
  temperature requirements, 187
  water relations, 42, 48
  world production, 82, 85
Pea, grass, 170
Peach, see also *Prunus* species, 195
  water relationships, 5, 34, 42
Peanut
  as oilseed crop
    area, yield, and production, 149
    characteristics of oil, 151
    composition, 159—161
    composition of oil, 153
    estimation of kernel age, 159
    protein and nitrogen, 155, 156
    uses, 150

world production, 145, 147, 148
  salt tolerance, 60, 70
  water relations, 18, 42
  weeds and, 99, 104
  world production, 77, 83, 145, 147, 148
Pear, 195
  horticultural features and uses, 205
  salt tolerance, 65
  utilization, 198
  world production, 82
Pear, evergreen, 66
Pecan, 70
Pelargonium × *hortorum*, 73
Pensacola bahia grass, 241
Pepper (*Capsicum*), 185—186
  composition, 192
  salt tolerance, 62, 67, 70
  soil pH limits, 191
  temperature requirements, 187
  water relations, 18
Pepper, chili, 81
Pepper, hot, 195
Pepper, red, 70
Pepper, sweet, 185
Perennial ryegrass, see Ryegrass
Perennial species
  forage crops, 217
  vegetables, 186
  weeds, 100
Persimmon
  horticultural features and uses, 205
  salt tolerance, 65, 70
  utilization, 198
Petroselinum crispum, 71, 184, 187
Pe-tsai, 184
pH, 191, 219
Phalaris arundinacea, see Reed canary grass
Phaseolus species, see also Bean crops, 184
  pulse crops, 169
  salt tolerance, 70
Phloem, 6
Phoenix dactylifera, see Date
Phormium tenax, 180
Phosphate, 48, 49
Photinia × *fraseri*, 66, 72
Photosynthesis, see also Dry matter accumulation
  C_3 and C_4 plants, 41, 46, 222
  chemicals inhibiting, 108
  drought stress response, 33—37
  dry-matter production, 89—90
  leaf orientation and, 38
  water and, 3
Phytin, 171
Phytohemmaglutinins, 170
Piassava, 180
Pigeon pea, 168, 184, 187
Pigweed, redroot, 102
Pigweed, smooth, 104
Pine, see specific *Pinus* species
Pineapple
  horticultural features and uses, 205—206
  salt tolerance, 65
  utilization, 198
  world production, 82
Pineapple guava, 66, 72
Pinus, water relations, 35
Pinus halepensis, 66
Pinus pinea, 66
Pinus thunbergiana, 66
Pistachio, 86
Pisum sativum, see Pea
Pittosporum tobira, 66, 72
Plantain, 82, 183, 198, 206
Platycladus orientalis, 66, 73
Plum, see *Prunus* species
Plum, cherry, 66
Plum, Natal, 66, 73
Poa pratensis, see Kentucky bluegrass
Podocarpus macrophyllus, 66, 73
Poi, 142
Poinsettia, 73
Polymnia sonchifolia, 139
Polyphenols, 170
Pomegranate, 65, 198, 206
Poncirus, 72, 201
Poppy, California, 73
Potato
  botanical classification, 185
  characteristics, 186
  composition, 192
  dry-matter production, 92, 94, 96
  economics of production, 186
  production levels, 80, 187
  salt tolerance, 62, 67, 70
  soil pH limits, 191
  as starch crop, 137—139
  temperature requirements, 187
  water relations
    evapotranspiration ratio, 48
    moisture stress, symptoms of, 43
    roots, 18
    stomates, number of, 34
    tissue content, 5
    yield losses, causes of, 4
  weeds and, 99
Potato, sweet, 183, 185
  botanical classification, 184
  characteristics, 186
  composition, 192
  production levels, 80, 187
  salt tolerance, 63, 70
  as starch crop, 138, 142
  temperature requirements, 187
  water relations, 47, 80
Potato bean, 169
Potato yam, 139
Potential evapotranspiration, 43, 44, 46
Precipitation, 3, 6
Preventive weed control, 105
Prickly sida, 103
Privet, wax-leaf, 72
Processing, protein crops, 168, 170

Production, see World production
Productivity, see Dry-matter production; Weeds, Yield
Prohibited noxious, defined, 105
Prolene, 30
Protease inhibitors, 170
Protein
 cassava, 140
 forages, 234—236, 239
 oilseed crops, 155—158, 164
Protein crops, see also Cereal crops; Legume crops; Nut crops; Oilseed crops
 processed components, 168, 170
 toxins, 170—171
 utilization of pulse crops, 167—169
 world production, 77, 168
Prunus species, 195
 horticultural features and uses, 206
 salt tolerance, 64, 65, 67—70, 72
 utilization, 196
 water relations, 42
 world production, 82
Prunus cerasifera, 66
Psidium guajava, 197
Psophocarpus tetragonolobus, 169
Pueraria lobata, see Kudzu
Pulse crops, see also Forage crops; Protein crops; Vegetable crops
 utilization of, 167—169
 world production, 77, 78, 85
Pummelo, see *Citrus* species
Pumpkin, 62, 82, 185
Punica granatum, 198, 206
Purple ice plant, 66
Purple nutsedge, 102
Pyracantha fortuneana, 66
Pyrenees cotoneaster, 66
Pyrethrum, 215
Pyrus communis, see Pear
Pyrus kawakamii, 66

Q

Quackgrass, 102, 217
 characteristics, 224
 dry matter and protein yields, 236
 seedling vigor, 226
Queensland arrowroot, 138
Quercus, 5, 35
Quince, 198, 206
Quinine, 213

R

Radish, 184, 186
 salt tolerance, 63, 70
Raffinose, 171
Rambutan, 198, 207
Ramie, 176, 177
Rape/rapeseed
 dry-matter accumulation, 94, 96
 as oilseed crop, 146, 147
  area, yield, and production, 149
  characteristics of oil, 151
  composition of oil, 153
  composition of seed, 159
  protein and nitrogen, 155
  uses, 150
  world production, 145, 148
 salt tolerance, 61
 world production, 83
Raphanus sativus, see Radish
Raphia, 180
Raphiolepsis indica, 66, 73
Raspberry, see *Rubus* species
Red clover, see Clover, red; Clover crops
Red pine, 35
Red raspberry, see *Rubus* species
Redroot pigweed, 102
Reed canary grass
 characteristics, 224
 composition, 239, 245
 digestibility, 238
 dry matter and protein yields, 236
 fertilization effects, 232, 233
 grass tetany and, 222
 harvest frequency effects, 234
 salt tolerance, 60
 seedling vigor, 226
 temperature effects on growth, 227
 temperature sensitivity, cold hardiness, 229
 toxins and antiquality constituents, 244
 weeds and, 105
Reflectance, 40
Relative humidity, 16, 20
Rescue grass, 61
Resistance, 32—34
Respiration, 3, 34, 35
Restricted noxious weed, defined, 105
Rhizobium, 167, 170, 219—220
Rhizopus chinensis, 140
Rhodes grass, 61
Rhubarb, 184, 187
Ribes species, 65, 82, 195, 207
Rice, 183
 as cereal crop, 118
  uses, 121—123
  world production, 119—120
 dry-matter production, 92—95
 proteins, 167
 salt tolerance, 58, 60, 63
 water relations, 47
 weeds and, 99, 100, 104
 world production, 77, 79, 117
Rice, wild, 118, 123
Rice bean, 169
Ricinus communis, see Castor bean
Root crops, see also Starch crops, 78
Roots, see Water relations
Rosa, 66

Rose, 64, 66
Rosea ice plant, 66
Rose-apple, 65, 198, 207
Roselle, 178
Rosemary, 66, 73
Rosemarinus officinalis, 66, 73
Rubber, 78, 83
Rubus species
    horticultural features and uses, 200
    salt tolerance, 68
    utilization, 196
    world production, 82
Rumex acetosa, see Sorrel
Runner bean, 169
Russian thistle, 102
Rutabaga, 184, 186—187, 192
Rye
    as cereal crop, 118
        uses, 121, 123
        world production, 119—120
    dry-matter production, 92
    salt tolerance, 60, 61
    uses, 117
    water relations, 10, 13
    weeds and, 105
    world production, 79, 119—120
Rye, wild, see *Elymus* species
Ryegrass
    characteristics, 224
    digestibility, 239
    perennial, 61, 227
    salt tolerance, 61
    seedling vigor, 226
Ryegrass, Italian, 61

S

Saccharin, 128
Safflower
    as oilseed crop, 147, 148
        characteristics of oil, 151
        composition of oil, 153
        protein and nitrogen, 155, 156
        uses, 150
    salt tolerance, 58, 60, 67
    water relations, 18
    world production, 84
Sage, yellow, 66, 72
Sagittaria sagittifolia, 138
Sago palm, 138, 143
Sainfoin, 223, 230
    and bloat, 245
    insect and disease pests, 237
    seedling vigor, 226
Salsify, 185
Salsola kali, 102
Salt grass, desert, 61
Salt tolerance
    boron, 69—73
    chloride, 67—68
    criteria, 57
    factors influencing, 57—58
    forage crops, 219
    herbaceous crops, 59—63
    ornamentals, 64, 66—67, 72—73
    sodium, 68
    woody crops, 63—65, 72
Samuela, 180
Sapodilla, 198, 207
Saponins, 170, 244
Sapote, 198, 207
Sapote, white, 65
Saranac, 238
Seasonal water use, 44—46
Secale cereale, see Rye
Seed laws, 105
Seedling vigor, forage crop, 221, 226
Senecio vulgaris, 108
Senescence, water deficits and, 38
Sesame
    as oilseed crop, 147, 148
        characteristics of oil, 151
        composition of oil, 153
        protein and nitrogen, 155, 156
        uses, 150
    salt tolerance, 60, 67, 70
    world production, 84
Sesamum indicum, see Sesame
Sesbania, 61
Setaria faberii, 100, 104
Shallot, 184
Shoti, 138
Sicklepod, 104
Sida spinosa, 103
Siduron, 108
Signalgrass, broadleaf, 104
Silk cotton tree, 181
Simazine, 108
Simmondsia chinensis, see Jojoba
Sirato, 61
Sisal, 47, 85, 178—179
Smooth pigweed, 104
Smother crops, 105, 106
Snap bean, see Bean; *Phaseolus* species
Sod-forming forages, 220—221
Sodium tolerance, 68
Soil, see also Water relations
    forage crop adaptability, 219, 231—233
    forage crop conservation of, 220—221
    vegetable crop pH limits, 191
Solanum curtilobum, 138, 139
Solanum juzepezukii, 138, 139
Solanum melongena, see Eggplant
Solanum sarachoides, 104
Solanum tuberosum, see Potato
Solenostemon rotundifolius, 138
Solute potential, 32
Sorghum
    as cereal crop, 118
        uses, 121, 123
        world production, 119, 120

as C₄ plant, 41
dry-matter production, 92, 93
salt tolerance, 58, 60, 67, 70
as sugar crop, 128, 133—134
textile uses, 180
uses, 117
water relations
   leaf water potential-stomatal conductance relationship, 31
   moisture stress effects, 38, 42
   reflectance, 40
   roots, 18
   tissue content, 5
   transpiration, 47—48
   yield losses, causes of, 4
weeds and, 99, 104, 105
world production, 79, 117
Sorghum bicolor, see Sorghum
Sorghum halepense, 102
Sorrel, 184
Soursop, see *Annona*
Southern magnolia, 66
Southern yew, 66, 73
Soybean
  botanical classification, 184
  dry-matter production, 92—94, 96
  as oilseed crop, 146—149
   characteristics of oil, 151
   composition of, 157, 158
   composition of oil, 153
   protein and nitrogen, 155, 156
   uses, 150
  as protein crop, 167
  salt tolerance, 60, 63
  toxic constituents, 170
  water relations
   moisture stress, 42—43
   photosynthesis, 35, 36
   roots, 18
   seasonal evapotranspiration, 47
   yield losses, causes of, 4
  weeds and, 99, 100, 103—105
  world production, 77, 83
Spanish moss, 181
Spanish oak, 35
Sphaerophysa, 61
Sphenostylis stenocarpa, 138, 169
Spinach (*Spinacea olearacea*), 184, 186
  as C₃ plant, 41
  composition, 192
  salt tolerance, 63
  soil pH limits, 191
Spinach, New Zealand, 184
Spindle tree, 66, 72
Sprangletop, bearded, 100, 104
Spreading juniper, 66
Squash, 185—186
  composition, 192
  salt tolerance, 63, 70
  soil pH limits, 191
  temperature requirements, 187

Squash, winter, 192
Stachyose, 171
Star-apple, 198, 208
Starch crops, see also Tubers; Vegetable crops
  cassava, 139—141
  cocoyam, 142—143
  potatoes, 137, 139
  protein crops as, 168, 170
  sago palm, 143
  species used as, 138—139
  sweet potatoes, 142
  world production, 77, 80
  yams, 141
Star jasmine, 66
Stomata, drought stress response, 28, 30—35
Stone pine, Italian, 66
Strawberry, 195, 198
  horticultural features and uses, 207—208
  salt tolerance, 63, 64, 68, 70
  utilization, 198
  world production, 82
Strawberry tree, 66
Sudan grass
  characteristics, 225
  quality, forage, 242
  salt tolerance, 61
  seedling vigor, 226
  toxins and antiquality constituents, 244
Sugar-apple, see *Annona*
Sugarcane
  as C₄ plant, 41
  dry-matter accumulation, 94, 96
  dry-matter production, 92, 93
  salt tolerance, 60
  as sugar crop, 128—130
  water relations, 5, 18, 42, 47
  world production, 77, 83
Sugar crops
  by-products, 134, 135
  consumption trends, 126—127
  qualities and caloric value, 125—126
  species used as, 127—128
   corn, 130—132
   sorghum, 133—134
   sugar beet, 128—129
   sugarcane, 129—130
   sugar maple, 132—133
  world production, 77, 78, 83
Sunflower
  dry-matter accumulation, 94, 95
  as oilseed crop, 148
   area, yield, and production, 149
   characteristics of oil, 151
   composition of oil, 153
   composition of seed, 158
   protein and nitrogen, 155, 156
   uses, 150
   world production, 145
  salt tolerance, 60, 67, 70
  water relations
   growth, 38

of leaf, 27, 31, 35
    moisture stress, critical periods, 42
    reflectance, 40
    roots, 18
    wilting percentage, 9
  weeds and, 104, 105
  world production, 77, 83, 145
Sunn hemp, 178
Swamp taro, 138, 142, 143
Sweet clover, see Clover, sweet
Sweet gum, 66
Sweet pea, 73
Sweetsop, 198
Switchgrass, 217, 225—226, 235
Sword bean, 169, 170
Syzygium jambos, 65, 198, 207
Syzygium paniculatum, 66, 73
Syzygium samarangense, 197, 208

T

Tacca contopetaloides, 138
Tall fescue, see Fescue, tall
Tall larkspur, 102
Tall morning glory, 104
Tall oatgrass, 61, 222, 245
Tamarind, 208
Tamarindus indica, 198
Tangerine, see *Citrus* species
Tannia, 138, 142—143
Tannins, 170, 244
Tapioca, see Cassava
Taraxacum officinale, 185
Taro, 138, 183, 184
Taro, giant, 138, 142—143
Taro, swamp, 138, 142, 143
Tarwi, 169, 171
Tea, 77, 78, 86, 207
Temperature, see also Climate
  dry-matter production, 92
  forage crop sensitivity, 227—230
  and water relations, 20, 21
    evaporative demand, 16
    vapor pressure, 19
Tepary bean, 169
Tetany, grass, 222
Tetragonia expansa, 184
Theobroma cacao, see Cocoa
Thistle, Russian, 102
Thorny elaeagnus, 66, 72
Tillandsia usneoides, 181
Tillers, 43
Timothy
  characteristics, 225
  digestibility, 239, 243
  dry matter and protein yields, 236
  fertilization effects, 232, 233
  quality parameters, 240
  salt tolerance, 61
  seedling vigor, 226

temperature sensitivity, cold hardiness, 229
tissue types, 241
world production, 87
Tobacco, 213—214
  as C_3 plant, 41
  salt tolerance, 70
  water relations, 31, 42, 47
  world production, 77, 78, 86
Tomato, 185
  composition, 192
  economics of production, 186
  as forced crop, 189
  production levels, 77, 81, 187
  salt tolerance, 58, 63, 67, 71
  temperature requirements, 187
  water relations
    moisture stress, 42, 43
    photosynthesis, 37
    roots, 18
    seasonal evapotranspiration, 47
    stomates, 34, 35
    tissue content, 5
  weeds and, 99
Tomato, cherry, 185
Topee tambo, 138
Toxins, 139
  forages, 244
  protein crops, 170
  soil elements as, 219
Trachelospermum jasminoides, 66
Tragopogon porrifolius, 185
Translocation, moisture stress and, 36—37
Transpiration, see Water relations
Trefoil, see also Birdsfoot trefoil, 61
Triazine herbicides, 108
Trifoliate orange, 201
Trifolium species, see Clover crops
Trigonellum foenumgraecum, 169
Triticale, 117, 118
  salt tolerance, 60
  uses, 123
Triticum species, see Wheat
Tropaeolum tuberosum, 138
Trypsin inhibitor, 170
Tubers, see Starch crops; specific crops
Tulip tree, 66
Tung, 148
  composition of farm feeds, 156
  oil, 151, 154
  uses, 145, 151
Turgor pressure, 23, 27, 31
Turnip, 184
  salt tolerance, 63, 70
  temperature requirements, 187
Tylosema esculentum, 169

U

Ulluco, 139
Ullucus tuberosus, 139

Ulmus americana, 72
Urea herbicides, 108
Urena lobata, 178

V

Vaccinium species, see Blueberry
Vaccinium macrocarpum (Cranberry), 197, 201
Vapor pressure, temperature and, 16, 19—21
Vegetable crops, see also Legume crops; Pulse crops; Starch crops
  classification, 184
  climate and, 185—187
  composition, 191, 192
  production, 187, 188
  production economics, 183, 185, 186
  production units, 188—189
  research trends, 190—191
  salt tolerance, 62—63, 67, 70
  water relations, 18, 47
  weeds and, 99
  world production, 77, 78, 81, 82
Velvet bean, 169
Velvetleaf, 102—104
Venice mallow, 103
Veratrum californicum, 102
Verticillium wilt, 170
Vetch
  digestibility, 238
  insect and disease pests, 237
  nitrogen fixation, 233
  salt tolerance, 61
  seedling vigor, 221, 226
Viburnus tinus, 66, 72
Vicia faba, see also Bean, broad; Bean crops
Vicia species, see Vetch
Vigna, 167
Vigna sesquipedalis, 184
Vigna sinensis, see Cowpea
Vigna species, 169, 184
Vigna unguiculata, see Cowpea
Viola odorata, 73
Viola tricolor, 73
Violet, 73
Vitis, see Grape
Voandzeia subterranea, 169

W

Walnut, 47, 70, 86
Walnut, black, 34
Wandering Jew, 35
Water relations
  crop water use
    seasonal, 43—44
    water-use efficiency, 46—48, 49
  deficits, responses to, 28, 30—37
    osmotic adjustment, 32—33
    photosynthesis, 33—37
    stomatal closure, 28, 30—35
  drought resistance, 48, 50—51
  dry-matter production, 92
  energy system, plant, 23, 25—29
  evaporative demand, 16—24
  forage crops, 219, 230, 231
  hydrologic cycle, 6
  importance of, 3—6
  modeling of soil-water relations, 51—53
  moisture stress, responses to, 37—41
    flowering habit and, 37
    growth, 37—40
    leaf orientation, 38, 40
    roots, 40—41
  moisture stress, yield effects, 41—43
  roots and, 10, 13—17
  soil, water storage in, 6—12
  soil-plant-atmosphere continuum, 23, 25
Water chestnut, Chinese, 138
Watercress, 184
Watermelon
  botanical classification, 185
  characteristics, 186
  horticultural features and uses, 208
  salt tolerance, 63
  utilization, 198
  water relations, 42
  world production, 81
Water potential, 27
Wax-apple/wax-jumbo, 198, 208
Wax-leaf privet, 72
Weeds
  adverse effects, 100
  classification and propagation, 100
  competition, 103—104
  contratoxicant, 109
  control, 104—109
  definitions, 99
  dispersal, 102
  economic losses, 99
  herbicide application equipment, 107—108
  herbicide interactions, 108
  herbicide resistance, 108—109
  introduction of, 100—101
  presence within croplands, 101—102
  production and longevity, 102
  yield loss and, 4
Weeping bottlebrush, 66
Western false hellebore, 102
West Indian cherry, 196, 200
Wheat, 183
  as cereal crop, 118
    uses, 121, 123
    world production, 119
  dry-matter production, 93—95
  salt tolerance, 58, 60, 61, 63, 70
  uses, 117
  water relations
    moisture stress, critical periods, 42

roots, 10, 14, 15, 18
  stomates, 34, 35
  transpiration ratio, 48
  yield losses, causes of, 4
 weeds and, 99
 world production, 77, 79, 117, 119
Wheatgrass
 characteristics, 224
 composition, 245
 growth, 227, 230
 seedling vigor, 226
Wheatgrass, crested, 226
Wheatgrass, intermediate, 230
Wheatgrass, pubescent, 227
White clover, see Clover, white; Clover crops
White ice plant, 66
White sweet clover, see Clover, sweet
Wild oat, 104
Wild rye, see *Elymus* species
Willow oak, 34
Wilting, permanent, 40
Wilting percentage, 9, 23
Wind, 33, 101
Winged bean, 169
World production, see also specific crop classes
 beverages and tobacco, 86
 cereal crops, 79, 119
 dry matter production, 95—96
 fibers, 85
 fruits, 82
 hay and alfalfa, 87
 nut crops, tree, 86
 oilseed crops, 83—84, 145—148
 principal groups, 78
 pulses, 85
 sugar, 83
 tubers, 80
 vegetables, 81, 82
Wormwood, biennial, 102

X

Xanthium pensylvanicum, 104
Xanthosoma, 138
Xylem, 6, 21, 22
Xylosma congestum, 66, 72

Y

Yacon, 139
Yam, 183, 185
 botanical classification, 184
 as starch crop, 138, 139, 141
 temperature requirements, 187
Yam, Elephant, 138
Yam, false, 138
Yam bean, 139, 169
Ye-eb, 169
Yellow sage, 66, 72
Yellow sweet clover, see also Clover, sweet, 224
Yew, southern, 66, 73
Yield, see also Dry-matter production
 dry matter production, 93—94
 forage crops, 235, 242
 limiting factors, 4
 water relations and, see also Water relations
  moisture stress and, 42—43
  water use efficiency and, 46
 weeds and, see Weeds
Yucca, 180

Z

Zea mays, see Corn
Zinnia, 73
Ziziphus jujuba, see Jujube